Ecological Studies
Analysis and Synthesis

Edited by
W.D. Billings, Durham (USA)　F. Golley, Athens (USA)
O.L. Lange, Würzburg (FRG)　J.S. Olson, Oak Ridge (USA)
H. Remmert, Marburg (FRG)

Volume 79

Ecological Studies

Volume 62
Frost Survival of Plants (1987)
By A. Sakai and W. Larcher

Volume 63
Long-Term Forest Dynamics of the Temperate Zone (1987)
By P.A. Delcourt and H.R. Delcourt

Volume 64
Landscape Heterogeneity and Disturbance (1987)
Edited by M. Goigel Turner

Volume 65
Community Ecology of Sea Otters (1987)
Edited by G.R. van Blaricom and J.A. Estes

Volume 66
Forest Hydrology and Ecology at Coweeta (1987)
Edited by W.T. Swank and D.A. Crossley, Jr.

Volume 67
Concepts of Ecosystem Ecology: A Comparative View (1988)
Edited by L.R. Pomeroy and J.J. Alberts

Volume 68
Stable Isotopes in Ecological Research (1989)
Edited by P.W. Rundel, J.R. Ehleringer, and K.A. Nagy

Volume 69
Vertebrates in Complex Tropical Systems (1989)
Edited by M.L. Harmelin-Vivien and F. Bourliere

Volume 70
The Northern Forest Border in Canada and Alaska (1989)
By J.A. Larsen

Volume 71
Tidal Flat Estuaries: Simulation and Analysis of the Ems Estuary (1988)
Edited by J. Baretta and P. Ruardij

Volume 72
Acidic Deposition and Forest Soils (1989)
By D. Binkley, C.T. Driscoll, H.L. Allen, P. Schoeneberger, and D. McAvoy

Volume 73
Toxic Organic Chemicals in Porous Media (1989)
Edited by Z. Gerstl, Y. Chen, U. Mingelgrin, and B. Yaron

Volume 74
Inorganic Contaminants in the Vadose Zone (1989)
Edited by B. Bar-Yosef, N.J. Barnow, and J. Goldshmid

Volume 75
The Grazing Land Ecosystems of the African Sahel (1989)
By H.N. Le Houérou

Volume 76
Vascular Plants as Epiphytes: Evolution and Ecophysiology (1989)
Edited by U. Lüttge

Volume 77
Air Pollution and Forest Decline: A Study of Spruce *(Picea abies)* **on Acid Soils** (1989)
Edited by E.-D. Schulze, O.L. Lange, and R. Oren

Volume 78
Agroecology: Researching the Ecological Basis for Sustainable Agriculture (1990)
Edited by S.R. Gliessman

Volume 79
Remote Sensing of Biosphere Functioning (1990)
Edited by R.J. Hobbs and H.A. Mooney

R.J. Hobbs H.A. Mooney
Editors

Remote Sensing of Biosphere Functioning

With 85 Illustrations and 6 Color Plates

Springer-Verlag
New York Berlin Heidelberg
London Paris Tokyo Hong Kong

R.J. Hobbs
Division of Wildlife and Ecology
CSIRO
Midland WA 6056
Australia

H.A. Mooney
Department of Biological Sciences
Stanford University
Stanford, CA 94305-5020
USA

Library of Congress Cataloging-in-Publication Data
Remote sensing of biosphere functioning/edited by Richard J. Hobbs
  and Harold A. Mooney.
      p.   cm.—(Ecological studies: vol. 79)
    ISBN 0-387-97098-3 (alk. paper)
    1. Biosphere—Remote sensing.   2. Biology—Remote sensing.
  I. Hobbs, Richard J.   II. Mooney, Harold A.   III. Series: Ecological
  studies; v. 79.
  QH501.R46   1989
  574.5—dc20   89-11524

Printed on acid-free paper.

© 1990 Springer-Verlag New York Inc.
All rights reserved. This work may not be translated or copied in whole or in part without the written permission of the publisher (Springer-Verlag New York, Inc., 175 Fifth Avenue, New York, NY 10010, USA), except for brief excerpts in connection with reviews or scholarly analysis. Use in connection with any form of information storage and retrieval, electronic adaptation, computer software, or by similar or dissimilar methodology now known or hereafter developed is forbidden.
The use of general descriptive names, trade names, trademarks, etc., in this publication, even if the former are not especially identified, is not to be taken as a sign that such names, as understood by the Trade Marks and Merchandise Marks Act, may accordingly be used freely by anyone.

Typeset by Asco Trade Typesetting Ltd., Hong Kong.
Printed and bound by Edwards Brothers, Inc., Ann Arbor, Michigan.
Printed in the United States of America.

9 8 7 6 5 4 3 2 1

ISBN 0-387-97098-3 Springer-Verlag New York Berlin Heidelberg
ISBN 3-540-97098-3 Springer-Verlag Berlin Heidelberg New York

Contents

Contributors	vii
1. Introduction HAROLD A. MOONEY and RICHARD J. HOBBS	1
2. Remote Sensing of Terrestrial Ecosystem Structure: An Ecologist's Pragmatic View R. DEAN GRAETZ	5
3. Measurements of Surface Soil Moisture and Temperature THOMAS SCHMUGGE	31
4. Estimating Terrestrial Primary Productivity by Combining Remote Sensing and Ecosystem Simulation STEVEN W. RUNNING	65
5. Remote Sensing of Litter and Soil Organic Matter Decomposition in Forest Ecosystems JOHN D. ABER, CAROL A. WESSMAN, DAVID L. PETERSON, JERRY M. MELILLO, and JAMES H. FOWNES	87

Contents

6. **Water and Energy Exchange** — 105
   Robert E. Dickinson

7. **Evaluation of Canopy Biochemistry** — 135
   Carol A. Wessman

8. **Remote Sensing and Trace Gas Fluxes** — 157
   Pamela A. Matson and Peter M. Vitousek

9. **Satellite Remote Sensing and Field Experiments** — 169
   Piers J. Sellers, Forrest G. Hall, Don E. Strebel,
   Ghassem Asrar, and Robert E. Murphy

10. **Remote Sensing of Spatial and Temporal Dynamics of Vegetation** — 203
    Richard J. Hobbs

11. **Remote Sensing of Landscape Processes** — 221
    Geoff Pickup

12. **Synoptic-Scale Hydrological and Biogeochemical Cycles in the Amazon River Basin: A Modeling and Remote Sensing Perspective** — 249
    Jeffrey E. Richey, John B. Adams, and Reynaldo L. Victoria

13. **Remote Sensing of Marine Photosynthesis** — 269
    John S. Parslow and Graham P. Harris

14. **Analysis of Remotely Sensed Data** — 291
    Jeremy F. Wallace and Norm Campbell

15. **Remote Sensing of Biosphere Functioning: Concluding Remarks** — 305
    Richard J. Hobbs and Harold A. Mooney

Index — 307

Contributors

ABER, JOHN D.
Institute for the Study of Earth, Oceans, and Space, University of New Hampshire, Durham, NH 03824, USA

ADAMS, JOHN B.
Department of Geological Sciences, University of Washington, Seattle, WA 98195, USA

ASRAR, GHASSEM
NASA Headquarters, Code EE, 600 Independence Avenue, S.W., Washington, DC 20456, USA

CAMPBELL, NORM
Division of Mathematics and Statistics, CSIRO, Wembley WA 6014, Australia

DICKINSON, ROBERT E.
National Center for Atmospheric Research, P.O. Box 3000, Boulder, CO 80307, USA

FOWNES, JAMES H.
Department of Agronomy and Soil Science, University of Hawaii, Honolulu, HI 96822, USA

GRAETZ, R. DEAN	Division of Wildlife and Ecology, CSIRO, P.O. Box 84, Lyneham ACT 2602, Australia
HALL, FORREST G.	NASA–Goddard Space Flight Center, Code 623, Greenbelt, MD 20771, USA
HARRIS, GRAHAM P.	Division of Fisheries, CSIRO Marine Laboratories, GPO Box 1538, Hobart 7000, Tasmania, Australia
HOBBS, RICHARD J.	Division of Wildlife and Ecology, CSIRO, LMB 4, P.O., Midland WA 6056, Australia
MATSON, PAMELA A.	Ecosystem Science and Technology Branch, NASA–Ames Research Center, Moffett Field, CA 94305, USA
MELILLO, JERRY M.	Ecosystem Center, Marine Biological Laboratory, Woods Hole, MA 02543, USA
MOONEY, HAROLD A.	Department of Biological Sciences, Stanford University, Stanford, CA 94305, USA
MURPHY, ROBERT E.	NASA Headquarters, Code EE, 600 Independence Avenue, S.W., Washington, DC 20456, USA
PARSLOW, JOHN S.	Division of Fisheries, CSIRO Marine Laboratories, GPO Box 1538, Hobart 7000, Tasmania, Australia
PETERSON, DAVID L.	M/S 242-4, NASA–Ames Research Center, Moffett Field, CA 94035, USA
PICKUP, GEOFF	Division of Wildlife and Ecology, CSIRO, Centre for Arid Zone Research, Box 2111, Alice Springs NT 5750, Australia

RICHEY, JEFFREY E.	School of Oceanography, University of Washington, Seattle, WA 98195, USA
RUNNING, STEVEN W.	School of Forestry, University of Montana, Missoula, MT 59812, USA
SCHMUGGE, THOMAS	USDA/ARS Hydrology Laboratory, Beltsville, MD 20705, USA
SELLERS, PIERS J.	COLA, Department of Meteorology, University of Maryland, College Park, MD 20742, USA
STREBEL, DON E.	Science Applications Research, 4400 Forbes Road, Lanham, MD 20706, USA
VICTORIA, REYNALDO L.	Centro de Energia Nuclear na Agricultura and Escola Superior de Agricultura "Luiz de Queroz," 13400 Piracicaba SP, Brazil
VITOUSEK, PETER M.	Department of Biological Sciences, Stanford University, Stanford, CA 94305, USA
WALLACE, JEREMY F.	Division of Mathematics and Statistics, CSIRO, Wembley WA 6014, Australia
WESSMAN, CAROL A.	Center for the Study of Earth from Space (CSES)/CIRES, University of Colorado, Campus Box 449, Boulder, CO 80309, USA

1. Introduction

Harold A. Mooney and Richard J. Hobbs

At present there is enormous concern about the changes that are occurring on the surface of the earth and in the earth's atmosphere, primarily as a result of human activities. These changes, particularly in the atmosphere, have the potential for altering the earth's habitability. International programs unprecedented in scope, including the International Geosphere-Biosphere Program, have been initiated to describe and understand these changes. The global change program will call for coordinated measurements on a global scale of those interactive physical and biological processes that regulate the earth system. The program will rely heavily on the emerging technology of remote sensing from airborne vehicles, particularly satellites. Satellites offer the potential of continuously viewing large segments of the earth's surface, thus documenting the changes that are occurring. The task, however, is not only to document global change, which will be an enormous job, but also to understand the significance of these changes to the biosphere. Effects on the biosphere may cover all spatial scales from global to local. The possibility of measuring biosphere function remotely and continuously from satellite imagery must be explored quickly and thoroughly in order to meet the challenge of understanding the consequences of global change. Initial guidelines and approaches are currently being formulated (Dyer and Crossley, 1986; JOI, 1984; NAS, 1986; Rasool, 1987).

There are many conceptual and technical issues that must be resolved before we can confidently monitor biosphere functioning. In July 1988 a joint U.S.–Australia workshop was held at the East-West Center, Honolulu, Hawaii, funded under a bilateral agreement through the U.S. National Science Foundation and the Australian Department of Industry, Technology and Commerce. Scientists from both countries met to address a number of specific problems, focusing on the kinds of functional properties that can now be monitored from aircraft and from space, the system generality of currently available sensing information, and, finally, the kinds of experiments, measurements, or sensors that will be required to improve our capability to measure continuously biosphere function. The aim of the meeting was to examine the monitoring of individual environmental factors driving biological processes (e.g., leaf area index, cover) and chemical features of the biotic system (e.g., protein and lignin content) that control function. Further, we examined the remote sensing of the actual processes central to biosphere functioning; namely, the exchanges of carbon, water, and trace gases. Finally, the meeting tackled the remote sensing of vegetation and landscape processes. In other words, we covered all levels of biosphere functioning, from the remote sensing of the underlying mechanisms to the detection of structural changes at the vegetation and landscape levels.

At present, imagery is available from a number of satellites that offer differing sampling frequencies (every 16 days for the current Landsat series versus daily for the National Oceanic and Atmospheric Administration (NOAA) Advanced Very High Resolution Radiometer (AVHRR) satellite), differing swath widths (2700 km in the case of the NOAA satellite to only 60 km for the French National Space Program SPOT satellite), and differing sampling resolutions. Further, the spectral regions scanned differ, as does the availability of photographic modes. Finally, the length of record available varies from the Landsat series, which commenced in 1972, to a few years in the case of SPOT (Greegor, 1986). Future development, including the Earth Observing System (EOS), offer increasing diversity and volumes of data.

The basis for assessments of the structure and function of biological features of the earth's surface is the information contained in the reflectance of radiation of specific spectral regions. Pigments differentially absorb visible radiation, as does water in the short-wave infrared region. Reflectance in the near infrared is related to leaf structural characteristics. Thus spectral information can yield correlative assessments of structural and functional features of vegetation relating to these characteristics, such as plant productivity and stress status (Rock et al., 1986). Good correlations have been made between absorptance in various visible bands and phytoplankton biomass and primary productivity. Developments utilizing satellite-derived measures of surface sea temperature and wind-induced mixing hold promise of refining these correlations. As Perry (1986) has noted,

"Satellite remote sensing. . . . is the only tool that can provide information on marine primary production on a global scale." Recently, approaches have also been developed that have yielded good correlations between spectral information and regional-scale terrestrial primary productivity (Tucker et al., 1985) and water balance (Sellers, 1985). In addition, satellite imagery has been used for structural surveys of large areas of terrestrial vegetation (Roller and Colwell, 1986), including monitoring tropical forest clearing (Tucker et al., 1984).

The assessment of most of the phenomena described here is by correlation, which in many cases may be unique to specific systems. This means that we will have a continuing need to relate remotely sensed data with ground- or ocean-based measurements. Also, and importantly, new experiments must be performed that would alter independently structural and functional properties of various systems in conjunction with new sensor development (Greegor, 1986). In this regard, experiments are now in progress that are relating ground-based with aircraft- and satellite-sensed measures of energy and water fluxes of selected ecosystems (Rasool and Bolle, 1984; Sellers et al., this volume).

In summary, the technology is now becoming available to make truly global assessments of the earth's changing structural and functional properties. Although the promise is great, considerably more work is needed to make this a reality. In this volume we aim to review current progress and point to new directions for the future.

References

Byer, M.E., and Crossley, D.A. Jr. (eds.) (1986). Coupling of ecological studies with remote sensing: potentials at four biosphere reserves in the United States. Man and the Biosphere (MAB) Program, U.S. Dept. of State, Washington, D.C.

Greegor, D.H. Jr. (1986). Ecology from space. *BioScience* 36:429–432.

JOI (Joint Oceanograph. Inst. Inc.). (1984). *Oceanography from Space. A Research Strategy for the Decade 1985–1995*. JOI, Washington, D.C.

NAS (Nat. Acad. Sci.). (1986). *Remote Sensing of the Biosphere*. NAS, Washington, D.C.

Perry, M.J. (1986). Assessing marine primary production from space. *BioScience* 36:461–467.

Rasool, S.I. (ed.) (1987). *Potential of Remote Sensing for the Study of Global Change*. COSPAR (Committee on Space Res.) Rep. to Int. Council of Scientific Unions (ICSU). Advances in Space Research 7(1). Pergamon, Oxford, England.

Rasool, S.I., and Bolle, H.J. (1984). ISLSCP: International satellite land-surface climatology project. *Bull. Amer. Met. Soc.* 65:143–144.

Rock, B.N., Vogelmann, J.E., Williams, D.L., Vogelmann, A.F., and Hoshizaki, T. (1986). Remote sensing of forest damage. *BioScience* 36:439–445.

Roller, N.E.G., and Colwell, J.E. (1986). Course-resolution satellite data for ecological surveys. *BioScience* 36:468–475.

Sellers, P.J. (1985). Canopy reflectance, photosynthesis and transpiration. *Int. J. Remote Sens.* 6:1335–1372.

Tucker, C.J., Holben, B.N., and Goff, T.E. (1984). Intensive forest clearing in Rondonia, Brazil, as detected by satellite remote sensing. *Remote. Sens. Envir.* 15:255–261.

Tucker, C.J., Vanpraet, C.L., Sharman, M.J., and Van Ittersum, G. (1985). Satellite remote sensing of total dry matter production in the Senegalese Sahel: 1980–1984. *Remote Sen. Envir.* 17:233–249.

2. Remote Sensing of Terrestrial Ecosystem Structure: An Ecologist's Pragmatic View

R. Dean Graetz

This chapter reviews the scientific concepts involved in the application of remote sensing technology to current and future problems in terrestrial ecology. The approach is pragmatic, being decisively user oriented, and is based on the proposition that currently available technology far exceeds the scientific capability of interpreting and applying it. For most terrestrial ecological problems of current and future concern, data types and volumes are not immediately limiting. Rather it is the understanding of the ecological significance of what has already been acquired that fetters the wider, more constructive use of remote sensing technology.

This situation must change! The problems that face ecologists are global in scale, yet require solutions and management that may be local or highly site specific. For example, there are global phenomena that involve the 'commons' of atmosphere and ocean, such as the greenhouse effect, El Niño, and acid rain. These jostle for attention along with the rapidly proliferating, small-scale problems of deforestation, soil erosion, and the like, which are locally severe and may in aggregate have global consequences.

It therefore can be argued that global understanding is required to act rationally at local scales; that is, to address local-scale problems that may have global ramifications. Global understanding is an impossible task for the discipline of ecology to achieve without extensive and intensive use of remotely sensed data. Remote sensing is the tool to use because it provides the only data sets that span the temporal and spatial scales of local systems aggregated to global systems.

Ecologists have been remiss by their diminutive input to the development and application of remote sensing to biosphere-scale problems. Too many data have been collected by those who are not operational users. Left in the hands of technocrats, remote sensing has frequently become an end in itself, rather than the powerful tool it can be. Given the current and widely disseminated capability to image the earth on a daily basis, few doubt the power of remote sensing. Yet in spite of this great power, the tool is blunt and the potential unrealized. The structure of this chapter is to indicate ways in which the tool can be sharpened. The prescription is not a technological "fix." Instead an increased input is required from field ecologists to the most critical stage in the use of remote sensing, the formulation of realistic scene models. Second, a change is required in ecologists' world view, with critical appraisal of the consequences for ecological description and analysis of accepting remotely sensed data as the primary data set.

Determining Future Requirements

The requirements for assessment and monitoring the structure of terrestrial ecosystems can be defined by four question: Why? What? Where? How? The methodology, the how, is the focus of this chapter, but it is best approached by considering the other three questions first.

The Why

Three widely accepted observations of the condition of the biosphere provide the motivation for global-scale studies. These are a recognition that biotic and abiotic components of the biosphere are inextricably linked, that human impacts on the earth now approach the global scale of biosphere processes, and that there are limits of habitability of the earth. This is the nature of the problem. It is global in scale with the overall objective of understanding the interactive role of the biosphere in global change.

The Where

Although the scale is ultimately global, the scope of the task varies greatly with geographic location. For example, one of the dominant issues of global change will be global habitability. Global habitability has several components, each of which is driven by forces with dissimilar response times and information requirements. The component of resource depletion, for example, will be forced by a burgeoning human population geographically located in the tropics or subtropics where we have the most extensive and rapid ecological transformation yet known on the earth. Of the total world population of 5 billion, more than 60% live in the tropical and subtropical belts, where, through rapidly expanding subsistence land use, the earth's most productive ecosystems are being destroyed at an ever-increasing rate. Therefore, the assessment and monitoring of tropical and subtropical terrestrial ecosystems are now of major concern to national and international agencies (Malingreau, 1986; Matson et al., 1987; Matson and Holben, 1987). The difficulty of remotely sensing these cloud-covered landscapes is well known.

The What

The biosphere, or even terrestrial ecosystems, is a somewhat abstract concept. Such concepts can be made more concrete by considering the space mission image of the earth, a blue-green planet laced with swirls of clouds. The planet is blue because 71% of its surface area is ocean. At the global scale, the oceans are far more important than the land in the exchange of solar energy, the process that ultimately drives the atmospheric engine. Thereore, any studies of the terrestrial contribution to global climate change must be comparable to, and in sympathy with, those for the oceans.

A closer examination of the green parts of the globe, the terrestrial ecosystems, reveals that they are not all green. In fact, it is only the forests and crops, approximately 25% of the land area, that are green. The land is only green where the projected foliage cover (PFC) of the vegetative cover is greater than 1. The remaining 75% of the earth's surface is covered by vegetation that is sparse and shows the color of the soil beneath. Therefore, although the green, highly productive areas of the earth's surface are of considerable interest in studies of carbon budgets, global monitoring requires that remote sensing technology be applied to very large areas of the land surface where the vegetative cover may be either a relatively minor or a fluctuating component. However thin or variable, this film of vegetation is the driving force of all terrestrial biological processes and is the interface of 29% of the global surface with the atmosphere.

The what, therefore, becomes the properties of vegetation that determine the exchange of energy and matter between terrestrial ecosystems and the atmosphere.

The Vegetation Component of Ecosystems

The characteristics of ecosystems are determined by the primary trophic level, the vegetation. Therefore vegetation can be taken as the functional, tangible equivalent of terrestrial ecosystems. Vegetation is the cover of plants in an area and can be characterized by three measures. The first is physiognomy, where the physiognomy of individual plants includes the attributes of leaf shape, growth form, phenology, and so on, and for assemblages of plants is more usually called vegetation structure. Vegetation structure has micrometeorological significance because it determines the magnitude and direction of the exchange of matter, momentum, and radiation between the earth's surface and the atmosphere (Running, 1986; Sellers and Dorman, 1987; Wilson et al., 1987; Taconet et al., 1986).

The second characteristic of vegetation is its dynamics. Vegetation changes in space in response to climatic and landscape factors, and, at any one location, alter in time. These temporal changes include the rhythmic phenological changes of growth and flowering, as well as the irregular, episodic alterations of disturbance (see Hobbs, this volume).

Disturbance is an intrinsic determinant of vegetation structure and composition. All vegetation is shaped by the disturbance regime it experiences. Disturbance regimes are the patterns of change and recovery that have characteristic impacts, frequencies, and spatial scales. For example, the time scales of disturbance may range from a storm (days), to the ploughing of crops (years), to clearing of forests (centuries); for example, see Nelson et al. (1987). Spatial scales range from the fine grain of grasslands (square meters), to the gap dynamics of forests (hectares), to the extensive (10^5 ha) wildfires of tropical savannas (Matson et al., 1987; Matson and Holben, 1987).

The disturbance regime and its consequence, the successional recovery, are an important characteristic of all plant communities because they determines the structure and structural dynamics of vegetation. The biotic response to disturbance generates the observable vegetation patterns (Delcourt et al., 1983). The space/time domains of these processes in, and properties of, vegetation are summarized as Figure 2.1.

Experimental terrestrial ecology is limited to the space/time domains of human experience. Remote sensing in general, and aerial photography in particular, has not yet facilitated an understanding and mapping of vegetation patterns of the entire globe. Satellite remote sensing is the only data source that can be used to assess and monitor this basic renewable resource on time and space scales that are comparable to those of the human transformation of the resource.

The last relevant characteristic of vegetation is its taxonomic composition. The taxonomic description of vegetation is based on its phylogenetic or evolutionary affinities and has been the focus of botanical science for several centuries. Thus, a taxonomic description has importance because the name of any species is the key by which data on such characteristics as physiognomy, phenology, productivity, and longevity are stored. However, in this functional context, the traditional taxonomy, the phylogenetic affinities of vegetation, has little value and will probably prove a hindrance. In the problems facing ecologists, function is all important and simplification is needed. The myriad of species and community classifications must be grouped by functional equivalence in terms of response to disturbance,

2. Remote Sensing of Ecosystem Structure

DISTURBANCE AND CHANGE + BIOTIC RESPONSE = VEGETATION PATTERNS

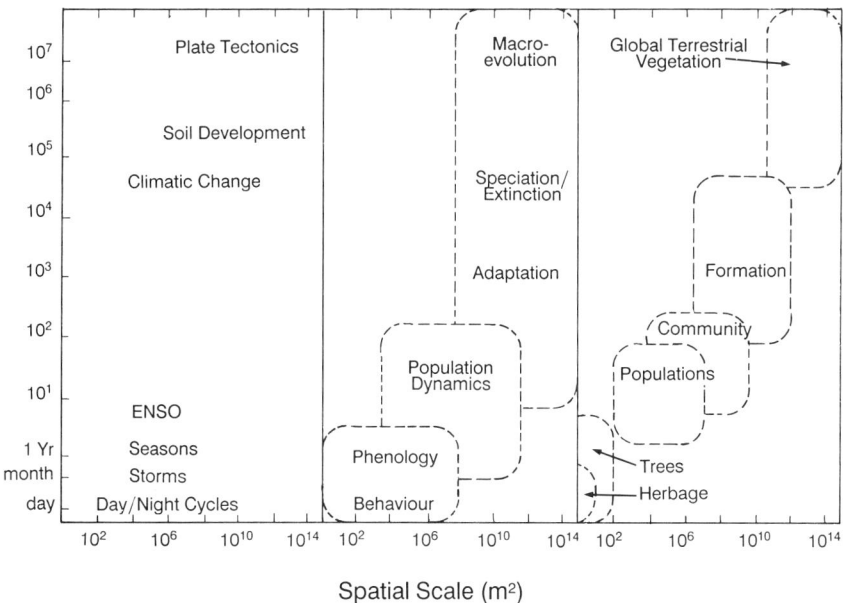

Figure 2.1. A diagrammatic representation of the space/time domains of disturbance, biotic response, and the resultant observable patterns of vegetation. Most ecological research and understanding is crowded into the space/time domains scaled to human experience; the episodic or rhythmic cycles of disturbance and the biotic response are studied as population dynamics and mapped as communities. The larger- and longer-scaled phenomena have largely been inferred from palaeo studies.

be it by humans, fire, or climatic change. Thus, of the three dimensions of vegetation discussed, structure and structural dynamics are fundamental for global-scale questions, with taxonomy the least important of the three.

The Utility of Three Dimensions of Vegetation

All three dimensions of vegetation description are applicable to global studies. For example, one of the principal effects of predicted climate change will be on vegetation. Potentially, this interaction will be reciprocal; climate-induced vegetation changes may well generate feedback on further climate change. The simplest approach would be to model vegetation as a structureless film of variable thickness across the continents. Such an approach might satisfactorily predict some processes—CO_2 exchange, for example—but it does not incorporate the dynamic properties of vegetation that are critical to predicting the response of vegetation as a whole to change over time.

Vegetation will not react as a unit to change or disturbance. Rather the outcome is the aggregate of reactions by component species. Ecological information is encoded, in part, in the physiological and taxonomic characteristics of plants. To limited extents, these include tolerances of environmental variables, or behavior in competitive conditions and under different disturbance regimes. For short-term or equilibrium conditions, it may be acceptable to model vegetation as a structureless green film of varying thickness. However, for longer term analysis, such as climate change, vegetation must be viewed as structured by differentially responsive elements. This view can be supported only by the inclusion of all three dimensions of vegetation—structure, dynamics, and taxonomy.

The Difficult Dimension: Taxonomy

The remainder of this chapter concentrates on the use of remotely sensed data to assess and monitor vegetation structure and dynamics. Fortunately, although these are the most important dimensions of vegetation, they are also the two most amenable to measurement of remote sensing. In contrast, the taxonomic composition of vegetation has proved quite elusive and cannot be done on regional scales with any reliability using satellite data (e.g., Badhwar et al., 1986a, 1986b). On continental scales, the use of spectral-temporal models of phenological behavior ("greening") does offer some hope of classifying broad-scale communities (Tucker et al., 1985). Even the higher spatial resolution data sets of aerial photography are usually inadequate to determine botanical composition, and taxonomic detail is most accurately determined by ground survey (e.g., van Gils and van Wijngaarden, 1984). It is possible, but unlikely, that newer space-borne sensors of high spectral and spatial resolution will dramatically improve the capability to determine botanical composition. Far greater promise is offered by the process of context modeling whereby the (structure and) taxonomy of vegetation is inferred from the context of other landscape variables such as soil type and elevation, modeled within a geographic information system (GIS) (Franklin et al., 1986; Strahler, 1981).

The GIS Imperative

Remotely sensed data are not an end in themselves and should be more correctly regarded as just one data set within a GIS. It is important to note that remotely sensed data become the dynamic information about the biosphere, in contrast to the ancillary data sets, such as elevation and soils, which are static. Numerous studies have demonstrated the value of ancillary data in increasing the information content of satellite data (Franklin et al., 1986; Graetz et al., 1986; Graetz and Pech, 1988; Pech et al., 1986a, 1986b; Strahler, 1981).

The Structure of Vegetation

A Framework

The key dimension of vegetation is structure. Over the decades, a wide variety of methods for describing and classifying vegetation structure have been developed (Mueller-Dombois 1984). The greatest difficulty in compiling global data sets of vegetation is the diversity and disparity of the classification systems used (Mathews, 1983; Olson et al., 1983). Because remote sensing is the only tool that ecologists have to work with on local to global scales, they must adapt their traditional methods of description and classification to be compatible with the nature of remotely sensed data. The first consequence of having to use remotely sensed data is the unfamiliar vertical perspective and scale of aggregation.

A useful framework within which to classify structure is a two-dimensional framework based on projected foliage cover (PFC or cover) and the life form of the tallest stratum; see Figure 2.2. This classification (e.g., Specht, 1981; Gillison and Walker, 1981; Walker and Hopkins, 1984) captures both abundance (cover) and structure (vertical distribution of biomass) and is directly compatible with remote sensing because a major axis utilizes the vertical view. The description of any vegetation type is determined by its location within the two-dimensional space of cover and height and a further qualification by botanical composition (Figure 2.2). Other classifications (e.g., FAO, 1973; Olson et al., 1983; van Gills and van Wijngaarden, 1984) are not readily transposed to the vertical perspective of satellite viewing.

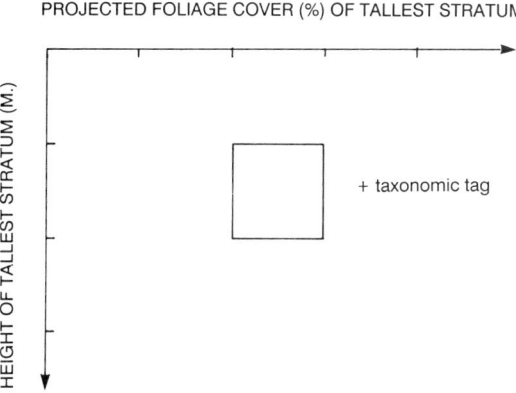

Figure 2.2. A very simple classification of vegetation based on the quantitative attributes of projected foliage cover and height of the tallest stratum. Systematic descriptive names could be assigned to any location within the table, for example, tall, open woodland. Further prescription could be attached by including a taxonomic tag, such as tall, open Acacia woodland .

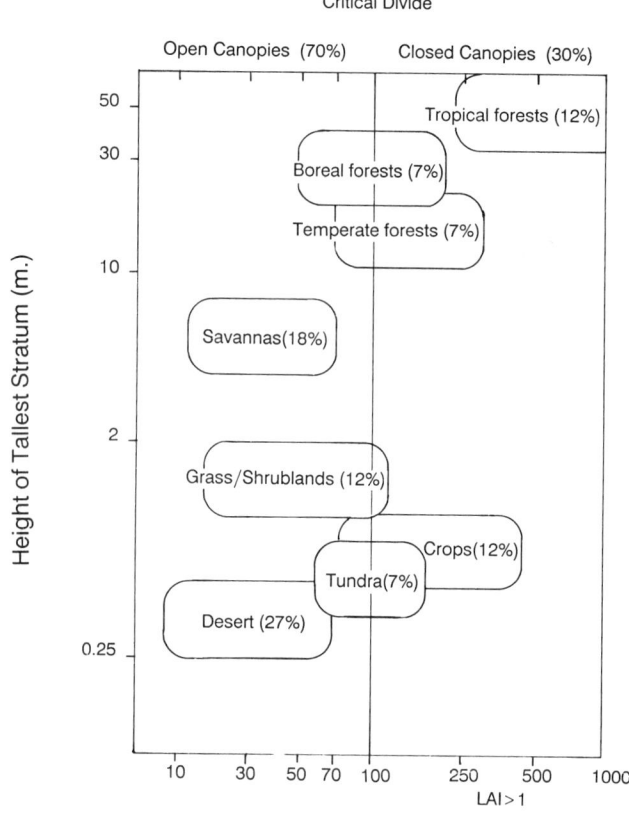

Figure 2.3. The two-dimensional framework of vegetation structure with domains representing a simplified grouping of the world's vegetation. The figures in the domain are the relative proportions of terrestrial vegetation in each domain. [Derived from the data of Olson et al. (1983).]

The most important characteristic of the classification presented in Figure 2.2 is that this system uses two (structure and taxonomy) of the three dimensions of vegetation defined in the foregoing. Also, the location and thus identification of any vegetation structural type are determined by two variables, cover and height, both of which can be determined by remote sensing (e.g., Nelson et al., 1988).

There is convergence of vegetation structure. At climatically similar sites on different continents, unrelated taxonomic groups of plants converge to similar morphologies and community structure. This is one of the key findings of plant biogeography. Given that community structure represents an integrated response of physiology to climate, then it follows that there is functional equivalence underlying vegetation structure. A recognition, grouping, and description of these functional equivalents—as vegetation functional types (VFTs), for example—would simplify the process of relating remotely sensed data to vegetation structure. This argument conforms to a trend among plant ecologists to group like-functioning species into guilds or plant functional types (PFTs). The emphasis is on function rather than taxonomy. Characteristics such as evergreen versus deciduous and C_3 versus C_4 metabolism have greater importance here than phylogenetic association.

Based on the data provided by Olson et al. (1983), a distribution of global terrestrial vegetation within the two-dimensional classification of structural types is suggested (Figure 2.3). Apart from rain forests and crops, the largest proportion of vegetation has an open canopy; that is, the projected foliage cover is less than 100% (Table 2.1). Unfortunately, almost all of the remote sensing literature has dealt with crops that are man-made, closed canopies. In contrast, there is relatively little collective experience in applying remote sensing techniques to the sparse grasslands, shrublands, and woodlands that together cover more than 60% of the land surface.

The point at which the PFC is less than 1 determines the type of scene models appropriate to relating the spectral data and ecological variables. This dependency is twofold. First, where PFC > 1, the interaction of radiation with plants (reflection, absorption, transmission) can be modeled as a simple one-dimensional (depth of canopy) process. Conversely, where the PFC is less than 1, the vegetation occupies discrete volumes in space and

Table 2.1. Areas of Terrestrial Vegetation Types Divided Into Closed (PFC > 1) or Open (PFC < 1) categories. [From Olson et al (1983)]

Type	Area (10^6 km^2)	Proportion (%)
Closed (LAI > 1)		
Tropical forest	15	12
Temperate forest	9	7
Cultivated cropland	15	12
		Subtotal ~ 30
Open (LAI < 1)		
Boreal forest	9	7
Savanna	23	18
Temperate grass-lands/		
shrub lands	15	12
Tundra	9	7
Desert	35	27
	130	Subtotal ~ 70

the interaction of radiation must be modeled as a two- or three-dimensional process. Second, as the relative proportion of vegetation in the scene declines, so does the associated spectral signal compared with that of the soil surface. For these plant communities, the soil surface itself should be as much an object of attention as is the vegetation. The ecological processes of vital importance to the functioning of sparse plant communities—the partitioning of rainfall and incoming radiation and the redistribution of water, organic matter, soil, and nutrients—all increasingly occur at the soil surface. Some of these processes are dealt with by Pickup (this volume). The role of the soil surface relative to vegetation in the functioning in sparse communities needs to be assessed and the ability of remotely sensed data to capture these processes evaluated more fully. The sparse communities are important. The savannas and grasslands, like the tropical rain forests, are under increasing pressure from destructive land use.

The Key Variables

Cover, height, and taxonomy are the key variables by which to describe vegetation types. Of these, cover emerges as the key variable with which to characterize the structure of vegetation canopies, not only because the largest proportion of the global terrestrial vegetation occurs as open structure, but also because cover will be the prime determinant of the remotely sensed signal in the visible to near-infrared (IR) regions. It is in these widely used optical wave bands that the spectral contrast between vegetation and soil is the greatest; see Figure 2.4. Here, the relative cover of vegetation and that of the soil within the field of view of the sensor determine the signal strength. If this can be modeled, and the model inverted, then these wave bands can provide accurate estimates of the relative cover of vegetation and soil in sparse communities (e.g., Foran, 1987; Graetz and Pech, 1988; Pech et al., 1986b).

Cover can also be related to other vegetation structural parameters such as height, biomass, and density, and for a given structural vegetation type these allometric relationships are usually adequate.

In the short term, much of the concern associated with global habitability will be related to the diminution of the productive capacity of terrestrial ecosystems, desertification, deforestation, soil erosion, and so on, and the sustainability of land use under a burgeoning human population. In the longer term, and in the unlikely event that the human population growth rate is checked, the cumulative influence of greenhouse-driven global climate change on human habitability may become of prime importance. Therefore, an ability to forecast the reciprocal interaction of climate change with the functioning of the biosphere generally, and with the terrestrial vegetation in particular, becomes an immediate and major objective. It follows that it is critical to be able to monitor vegetation structure

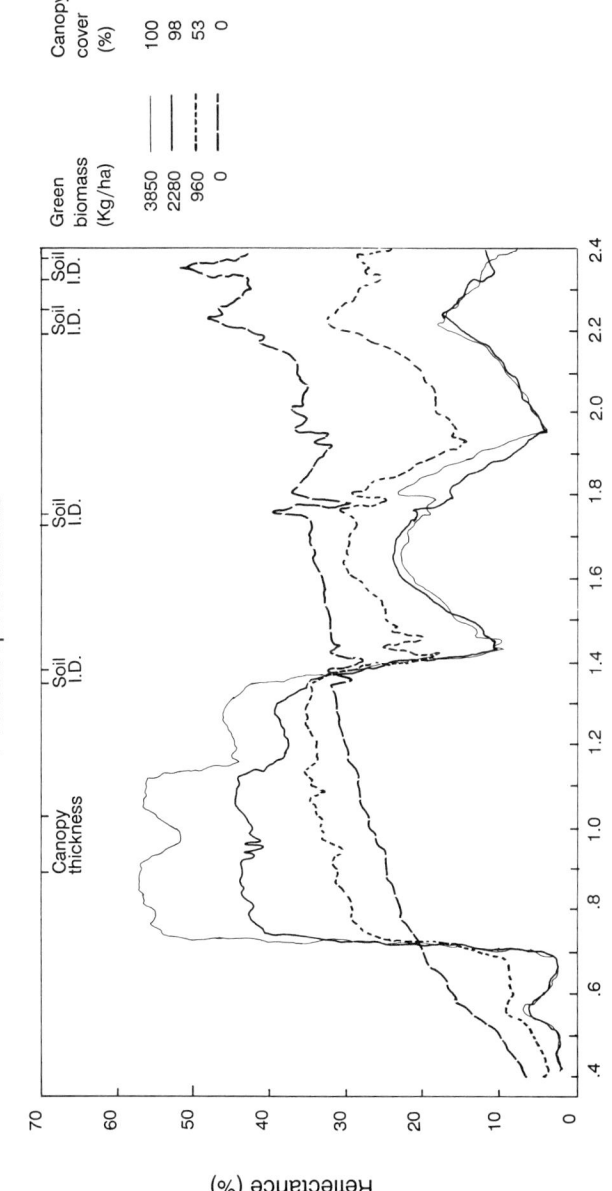

Figure 2.4. A diagram of reflectance as a function of wavelength for a bare agricultural soil and for three levels of plant cover and biomass superimposed. The sharp contrast in reflectance of vegetation and soil in the visible red, near-IR, and mid-IR wavelengths is the basis of most vegetation indices. Windows of maximal spectral contrast are noted. It is simplistic to translate experimental spectral curves such as these directly to whole scenes.

because it can be related to function. And it is the understanding of the functioning of the biosphere that is the ultimate goal. Therefore, with our concern for climate change we should concentrate on the attributes of structure that can be most closely tied to functioning—that is, to ecosystem processes—and on modeling the relationships of these with climate.

The processes of particular interest in climate, and thus biosphere change, are the radiation, heat, mass, and momentum exchange with the atmosphere, and carbon storage in the biospheric and lithospheric pools. Most, but not all, of these exchange processes are between plant surfaces and the atmosphere, driven by the incoming energy from the sun and controlled by both surface factors (stomata) and the structural characteristics of the canopy itself. Structural characteristics, such as albedo and aerodynamic roughness, determine the magnitude of the potential exchanges that can occur. Surface physiological factors, stomata, leaf orientation, and so forth respond to ambient environmental conditions and soil-water potential and act in sequence with the structural controls to reduce the potential to the actual.

The relative importance of these two levels of control, structural versus physiological, is evident from the initial prerequisites of Global Circulation Models (GCMs). The parameters required for a vegetated land surface interacting with the atmosphere are measures of albedo (A), evapotranspiration (E_T), surface roughness (Z_0), and net CO_2 flux in the case of Global Biosphere Models (GBMs). All of these parameters are determined by the structure of vegetation. More advanced modeling has incorporated the surface internal resistances to capture the "interactive" nature of the exchange of energy and matter; for example, the Biosphere Atmospheric Transfer System (BATS) model (Wilson et al., 1987) and the Simple Biosphere (SiB) model (Sellers and Dorman, 1987).

In summary, the capacity to assess and monitor the structure of terrestrial vegetation using remote sensing is important because structure can be related to functioning, that is, to ecosystem processes that are ultimately aggregated up to the functioning of the biosphere-atmosphere-geosphere at global scales. However, it is important to go further and explore the nature of the relationships between structure and function to establish what the key variables are and how accurately they must be measured.

Problems in the Collection and Interpretation of Remotely Sensed Data

The foregoing argument is basic and has been expressed for at least a decade. Why, then, is it necessary to reiterate it here? In the past 20 years, there has been an enormous increase in scientific understanding of the functioning and behavior of the earth's surface through the application of remotely sensed data. In some areas of application (e.g., terrestrial hydrology, oceanography), significant advances are still to be made as a result of new sensors, as in the microwave region, or of the availability of higher spatial- and, more important, temporal-resolution data sets. In contrast, for vegetated landscapes the spectrum of sensors has been thoroughly evaluated and no breakthroughs as a result of technological innovation are likely.

The basic life processes and functioning of plant canopies are best detected and contrasted with the nonliving soil background in the visible and near- or mid-IR wave bands, the area of most research and development. Success in the use of microwave wavelengths has been slow in coming and application studies are relatively primitive. Combining microwave with optical wavelengths appears to offer more promise (Paris and Kwong, 1988).

Significant advances in remote sensing have come not through direct observation but rather by inference, utilizing models that relate the spectral-temporal or spectral-spatial behavior of land surfaces to such processes as plant growth (Honey and Tapley, 1981; Pickup and Foran, 1987) or to such phenomena as soil erosion (Pickup, this volume). Innovative "lateral" thinking has been at a premium and there have been but few cases in the past decade or so. The most striking has been the application of data from the NOAA series of satellites to the assessment and monitoring of the productivity of terrestrial ecosystems by the NASA group led by C. J. Tucker. Utilizing the computation of a relatively simple but robust vegetation index, the normalized difference vegetation index or NDVI, and its demonstrated relationship to surface vegetation parameters (e.g., Holben et al., 1980) and insensitivity to atmospheric conditions (Holben and Fraser, 1984), it has been possible to monitor the phenological behavior of regions, such as the Sahel of Africa (Tucker et al., 1986b), continents (Townshend and Justice, 1986), and global vegetation (Justice et al., 1985). By exploring the relationship between the vegetation index and the radiometric characteristics of canopies (Sellers, 1985, 1987), it has been possible to estimate the net aboveground primary productivity (NPP) of whole biomes (Goward et al., 1985, 1987) and to map the vegetation types of entire continents (Tucker et al., 1985). These and other contributions have catalyzed the widespread application of NOAA data. The contribution of these high temporal frequency, low spatial and spectral resolution data to the understanding of the functioning of the earth's surface has far exceeded that of the high spectral and spatial resolution but low temporal frequency Landsat MSS and TM data sets.

Present Methods and Their Limitations

Remote sensing is the process of extracting information from a data stream that involves the earth's surface, the atmosphere that lies between it and the spacecraft, and the image-forming sensors on board the spacecraft. Information extraction can proceed only by the derivation and use of models to account for the transformations that take place at each of these three steps and at the processing stage (Strahler et al., 1986). These modeling stages are absolutely critical to the information-extraction process and are discussed at length.

Image Processing Models

Image processing models are those used to extract information from the data; that is, to "infer the order in the properties and distributions of matter and energy in the scene from a set of measurements comprising the image" (Strahler et al., 1986). Explicitly or otherwise, scene inference requires the construction and application of a remote sensing model that has three submodels: for the scene, for the atmosphere, and for the sensor. The central problem of information extraction or scene inference is a problem of model inversion.

Scene Models

A scene model quantifies the relationships among the type, number, and spatial distribution of objects and backgrounds in a scene (pixel), and their interactions with radiation (reflectance, transmittance, emittance) and illumination geometry; see Figure 2.5. The scene model quantifies the analyst's conceptual understanding of the target. For terrestrial vegetation, we can identify the two extremes of model type: the discrete model and the continuous model. The discrete model is intuitively the most appealing, containing within each resolution cell discrete objects, such as trees, shrubs, and grasses, and backgrounds, such as soil, rocks, or snow. Because of similarities in electromagnetic properties (reflectance, radiative temperature, etc.), individual objects may be abstracted into a (usually) smaller number of scene elements of uniform properties, such as sunlit vegetation or shadowed soil. These elements have a spatial distribution within a resolution cell that can be statistically characterized (random, clumped, etc.) and, therefore, parameterized, as with texture models.

In the real world, within the more commonly used electromagnetic wave bands, discrete scene models are complex; there are many classes of elements and several classes of backgrounds. The second characteristic of discrete scene models is resolution, the size of the elements relative to that of the resolution cell. Strahler et al. (1986) identify H- and L-resolution models. The H-resolution model is applicable where the elements of the scene are larger than the resolution cells; with the L-resolution model, the converse is true. L-resolution models can be regarded as a type of (continuous) mixture model where the proportions are functions of the sizes and shapes of the elements (vegetation) in the scene model and their relative densities within the resolution cell (Strahler et al., 1986).

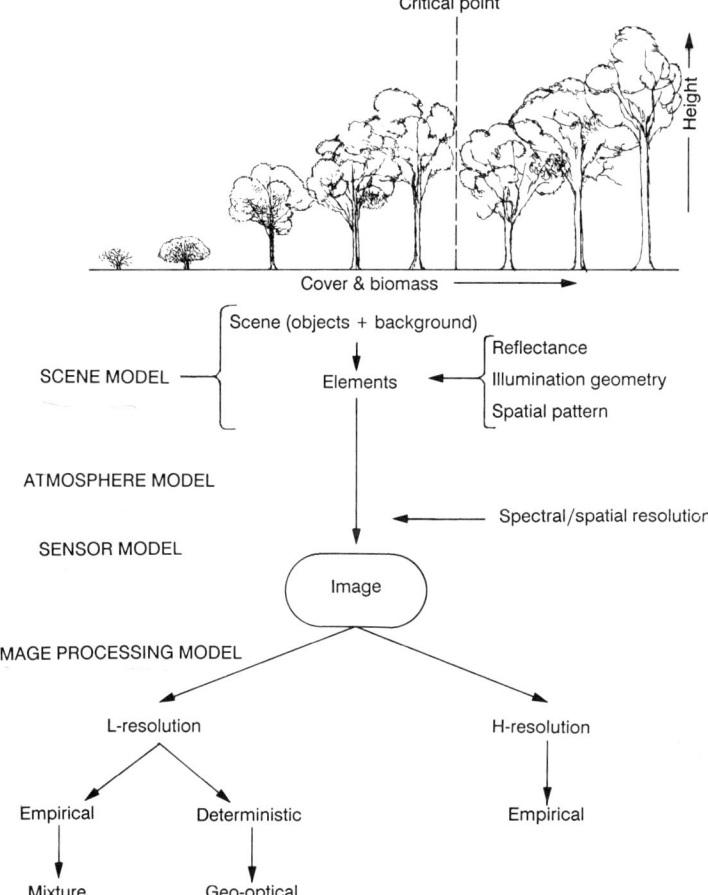

Figure 2.5. A schematic outline of the data flow of remote sensing and the four models that account for the transformations of the data stream. The scene model relates the interaction of objects (vegetation) and backgrounds (soils, etc.) with electromagnetic radiation, including the influence of illumination. The geometry of the scene is determined by the spatial patterning within the scene (pixel) and the complexity of the vegetation itself. The data stream is transformed by the atmosphere and the sensors, for which models are also required. The analysis or image processing model can be one of two types; the option is determined by the relative size of the elements and the scene (pixel). A gradation in complexity of the scene is provided. The critical point, when PFC = 1, influences the type of scene model that needs to be constructed.

H-resolution models underlie the classification image processing models, that is, where the purpose is to assign a pixel to a class or otherwise label a pixel. Even though classification has been the dominant remote sensing model for nearly two decades, it has a much reduced role to play in the inference of vegetation structure. In an ideal system of the future, classification or H-resolution scene models will operate in a subservient role to L-resolution models.

In contrast, the continuous scene model is suitable for resolution cells within which the objects need not be considered as separate. Instead, measurements of resolution cell can be modeled as being the sum of interactions among the various classes of scene elements weighted by their relative concentrations, such as layered atmospheric absorption models. Where a scene can be modeled by proportions, it is called a *mixture model*, and these have considerable applications to the remote sensing of vegetation (McLachlan and Basford, 1988). Both of these scene models can be inverted to infer vegetation structure. By changing the spatial resolution (i.e., the size of the resolution cell), the same vegetated landscape can be described by either model.

Atmosphere Models

An atmospheric model must explain any transformation of the stream of electromagnetic radiation between the surface and the spacecraft. For most applications, it has been convenient, though not necessarily appropriate, to ignore both atmosphere and sensor in the information-extraction process. The cost of this omission has been that the methodology and procedures used become site and time specific. They have no generality; processing parameters are not transferable. For other times or places, the entire procedure must be repeated. In some applications, such as classification, this cost is significantly less than for the development and incorporation of both atmosphere and sensor models in the analysis process.

Sensor Models

The sensor model quantifies how the radiation stream is converted into objective data and the two key parameters of this model are spatial and spectral resolution. Of the three models in the data-acquisition path, the sensor model has been regarded as a static engineering calibration and mostly of no concern to practitioners. This is not always the case. In some applications, where absolute values of a remotely sensed variable are required over a long time series, significant differences in sensor responses have been found for the same craft between times (Musick, 1986; Suits et al., 1988a) and between different craft at the same time (e.g., Gallo and Eidenshink, 1988). The interconversion of data from one spacecraft to another is also critical (Suits et al., 1988b). Thus a requirement for accurate estimates of an absolute variable, such as temperature or reflectance, will also demand an equally accurate model of the influence of the atmosphere at the time of acquisition to separate its contributions from those of the scene.

The spectral resolution of the sensor—that is, which part of the electromagnetic spectrum and with which bandwidths—is a critical choice. Unforunately, the past two decades have been dominated by the "curse of spectroscopy;" a naive, simplistic view that the spectra of isolated pure samples are a sufficient guide to deciding not only on the spectral bandwidth but also on the analysis model. The development of sensors with high spectral resolution without a concurrent development of scene models for analysis results in very limited qualitative results (e.g., Thomas and Ustin, 1987).

Agriculturalists have found that even scenes of spatially homogeneous crops differed from the spectral behavior of a single leaf acquired in the laboratory, being strongly determined by scene-specific factors such as soil color, leaf moisture content, and canopy geometry (Curran and Wardley, 1988). Crop canopy geometry is complex, volatile, and difficult to model, and yet crop canopies are man-made of annual grasses or forbs, designed to be uniform in space and time. By comparison, natural perennial vegetation appears to be almost intractable. Obviously, there are real limits to the accuracy of estimates of canopy variables of natural vegetation that can be derived from remotely sensed data.

The question of spatial resolution has been more recently addressed by Woodcock and Strahler (1987) and Townshend and Justice (1988). In particular, Woodcock and Strahler (1987) demonstrate that the spatial resolution of the sensor determines which analysis pathways are appropriate for the scene inference required and they illustrate how the spatial structure of the scene itself can be extracted using an analysis technique based on the local variance within the image. Further theoretical and experimental studies in this spectral-spatial scene modeling have been explored by Jupp et al. (1988a, 1988b) and Woodcock et al. (1988a, 1988b).

In summary, the analysis of all remotely sensed data involves models of the many processes wherein electromagnetic radiation is transformed (the scene, atmosphere, and sensor) and whereby inference is made about the scene from the image data (the image processing model). These models are an implicit part of the process but their importance has not always been recognized. This neglect is regarded as one of the major stumbling blocks to the further development and application of remote sensing technology to ecological problems at all scales. Without the explicit formulation and incorporation of scene models, application studies will continue to be left as a correlation table—statistically significant but functionally useless. Similarly, without explicit atmospheric models in the analysis, each application study will be unique and not transferable.

All of the submodels comprising the overall image processing model are shown in Figure 2.5. This framework will be used to discuss the analysis of two ecosystem structural variables: cover and LAI.

Vegetation Indices

By far the most common strategy for relating remotely sensed data to vegetation canopies has been via the correlation of vegetation indices with such variables as cover and biomass. This simple empirical approach has initially yielded substantial understanding of the structure and dynamics of vegetation at all scales. The most striking advance has been the first demonstration of the close coupling of the greening of terrestrial vegetation and the fluctuations of global atmospheric CO_2 (Tucker et al., 1986a).

The development and use of indices, with the normalized difference vegetation index (NDVI) being the most widely used, are an attempt to derive a robust index that contains a simplistic scene spectral model (near-IR–red) which is normalized (near-IR–red/near-IR + red), thereby including a simple sensor model. The scene model is L-resolution and empirical relationships are derived by calibration of the index with spatially averaged or smeared variables such as LAI and biomass (Figure 2.5).

However crude, the use of indices has permitted the remote assessment of canopy variables in crops and native grasslands (e.g., Aase et al., 1986, 1987; Wanjura and Hatfield, 1988) and the prediction of yields of crops (Wiegand et al., 1986; Wiegand and Richardson, 1987). The widely used NDVI of visible and near-IR wavebands can be correlated with a microwave equivalent (Becker and Choudhury, 1988) or transposed into the microwave region (Owe et al., 1988).

The limitations of the lack of scene models must be recognized in trying to empirically derive relationships among different sensors, scales, and target canopies. What is derived for NOAA satellite data and soybeans cannot be expected to apply to Landsat data and soybeans. The differences in spectral wave bands of the two spacecraft will have an effect (sensor model) but it will be secondary to the differences caused by disparities in the time of overpass (scene model). The same target canopy will, because of shadowing, have a very different spectral reflectance at 0930 hours, compared with 1430 hours (Curran and Wardley, 1988).

However valuable and utilitarian simple vegetation indices have been, it is unlikely that they will continue to be so in the future. Explicitly formulated scene models are required to facilitate the transfer of image processing models between spacecraft and between scales.

The Estimation of Vegetative Cover

The cover of vegetation in open, sparse communities is strongly correlated with brightness in most of the widely used visible and near-IR wave bands. Cover is the prime determinant of the spectral signal because of the con-

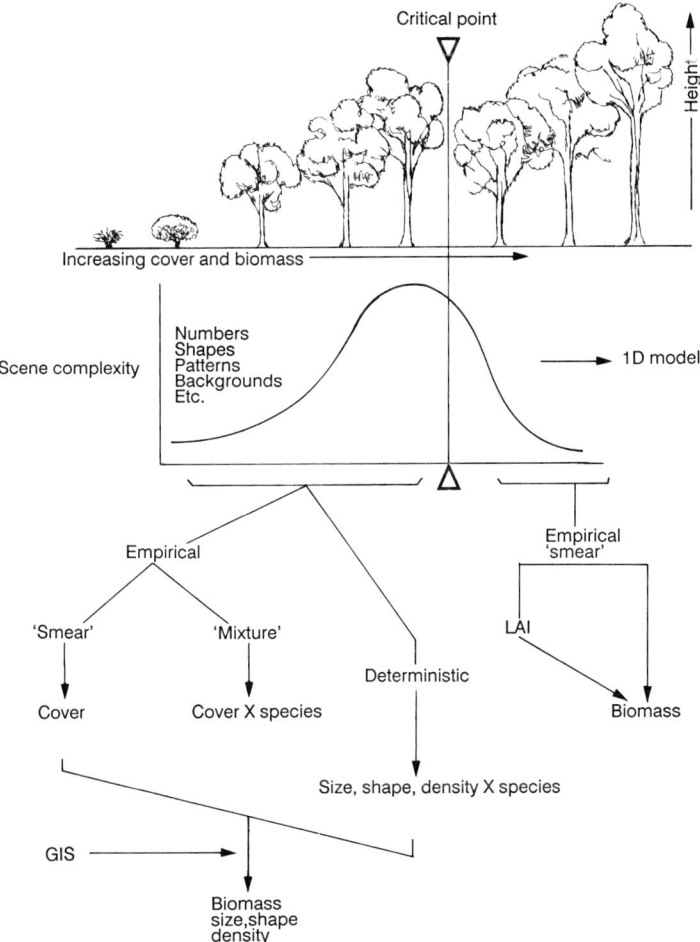

Figure 2.6. A schematic diagram of how the image analysis pathways are currently determined by the complexity of the real world scene and how this complexity is captured by scene models. Empirical models (i.e., those that are calibrated by synchronous measurements from space and the ground) predominate, and the estimates of cover or biomass vary in specificity. Where possible, the inclusion of ancillary data via a GIS will expand the range of variables that can be derived from remotely sensed data.

trast between vegetation and soil (Figure 2.4). Typically, this is analyzed using an invertible L-resolution empirical model. The flow path for the estimation of plant cover, including LAI, from the scene to the image processing model is represented in Figure 2.6. The complexity of the scene increases to a maximum just before canopy closure in response to increasing numbers, shapes, and patterns of the vegetation and their interaction with the background soil through shadowing. Past the critical point of canopy closure, the scene complexity rapidly declines as the interaction of radiation and vegetation approaches a simple one-dimensional system of absorption and reflection as a function of depth (LAI). The image analysis models that can be used differ on either side of the critical point. For LAI greater than 1, empirical regression models can be derived between canopy variables, such as LAI, and biomass, and reflectance or a vegetation index, either independently or sequentially.

Unlike the simple variable cover, LAI is structurally a more specific variable for any plant community (Bartlett et al., 1988). It is not, however, a very useful parameter since it is part of an idealized plane/parallel canopy model that holds in simple plant communities (Clevers, 1988). Indeed, it can be argued that LAI is really only meaningful in certain planophile crop canopies, and its use in these agricultural communities now seems superseded (Wiegand and Richardson, 1987).

The geometrical structure of vegetation canopies, of which LAI is a primitive descriptor, determines their reflectance characteristics in the visible and near-IR range (Bartlett et al., 1988; Curran and Wardley, 1988) and in the mid- and thermal-IR range (Williamson, 1988). This has been demonstrated for the widely different canopies of grassland (Asrar et al., 1986), forests (Peterson et al., 1987; Badhwar et al., 1986a, 1986b), and crops (Aase et al., 1986, 1987).

To the left of the critical divide where scene complexity is substantial, two approaches can be used. The most common one is to follow an empirical, invertible calibration path. The simplest of these ignores (smears) the scene complexity entirely to derive relationships between cover and reflectance (Figure 2.6). Examples of the use of such a simple correlative approach to estimate vegetation cover are for forests and woodlands (Peterson et al., 1986; Karaska et al., 1986; Vujakovic, 1987; Walker et al., 1986) and for sparse rangelands (Foran, 1987; Foran and Pickup, 1984). Foran (1987) included a sensor model (calibration to reflectance) and restricted observations to times at which atmospheric influence was minimal and constant between observations.

Of the six satellite-based studies listed here, only those that developed an empirical relationship between cover and Landsat (MSS) data (e.g., Foran, 1987) are transferable to other situations. The others are both site and time specific and demonstrate only the correlation between the two variables. The relationship was not inverted nor was its adequacy as a predictor demonstrated by the computation of 95% confidence limits over a reasonable range of expected values.

An alternative path is likewise empirical in that it relies on calibration by synchronous ground and satellite measurements. However, it uses a scene model that explicitly recognizes the contribution of the discrete elements to the reflectance of the overall scene (pixel) by inverting the model using mixture or multivariate calibration techniques (Figure 2.6). Such a model has been used for rangeland vegetation by Pech et al. (1986b). A simple intuitive application of this model based on the geometry of the Landsat MSS red/near-IR data space led to the development of indices of "cover" and "bare" that contained an empirical sensor and atmospheric model (Graetz et al., 1986) and could be calibrated against ground measurements (Pech et al., 1986a; Graetz and Pech, 1988). Pech et al. (1986b) list the interactions that can occur between objects and backgrounds within a scene. The importance of variations in shadows (Ranson and Doughtry, 1987) and spatial patchiness (Peterson et al., 1987) has also been noted. Pech and Davis (1987) have extended mixture modeling and multivariate calibration to woodlands.

The image processing path that utilizes the most explicit scene model is that labeled "deterministic" in Figure 2.6. Here the contributions of the size, shape, and spatial patterning of the scene elements are formulated in terms of their geometry and of the variance in reflectance that will be generated between adjacent scenes or pixels by the spatial patterning of the elements. Inversion of this model generates estimates of elements size, shape, and so forth. This geometric scene modeling has been pioneered by Li and Strahler (1985) and Franklin and Strahler (1988). Theoretically, this approach offers the greatest potential to extract information from remotely sensed data but in practice it seems that the intrinsic variability in soil background, object (tree, shrub, etc.) shape, and so on will impose restrictions.

Spectral Is but One Dimension

In almost all of the examples used herein, the scene models had only one dimension, the spectral dimension. The exception was the deterministic geometric scene models, which are based on the spectral variation in space; that is, they are spectral-spatial models (Pickup and Foran, 1987). Scene models that include the temporal dimension have been growing in number and power. In particular, the most famous, and the most basic, of these scene models was the *tasseled cap*, the spectral-temporal model of the phenological (greenness) development of crops that revolutionized the application and interpretation of Landsat data for agricultural landscapes. The explicit inclusion of the temporal dimension in the simple scene models changed the field of agricultural remote sensing almost overnight. Now these models of spectral trajectory through time are sufficiently well developed both to identify crops and to forecast yields on a global basis (Hall and Badhwar, 1987). So much information is carried in the spectral-temporal domain that it overrides the noise provided by differing spacecraft sensors and atmospheres. Surprisingly, there has been comparatively little application or development of these scene models to natural vegetation and examples are few (Honey and Tapley, 1981; Tucker et al., 1985; Justice and Hiernaux, 1986).

Calibration

The synchronous acquisition of spectral and ecological data to calibrate scene, atmosphere, and sensor models is an essential part of remote sensing. Current advances in radiometry and instrumentation are such that the technological aspects of data collection are no longer limiting. Sampling and experimental design remain the greatest challenge. A significant contribution has been made in this area by Curran and co-workers (Curran and Williamson, 1985, 1986, 1987a, 1987b, 1987c).

Conclusions

Because of the depredations of the human population and the greenhouse-driven climate change, assessment and monitoring of the biosphere now loom large in the priorities of mankind. Understanding the behavior of the biosphere and its interaction with the atmosphere is an ecological problem of some magnitude. It is not an impossible task, just difficult. To attack this problem, ecologists must arm themselves with tools commensurate to the task. The most powerful and readily available tool is remote sensing. Ecologists must come to grips with this tool, and the foremost question must be: "What are the consequences of having to use remotely sensed data as the basic data source for the terrestrial vegetation component of the biosphere?"

The answer is a fundamental reorientation of the methodology of describing and classifying landscapes from the traditional lateral, oblique view to the vertical perspective of spacecraft. Ecologists must come to terms with the way in which data are collected by remote sensing systems. They require a working familiarity with the spectral, spatial, and temporal dimensions of current and future earth-observing systems. Above all, they must recast their ecological problems into these three dimensions of remote sensing.

The science of global ecology should not be overwhelmed by the technology of the remote sensing industry. Science, unlike technology, is still driven by fundamental questions, which, because they are basic, can be simply expressed. The questions for global ecology, as for any other problem, are: What do I want to know and how well do I want to know it?"

The difficult but challenging and exciting task is then to devise the scene models, the spectral, spectral-temporal, or spectral-spatial models, to answer these questions.

References

Aase, J.K., Frank, A.B., and Lorenz, R.J. (1987). Radiometric reflectance measurements of northern great plains rangeland and crested wheatgrass pastures. *J. Range Manag.* 40:299–302.

Aase, J.K., Millard, J.P., and Brown, B.S. (1986). Spectral radiance estimates of leaf area and leaf phytomass of small grains and native vegetation. *IEEE Trans. Geosci. Remote Sens.* GE-24:685–692.

Asrar, G., Kanemasu, E.T., Miller, G.P., and Weiser, R.L. (1986). Light interception and leaf area estimates from measurements of grass canopy reflectance. *IEEE Trans. Geosci. Remote Sens.* GE-24:76–82.

Badhwar, G.D., MacDonald, R.B., Hall, F.G., and Carnes, J.G. (1986a). Spectral characterization of biophysical characteristics in a boreal forest: relationship between thematic mapper band reflectance and leaf area index for aspen. *IEEE Trans. Geosci. Remote Sens.* GE-24:322–326.

Badhwar, G.D., MacDonald, R.B., and Mehta, N. (1986b). Satellite-derived leaf-area-index and vegetation maps as inputs to global carbon cycle models—a hierarchical approach. *Int. J. Remote Sens.* 7:265–281.

Bartlett, D.S., Hardisky, M.A., Johnson, R.W., Gross, M.F., Klemas, V., and Hartman, J.M. (1988). Continental scale variability in vegetation reflectance and its relationship to canopy morphology. *Int. J. Remote Sens.* 9:1223–1241.

Becker, F., and Choudhury, B.J. (1988). Relative sensitivity of normalised difference index (NDVI) for vegetation and desertifiation monitoring. *Remote Sens. Envir.* 24:297–311.

Clevers, J.G.P.W. (1988). The derivation of a simplified reflectance model for the estimation of leaf area index. *Remote Sens. Envir.* 25:53–69.

Curran, P.J., and Wardley, N.W. (1988). Radiometric leaf area index. *Int. J. Remote Sens.* 9:259–274.

Curran, P.J., and Williamson, H.D. (1985). The accuracy of ground data used in remote-sensing investigations. *Int. J. Remote Sens.* 6:1637–1651.

Curran, P.J., and Williamson, H.D. (1986). Sample size for ground and remotely sensed data. *Remote Sens. Envir.* 20:31–41.

Curran, P.J., and Williamson, H.D. (1987a). Airborne MSS data to estimate GLAI. *Int. J. Remote Sens.* 8:57–74.

Curran, P.J., and Williamson, H.D. (1987b). GLAI estimation using measurements of red, near-infrared, and middle infrared radiance. *Photogram. Eng. Remote Sens.* 53:181–186.

Curran, P.J., and Williamson, H.D. (1987c). Estimating green leaf area index of grassland with airborne multispectral scanner data. *Oikos* 49:141–148.

Delcourt, H.R., Delcourt, P.A., and Webb, T. (1983). Dynamic plant ecology: the spectrum of vegetation change in space and time. *Quat. Sci. Rev.* 1:153–175.

FAO. (1973). *FAO Manual of Forest Inventory, with special reference to mixed tropical forests.* UN, FAO, Rome, Italy.

Foran, B.D. (1987). Detection of yearly cover change with Landsat MSS on pastoral landscapes in central Australia. *Remote Sens. Envir.* 23:333–350.

Foran, B.D., and Pickup, G. (1984). Relationship of aircraft radiometric measurements to bare ground on semi-desert landscapes in central Australia. *Austral. Rangelands J.* 6:59–68.

Franklin, J., Logan T.L., Woodcock, C.E., and Strahler, A.H. (1986). Coniferous forest classification and inventory using Landsat and digital terrain data. *IEEE Trans. Geosci. Remote Sens.* GE-24:139–149.

Franklin, J., and Strahler, A.H. (1988). Invertible canopy reflectance modelling of vegetation structure in semiarid woodland. *IEEE Trans. Geosci. Remote Sens.* GE-26:809–825.

Gallo, K.P., and Eidenshink, J.C. (1988). Differences in visible and near-IR responses, and derived vegetation indices, for the NOAA-9 and NOAA-10 AVHRRs: A case study. *Photogram. Eng. Remote Sens.* 54:485–490.

Gillison, A.N., Walker, J. (1981). Woodlands. pp. 177–197. In R.H. Groves (ed.), *Australian Vegetation.* Cambridge University Press, Cambridge, England.

Goward, S.N., Dye, D., Kerber, A., and Kalb, V. (1987). Comparison of North and South American biomes from AVHRR observations. *Geocarto Int.* 1:27–39.

Goward, S.N., Tucker, C.J., and Dye, D.G. (1985). North American vegetation patterns observed with Nimbus-7 Advanced Very High Resolution Radiometer. *Vegetatio* 64:3–14.

Graetz, R.D., and Pech, R.P. (1988). The assessment and monitoring of sparsely vegetated rangelands using calibrated Landsat data. *Int. J. Remote Sens.* 9:1201–1222.

Graetz, R.D., Pech, R.P., Gentle, M.R., and O'Callaghan, J.F. (1986). The application of Landsat image data to rangeland assessment and monitoring: the development and demonstration of a land image-based resource information system (LIBRIS). *J. Arid Envir.* 10:53–80.

Hall, F.G., and Badhwar, G.D. (1987). Signature-extendible technology: global space-based crop recognition. *IEEE Trans. Geosci. Remote Sens.* GE-25:93–103.

Holben, B.N., and Fraser, R.S. (1984). Red and near-infrared response to off-nadir viewing. *Int. J. Remote Sens.* 5:145–160.

Holben, B.N., Tucker, C.J., and Fan, C.J. (1980). Assessing soybean leaf area and leaf biomass with spectral data. *Photogram. Eng. Remote Sens.* 26:651–656.

Honey, F.R., and Tapley, I.J. (1981). A vegetation response model applied to range inventory using Landsat MSS data. *Proc. 15th Int. Symp. on Remote Sensing of the Environment*, Ann Arbor, MI.

Jupp, D.L.B., Strahler, A.H., and Woodcock, C.E. (1988). Autocorrelation and regularization in digital images I: Basic theory. *IEEE Tran. Geosci. Remote Sens.* GE-26,463–473.

Justice, C.O., and Hiernaux, P.H.Y. (1986). Monitoring the grasslands of the Sahel using NOAA AVHRR data: Niger 1983. *Int. J. Remote Sens.* 7:1475–1497.

Justice, C.O., Townshend, J.R.G., Holben, B.N., and Tucker, C.J. (1985). Analysis of the phenology of global vegetation using meteorological satellite data. *Int. J. Remote Sens.* 6:1271–1318.

Karaska, M.A., Walsh, S.J., and Butler, D.R. (1986). Impact of environmental variables on spectral signatures acquired by the Landsat thematic mapper. *Int. J. Remote Sens.* 7:1653–1667.

Li, X., and Strahler, A.H. (1985). Geometrical-optical modelling of a conifer forest canopy. *IEEE Trans. Geosci. Remote Sens.* GE-23:705–721.

Malingreau, J.P. (1986). Global vegetation dynamics: satellite observations over Asia. *Int. J. Remote Sens.* 7:1121–1146.

Mathews, E. (1983). Global vegetation and land use: new high-resolution data bases for climate studies. *J. Clim. App. Meteorol.* 22:474–487.

Matson, M., and Holben, B. (1987). Satellite detection of tropical burning in Brazil. *Int. J. Remote Sens.* 8:509–516.

Matson, M., Stephens, G., and Roinson, J. (1987). Fire detection using data from the NOAA-N satellites. *Int. J. Remote Sens.* 8:961–970.

McLachlan, G.J., and Basford, K.E. (1988). *Mixture Models: Inference and Applications to Clustering*. Marcel Dekker, NY.

Mueller-Dombois, D. (1984). Classification and mapping of plant communities: a review with emphasis on tropical vegetation. pp. 21–88. In G.M. Woodwell (ed.), *The Role of Terrestrial Vegetation in the Global Carbon Cycle: Measurement by Remote Sensing*, SCOPE 23. Wiley, NY.

Musick, H.B. (1986). Temporal change of Landsat MSS albedo estimates in arid rangeland. *Remote Sens. Envir.* 20:107–120.

Nelson, R., Horning, N., and Stone, T.A. (1987). Determining the rate of forest

conversion I. Mato Grosso, Brazil, using Landsat MS and AVHRR data. *Int. J. Remote Sens.* 8:1767-1784.
Nelson, R., Krabill, W., and Tonelli, J. (1988). Estimating forest biomass and volume using airborne laser data. *Remote Sens. Envir.* 24:247-267.
Olson, J.S., Watts, J.A., and Allison, L.J. (1983). *Carbon in Live Vegetation of Major World Ecosystems.* ORNL-5862, Pub. No. 1997, Envir. Sci. Div., Oak Ridge Nat. Lab., Oak Ridge, TN.
Owe, M.F., Chang, A., and Golus, R.E. (1988). Estimating soil moisture from satellite microwave measurements and a satellite derived vegetation index. *Remote Sens. Envir.* 24:331-345.
Paris, J.F., and Kwong, H.H. (1988). Characterization of vegetation with combined thematic mapper (TM) and shuttle imaging radar (SIR-B) image data. *Photogram. Eng. Remote Sens.* 54:1187-1193.
Pech, R.P., and Davis, W.A. (1987). Reflectance modelling of semi arid woodlands. *Remote Sens. Envir.* 23:365-377.
Pech, R.P., Davis, A.W., Lamacraft, R.P., and Graetz R.D. (1986a). Calibration of Landsat data for sparsely vegetated semi-arid rangelands. *Int. J. Remote Sens.* 7:1729-1750.
Pech, R.P., Graetz, R.D., and Davis, A.W. (1986b). Reflectance modelling and the derivation of vegetation indices for an Australian semi-arid shrubland. *Int. J. Remote Sens.* 7:389-403.
Peterson, D.L., Spanner, M.A., Running, S.W., and Teuber, K.B. (1987). Relationship of thematic mapper simulator data to leaf area index of temperate coniferous forests. *Remote Sens. Envir.* 2:323-341.
Peterson, D.L., Westman, W.E., Stephenson, N.J., Ambrosia, V.G., Brass, J.A., and Spanner, M.A. (1986). Analysis of forest structure using thematic mapper simulator data. *IEEE Tran. Geosci. Remote Sens.* GE-24:113-121.
Pickup, G., and Foran, B.D. (1987). The use of spectral and spatial variability to monitor cover change on inert landscapes. *Remote Sens. Envir.* 23:351-363.
Ranson, K.J., and Doughtry, C.S.T. (1987). Scene shadow effects on multispectral response. *IEEE Trans. Geosci. Remote Sens.* GE-25:502-509.
Running, S.W. (1986). Global primary production from terrestrial vegetation: estimates integrating satellite remote sensing and computer simulation technology. *Sci. Total Envir.* 56:233-242.
Sellers, P.J. (1985). Canopy reflectance, photosynthesis and transpiration. *Int. J. Remote Sens.* 6:1335-1372.
Sellers, P.J. (1987). Canopy reflectance, photosynthesis, and transpiration. II. The role of biophysics in the linearity of their interdependence. *Remote Sens. Envir.* 11:171-190.
Sellers, P.J., and Dorman, J.L. (1987). Testing the simple biosphere model (SiB) using point micrometeorological and biophysical data. *J. Clim. Appl. Meteorol.* 26:622-651.
Specht, R.L. (1981). Foliage projective cover and standing biomass. pp. 10-21. In A.N. Gillison, and D.J. Anderson (eds.), *Vegetation Classification in Australia.* Australian Nat. Univ. Press, Canberra, Australia.
Strahler, A.H. (1981). Stratification of natural vegetation for forest and rangeland inventory using Landsat digital imagery and collateral data. *Int. J. Remote Sens.* 2:15-41.
Strahler, A.H., Woodcock, C.E., and Smith J.A. (1986). On the nature of models in remote sensing. *Remote Sens. Envir.* 20:121-139.
Suits, G., Malila, W., and Weller, T. (1988a). The prospects for detecting spectral shifts due to satellite sensor aging. *Remote Sens. Envir.* 26:17-29.
Suits, G., Malila, W., and Weller, T. (1988b). Procedures for using signals from one sensor as substitutes for signals of another. *Remote Sens. Envir.* 25:395-408.

Taconet, O., Carlson, T., Bernard, R., and Vidal-Madjar, D. (1986). Evaluation of a surface/vegetation parameterization using satellite measurements of surface temperature. *J. Clim. Appl. Meteorol.* 25:1752–1767.

Thomas, R.W., and Ustin, S.L. (1987). Discriminating semiarid vegetation using airborne imaging spectrometer data: a preliminary assessment. *Remote Sens. Envir.* 23:273–290.

Townshend, J.R.G., and Justice, C.O. (1986). Analysis of the dynamics of African vegetation using the normalized difference vegetation index. *Int. J. Remote Sens.* 7:1435–1446.

Townshend, J.R.G., and Justice, C.O. (1988). Selecting the spatial resolution of satellite sensors required for global monitoring of land transforms. *Int. J. Remote Sens.* 9:187–236.

Tucker, C.J., Fung, I.Y., Kealing, D.C., and Gammon, R.H. (1986a). Relationship between atmospheric CO_2 variations and a satellite derived vegetation index. *Nature* 319:195–199.

Tucker, C.J., Justice, C.O., and Prince, S.D. (1986b). Monitoring the grasslands of the Sahel 1984–1985. *Int. J. Remote Sens.* 7:1571–1582.

Tucker, C.J., Townshend, J.R.G., and Goff, T.E. (1985). African land-cover classification using satellite data. *Science* 227:369–375.

Van Gils, H.A.M.J., and Van Wijngaarden, W. (1984). Vegetation structure in reconnaissance and semi-detailed vegetation surveys. *ITC J.* 13:213–218.

Vujakovic, P. (1987). Monitoring extensive "buffer zones" in Africa: an application for satellite imagery. *Biol. Conserv.* 39:195–208.

Walker, J., and Hopkins, M.S. (1984). Vegetation. pp. 44–67. In R.C. McDonald, R.F. Isbel, J.G. Speight, J. Walker, M.S. Hopkins (eds.), *Australian Soil and Land Survey: Field Handbook*. Inkata Press. Melbourne, Australia.

Walker, J., Jupp, D.L.B., Penridge, L.K., and Tian, G. (1986). Interpretation of vegetation structure in Landsat MSS imagery: a case study in disturbed semiarid Eucalypt woodland. Part 1. Field data analysis. *J. Envir. Manag.* 23:19–33.

Wanjura, D.F., and Hatfield, J.L. (1988). Vegetative and optical characteristics of four row crop canopies. *Int. J. Remote Sens.* 9:249–258.

Weigand, C.L., Richardson, A.J. (1987). Spectral components analysis: rationale, and results for three crops. *Int. J. Remote Sens.* 8:1011–1032.

Weigand, C.L., Richardson, A.J., Jackson, R.D., Pinter, P.J., Aase, J.K., Smika, D.E., Lautenschlager, L.F., and McMurtrey, J.E. (1986). Development of agrometeorological crop model inputs from remotely sensed information. *IEEE Trans. Geosci. Remote Sens.* GE-24:90–98.

Williamson, H.D. (1988). Evaluation of middle and thermal infrared radiance in indices used to estimate GLAI. *Int. J. Remote Sens.* 9:275–283.

Wilson, M.F., Henderson-Sellers, A., Dickinson, R.E., and Kennedy, P.J. (1987). Sensitivity of the biosphere–atmosphere transfer scheme (BATS) to the inclusion of variable soil characteristics. *J. Clim. Appl. Meteorol.* 26:341–362.

Woodcock, C.E., and Strahler, A.H. (1987). The factor of scale in remote sensing. *Remote Sens. Envir.* 21:311–332.

Woodcock, C.E., Strahler, A.H., and Jupp, D.L.B. (1988a). The use of variograms in remote sensing I: Scene models and simulated images. *Remote Sens. Envir.* 25:323–348.

Woodcock, C.E., Strahler, A.H., and Jupp, D.L.B. (1988b). The use of variograms in remote sensing II: Real digital images. *Remote Sens. Envir.* 25:349–379.

3. Measurements of Surface Soil Moisture and Temperature

Thomas Schmugge

The monitoring of the energy and moisture fluxes between the soil and the atmosphere as well as of the water budget of the root zone in the soil is recognized as important for applications ranging from the study of biospheric processes at local scales (10s of meters) to the modeling of atmospheric behavior at regional scales (10s of kilometers). To estimate these land surface fluxes, it is necessary to determine the following quantities:

1. The energy driving forces, that is, the incident solar energy, surface albedo, and resulting net radiation.
2. The moisture availability or status in the soil and the vegetation/soil interaction.
3. The capacity of the atmosphere to absorb the flux, which depends on the surface air temperature, vapor pressure gradients, and surface winds.

From remotely sensed data it is possible to estimate surface parameters related to the soil/vegetation system, such as vegetation indices and surface soil moisture; components of the radiation forcing, such as solar insolation and surface albedo; and indicators of the response to it or surface temperature.

Measurement of the thermally emitted radiation at various wavelengths from the earth's surface can yield useful estimates of surface soil moisture and temperature. This radiation, which is emitted by any surface with a temperature above absolute zero, is described by the Planck/Blackbody equation:

$$E = C_1 k^3/(\exp(C_2 k/T) - 1) \qquad (1)$$

where C_1 is 1.191×10^{-8} W/(m² sr cm^{-4}), C_2 is 1.439 cm K, and k is the wave number (cm^{-1}). This gives the radiation for a perfect emitter or blackbody with an emissivity of one ($e = 1$). For real surfaces, $e < 1$. To estimate surface temperatures, radiation at wavelengths around 10 μm (micrometers) is used because:

1. The peak intensity of Equation (1) occurs in this region for terrestrial temperatures ($\approx$300 K).
2. The atmosphere is relatively transparent in this region.

Here the variations in the intensity of radiation are assumed to be primarily due to surface temperature variations. To estimate surface soil moisture, radiation at much longer wavelengths (10s of centimeters) is used, and in this case, the changes in the intensity of the emitted radiation are attributable to the variation of the emissivity with the moisture content of the soil. This effect results from the large dielectric contrast between water and dry soils at these longer wavelengths.

In this chapter we will discuss the basic principles for the remote sensing of these two parameters and give examples of measurement results.

Thermal Infrared

In Figure 3.1 we have plotted Equation 1 from 5 to 20 μm for temperatures of 280, 290, and 300 K (Kelvin) (7 to 27C), that is, the low range of terrestrial temperatures. At these temperatures, the peak of the emission occurs in the 8- to-10-μm range of wavelength. In this figure we have also plotted the atmospheric transmission for cloud-free conditions calculated with the Lowtran-6 path radiance model (Kniezys et al., 1983) for the U.S. standard mid-latitude summer atmosphere, assuming that the radiometer is at satellite altitude. As can be seen, the atmosphere is also relatively transparent in the 8-to-12-μm range, but that there is still a significant effect, that is, there is only 60 to 70% transmission, with a major dip at about 9.5 μm as a result of ozone absorption. With the exception of this dip, water vapor is the dominant absorber in the 8-to-12-μm window and is due to what is called the water vapor continuum and not any individual absorption lines. Thus the magnitude of the atmospheric effect will depend on the water vapor content of the intervening atmosphere for clear sky conditions. And this unknown or uncertain atmospheric contribution is one of the problems in the remote sensing of surface temperature.

Several approaches have been developed for eliminating atmospheric effects in the estimation of sea surface temperature from space. Here the problem is simpler in that the temperature does not change rapidly with time and a week's worth of data can be used to estimate the surface

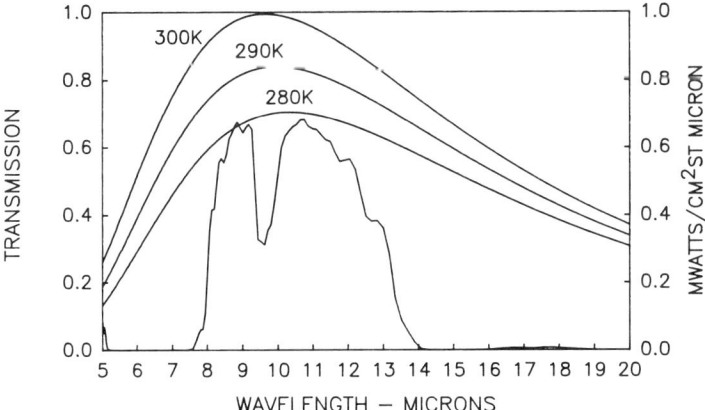

Figure 3.1. Calculated blackbody spectral radiance at three temperatures and atmospheric transmission for the U.S. standard mid-latitude summer atmosphere.

temperature. The technique used with the Advanced Very High Resolution Radiometer (AVHRR) data from the National Oceanic and Atmospheric Administration (NOAA) series of satellites involves the differential water vapor absorption in the 10-to-13-μm window, the so-called split-window technique (McClain et al., 1983). However, this assumes that the surface emissivity is constant over this spectral band, which, as we will see later, may not be the case for land surfaces (Price, 1984; Becker, 1987).

Examples of multispectral observations in the 8-to-12-μm region from an aircraft platform will be presented. These data were obtained with the National Aeronautics and Space Administration's (NASA) Thermal Infrared Multispectral Scanner (TIMS) during the HAPEX-MOBILHY experiment in 1986. The HAPEX-MOBILHY (Hydrologic Atmospheric Pilot Experiment—Modelisation du Bilan Hydrique) program is aimed at studying the hydrological budget and evaporation flux at the scale of a General Circulation Model (GCM) grid square, that is, 10^4 km^2 (Andre et al., 1986, 1988). The experiment was performed over a 100 by 100-km square in southwestern France in 1985 and 1986. As part of the program, several surface and subsurface networks were in operation from mid-1985 to early 1987, to measure and monitor soil moisture, surface energy budget, and surface hydrology, as well as atmospheric parameters. During the special observing period (SOP) from May 7 to July 15, 1986, additional intensive measurements were taken, including detailed measurements of atmospheric fluxes and remote sensing of surface properties, using two well-instrumented aircraft—the NCAR King-Air for flux measurements and the NASA C-130 for remote sensing observations. Results from the TIMS sensor on the latter aircraft will be discussed.

Thermal Infrared Multispectral Scanner (TIMS)

The TIMS is a six-channel scanner operating on the NASA C-130 aircraft in the thermal infrared (IR) (8 to 12 μm) region of the electromagnetic spectrum. The channels are at the following wavelengths (in micrometers): 8.2 to 8.6, 8.6 to 9.0, 9.0 to 9.4, 9.4 to 10.2, 10.2 to 11.2, and 11.2 to 12.2. The scan rate can be varied from 7.3 to 25 scans per second. The instantaneous field of view (IFOV) is 2.5 mrad; the detector analog signals are sampled every 2.08 mrad, yielding a scan of 638 pixels and covering 76.6 degrees (Kahle and Abbott, 1986; Palluconi and Meeks, 1985).

For calibration, the system is equipped with cold and warm reference sources or blackbodies, approximately covering the temperature range of the scene of interest and typically separated by 30 C. The temperatures of the references are known to better than 1 C (Palluconi and Meeks, 1985).

The scanner responds to the incident radiance (W/m^{2}sr cm^{-1}) and not to the temperature directly. The brightness temperature is related to the incident radiance via the Planck equation for blackbody radiation. However, if the observed temperatures are close to the calibration temperatures, say within 10 or 15 C, the error arising from using a linear temperature calibration is less than 1 C and that approximation will be used here (Schmugge and Janssen, 1988).

Sensitivity and Accuracy

In order to get a better understanding of the sensitivity and accuracy of the thermal data, the temperatures of a small reservoir (Lac de L'Uby) were determined. A water surface was selected for reasons of constant emissivity and a small temperature range. This lake was chosen because it was covered on almost every flight, for a total of ten passes. The data used were always at nadir, from a more or less square area of 110 to 930 pixels. The average temperature in each channel was determined. In general, the range of digital values over the lake is two or three counts, or about 0.6 to 1.0 C.

The TIMS temperatures represent the detected radiance at the aircraft altitude. In order to convert this observation to the actual surface temperature, atmospheric effects must be taken into account. These primarily include the absorption and emission by the atmospheric gases, mainly water vapor for this portion of the spectrum. Since the relationship between radiance and temperature can be approximated as linear over narrow temperature ranges (Price, 1984), Equation 2 can be used to correct for these atmospheric effects.

$$TIR = \tau T_s + (1 - \tau) T_a \quad (2)$$

Here, TIR is the temperature (degrees Kelvin) observed by the sensor, T_s is the surface temperature, T_a is the average air temperature for the atmospheric layer between the surface and the aircraft, and τ is the transmittance of this layer for a specific channel. For all the days, nearby radio soundings released within one hour of the pass were used to determine the transmittance and the air temperature. For calculating the average air temperature (T_a) the following equation was used:

$$T_a = \Sigma T_i^* W_i^* z_i / \Sigma W_i^* z_i \qquad (3)$$

in which T_i, W_i, and z_i are the temperature, water vapor content in kilograms per cubic meter (kg m³), and layer thicknesses of the atmosphere obtained from the radio soundings. This equation assumes that the principal absorption is by water vapor and so the temperatures are weighted by the vapor content of the layer. LOWTRAN-6, an atmospheric path radiance model developed by the Air Force Geophysics Laboratory (Kniezys et al., 1983), is used to calculate the transmittance for the different channels. The values of τ range from a low of 65% to a high of 91% and they were the greatest for channels 3 and 5 and lowest for channel 1. The values were generally correlated with the water vapor content (W) in the atmosphere as seen in Figure 3.2, where we have plotted the variation of τ versus the integrated water content for channels 1 (8.2 to 8.6 μm), 5 (10.2 to 11.2 μm), and 6 (11.2 to 12.2 μm) for the ten days on which we have coverage. The figure shows an approximate linear dependence of τ on W and the large difference between channels 1 and 5. Channel 1 is the most sensitive to the water vapor whereas channel 5 is one of the least sensitive, that is, most transparent. The TIMS channels 5 and 6 correspond approximately to channels 4 and 5 of the AVHRR, and thus Figure 3.2 demonstrates the differential sensitivity to water vapor that is used to compensate for the atmosphere in sea surface temperature determinations.

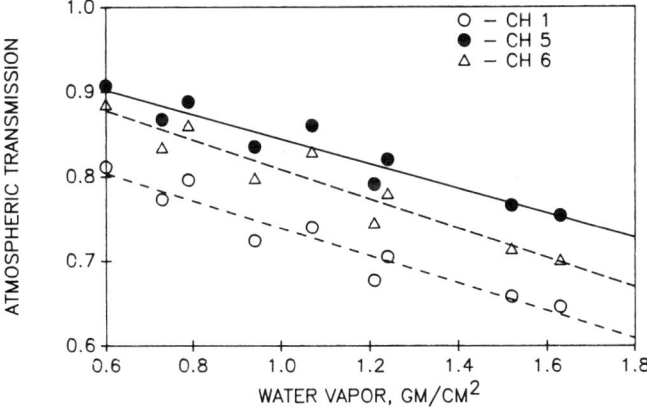

Figure 3.2. The variation of atmospheric transmission with water vapor for TIMS channels 1, 5, and 6.

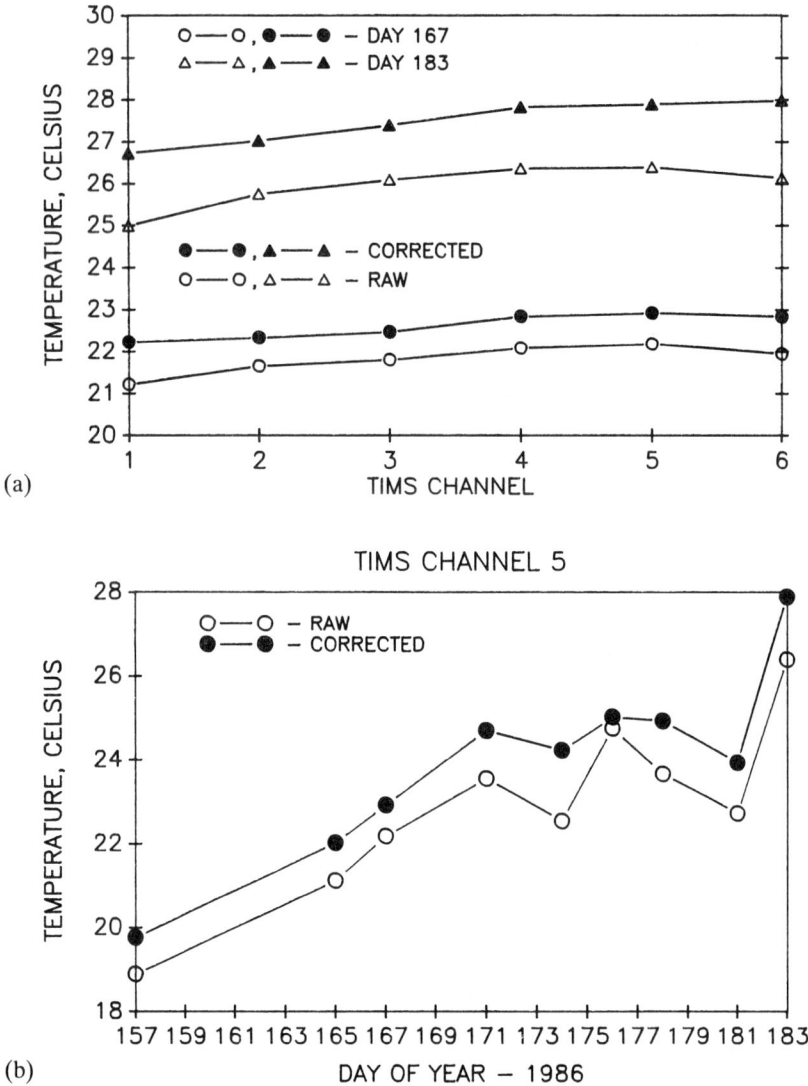

Figure 3.3. Water temperatures from TIMS for a small lake in HAPEX: (a) raw and atmosphere-corrected water surface temperatures for the six channels of TIMS on two days during HAPEX, and (b) raw and corrected temperatures for channel 5 for nine days.

Using Equation 2 and the calculated values of τ, the observed temperatures were corrected for the atmospheric effects. Because T_a was cooler than the lake surface, the effect of the correction is generally to increase the temperature for the water surface by about 0.7 to 2.9 C. The changes were the greatest for channel 1, which always had about a 10% lower transmittance than any other channel. Figure 3.3a presents plots of the raw and corrected temperatures for June 16 (day 167) and July 2 (day 183) as examples of the corrections. On day 167, T_a was 19 C, or about 3 C cooler that the water, and as a result the correction was small, less than 1 C. Also, the temperatures for the six channels were within 0.7 C of each other. On day 183, the difference between T_a and the water was greater, 5 C, and the transmittance was lower so the correction was larger, about 2 C, and the agreement among the channels was not as good, with a 1.2 C range. Another feature is that the corrected temperatures for channel 1 are usually significantly lower than those for the other channels, indicating that the Lowtran model is perhaps underestimating the atmospheric absorption for this channel. Figure 3.3b shows the variation of the water surface temperature before and after atmospheric correction for channel 5 for nine days between June 6 and July 2. The results show a more or less continuous increase with time, with the magnitude of the correction varying between 0.3 C and about 1.5 C. Unfortunately, there were no surface measurements available to verify the accuracy of these surface temperature estimates but the temperature trend is certainly in the proper direction.

Spatial Variations

For three days in June, the temperature distribution was determined for a 47-ha oat field. The field is located in the northern test site of HAPEX near the village of Lubbon. Two sites are defined for analysis—site A, which covers approximately half of the oat field, and site B, a smaller part of the oat field with a relatively high crop density and which generally appeared healthier than the rest. The approximate area of site A is 23 ha and that of site B is 0.6 ha. On June 7, the oat plants were measured and found to be about 80 cm high and on June 20 the canopy cover was estimated at 80%, with heights ranging from 90 to 130 cm; the crop was in good condition.

Multispectral data acquired by the NS001 scanner at the same time were used to determine the vegetation index (VI) of the oat.

$$VI = (\text{reflective IR} - R)/(\text{reflective IR} + R) \tag{4}$$

For the IR, the 0.767 to 0.910 μm band is used and for the red (R), the 0.633 to 0.697 μm band is used. On June 16, the VI at site A ranged from 0.12 to 0.63, while on June 23, the VI varied from 0.14 to 0.74. The large range in VI values indicates a nonhomogeneous crop cover for site A. At site B, the mean VIs on June 16 and 23 were 0.59 and 0.67, respectively, with a much smaller range of values.

Figure 3.4(a) is a histogram of the surface temperatures in the oat field (site A), as given by the scanner and not corrected for atmospheric effects. Figure 3.4(b) is the histogram for site B. In Table 3.1, the surface temperatures for both sites on the three days are listed together with the wind speed (meters per second) and (surface) air temperature obtained from a

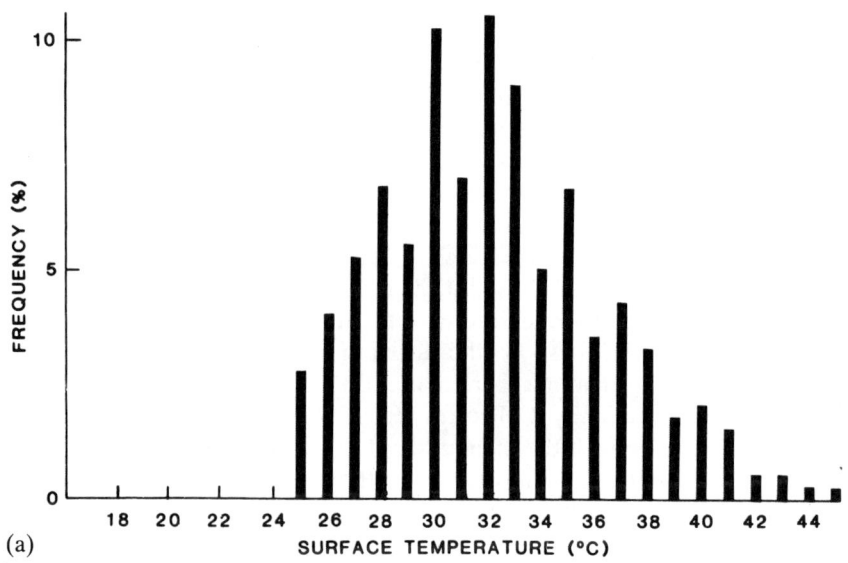

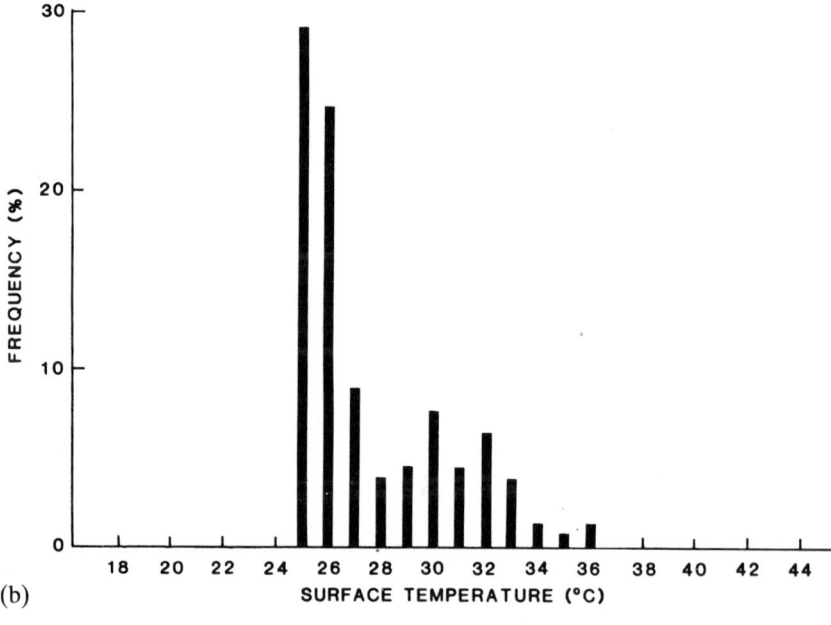

Figure 3.4. Histograms of TIMS channel 5 temperatures for the oat field at the central site of HAPEX: (a) for the larger portion of the field, and (b) for the small cool portion. Note that these temperatures were not corrected for atmospheric effects, so the high temperatures would actually be higher than the values shown.

3. Measurements of Surface Soil Moisture and Temperature

Table 3.1. Surface Temperatures (C) of Sites A and B

Day	T_s Mean	Site A Range	T_s Mean	Site B Range	Wind Speed at 3.2 m	T_{air} at Surface
June 14	22.1	18.3/36.7	21.7	18.3/26.6	0.8	22.0
June 16	32.2	25.1/48.0	21.7	25.5/36.2	3.5	28.5
June 23	30.0	21.3/45.1	25.5	21.3/33.1	5.7	23.5

surface flux station (Itier, 1982) that was placed in the oat field. The data are taken every 15 minutes and these values are valid for the time of the aircraft overpass. The temperature ranges for both sites A and B in the oat field are very large. As seen in Table 3.1, the temperature range for site A varies from 18.4 K on June 14 to 23.8 K on June 23. During this period, the canopy cover was estimated to be 80% but not very homogeneous from ground observations. On false color aerial photos, taken on June 16, most of the crop appeared healthy but there were portions with little green vegetation. Also, the VI as measured with the NS001 indicated a non-homogeneous crop distribution. On June 16 (at around 10:30 a.m.), the soil surface temperature was measured and found to range from 42 to 45 C in a bare portion of the field. It can be concluded that the temperatures of the pixels do not always represent the temperature of the crop. In particular, for the cases where the canopy cover is not 100%, the soil surface influences the TIMS data, and the remotely sensed surface temperatures cannot be taken for crop temperatures.

For site B, the ranges of temperatures were from 8.3 to 11.8 C for the three days, which is still a fairly large temperature range for a crop considered to be healthy. On June 18, for this organic sandy soil, the gravimetric soil moisture ranged from 5% to 8% for depths down to 50 cm, indicating that a shortage of moisture may have been a cause of the high temperatures. The histograms in Figure 3.4 indicate that there can be significant variations in surface temperatures for what had been considered a reasonably uniform field. Therefore, it would not be wise to extrapolate the flux measurements made in a healthy (cool) part of the field to the whole field without taking into account the variations in surface temperature using approaches such as those proposed by Nieuwenhuis et al. (1985), Sequin and Itier (1983), or Riou et al. (1988).

Spectral Variations

As we noted earlier, one of the problems in estimating land surface temperatures using remotely sensed data is the spectral variation of land surface emissivity, particularly for bare soils or exposed rocks. For vegetation, with its multiple surfaces of an absorbing material, we would expect that its emissivity would be close to one and uniform. Palluconi in (1988) reporting at the first FIFE results workshop, verified this conclusion.

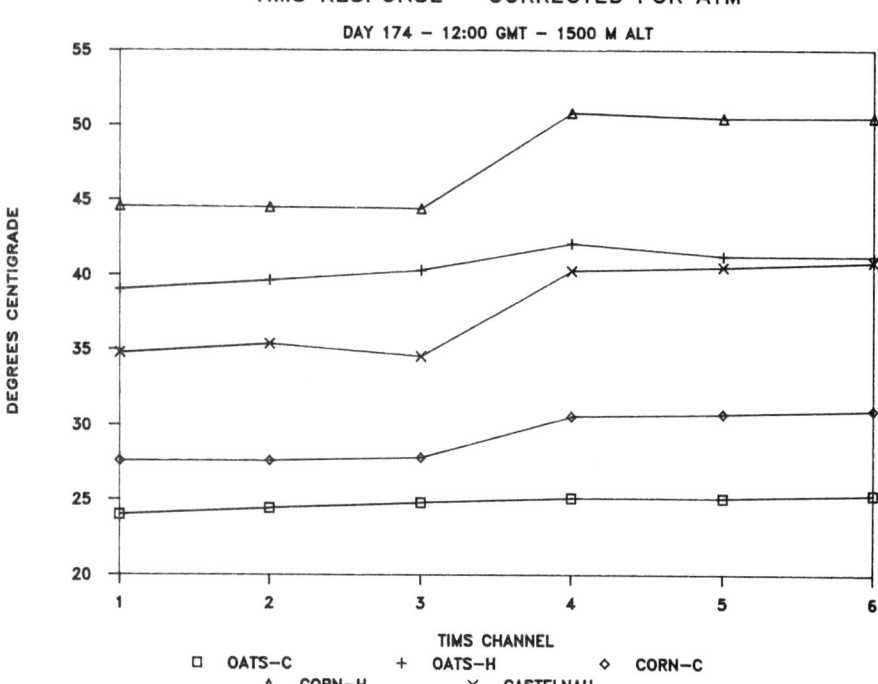

Figure 3.5. TIMS spectra for several surfaces from HAPEX showing the corrected temperatures for an oat and corn fields. The data for both fields have results from both a cold and a hot spot in each field.

Figure 3.5 is an example of the difficulty that might be expected when bare soils are present in the field of view. These are data from several fields in the HAPEX experiment. For the cool portion of the oat field discussed earlier, there is essentially no variation in the surface temperature across the six channels. However, for the hot portion of an adjacent cornfield, there is a 6 or 7 C increase for channels 4, 5, and 6 over the lower three channels. This field was planted in corn but had much exposed soil between the rows of plants, which were separated by 1 m. These data were from an altitude of 1.5 km with an IFOV of ≈4 m, thus the individual rows were not resolved. This spectral variation arises from the absorption of IR radiation owing to the stretching vibrations of the silicon-oxygen bonds of the silicates. These are the restrahlen bands and are most pronounced in quartz. Christianson et al. (1986) and Gillespie (1986) report variations in emissivity from <0.8 for channels 1, 2, and 3 of TIMS to >0.9 for the higher three channels for quartzite and quartz sand. This is qualitatively the behavior observed in Figure 3.5 for fields that have much exposed soil. Because of this strong spectral difference between the soil and vegetation, the TIMS data may be useful for estimating the amount of vegetation cover present.

Since there was no difference observed in emissivity for channels 5 and 6, these results indicate that it may be possible to use differential absorption by water vapor for these channels, seen in Figure 3.2, to compensate for atmospheric effects in surface temperature determinations for this region.

Discussion

In this section, we have presented results from aircraft observations of surface radiometric temperatures and briefly discussed their possible use. There are two major problem areas: atmospheric effects for converting the aircraft observation radiance to a surface radiant temperature and the uncertainty of surface emissivity for converting the radiant temperature to a physical temperature. By using either the split-window approach or some ancillary knowledge of the atmospheric profiles, it should be possible to estimate the surface radiant temperature to an accuracy of 1 or 2 C (Price, 1984). A possible problem with split-window technique is the variation of the surface emissivity with wavelength over the two bands. For vegetation, we have seen that there is litle or no emissivity variation, but for exposed soil, some variation was observed, however, not in the 10-to-12-μm range. In general, this may be a problem.

At present, there is a continuing source of thermal IR data from space, namely, the AVHRR instrument on board the NOAA polar orbiting satellites. This sensor has two channels approximately equivalent to the TIMS channels 5 and 6 with a spatial resolution of 1 km. There are usually two satellites in orbit, one with an equator crossing of mid-morning (9:30) and the other at mid-afternoon (14:30). The latter time is near the maximum in surface temperature and, therefore, is the most useful for detecting moisture stress in vegetation. There is almost daily coverage with this sensor.

Microwave Sensing of Soil Moisture

In the introduction, we noted that the emission of microwave radiation from a soil depends on its moisture content because of the large contrast between the dielectric constant of water ($\approx$80) and that of dry soil (3.5). This arises from the ability of the electric dipole of the water molecule to align itself in response to the oscillating electric field in an electromagnetic wave. Figure 3.6 is a plot of the real and imaginary parts of the dielectric constant for water in the microwave portion of the spectrum. It shows a large value ($\approx$80) of the real part at low frequencies, with a transition to a much smaller value at the high frequencies. At the higher frequencies, the

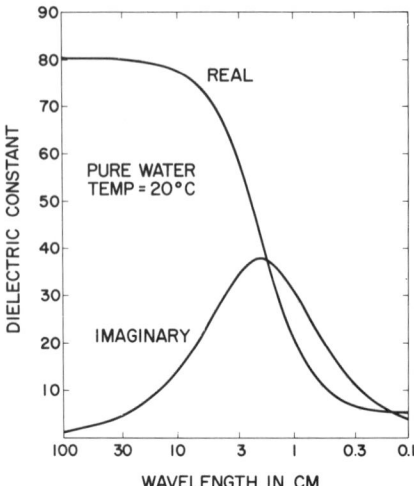

Figure 3.6. The real and imaginary parts of the dielectric constant for water in the microwave frequency range.

fields oscillate so rapidly that the dipole of the water molecule cannot follow, and thus its contribution to the dielectric constant decreases. For ice, the dielectric constant is large at low frequencies, but because of the tighter binding of the water molecule in the solid, the motion is inhibited at about 10^4 Hz (hertz). As seen in Figure 3.6, for liquid water this reduction in the molecule's ability to rotate does not occur until about 10^{10} Hz. Since the first water added to a soil is tightly bound, its rotation will be inhibited in a fashion similar to the behavior in ice.

The dielectric constant is of importance here because it describes the propagation characteristics of an electromagnetic wave in the medium. These characteristics include the velocity of propagation, the wavelength in the medium, and the absorption of energy in the medium. The square root of the dielectric constant is the index of refraction (n) for the material, and it is the contrast in n at the boundary between two media that determines the reflection and transmission coefficients of electromagnetic waves at such a boundary.

In this chapter, we will consider the frequency range from 1 to 100 GHz (gigahertz) (1 GHz = 10^9 Hz; wavelengths between 30 and 0.3 cm) and show that frequencies below 5 or 6 GHz are most effective for soil moisture sensing. This variation of the soil's dielectric constant with moisture produces a variation in emissivity from 0.95 for dry soils to 0.6 or less for wet soils, with changes of a corresponding magnitude for the soil's reflectivity. These variations have been observed by both passive and active microwave sensors. The former are radiometers that observe the variations in the thermal emission from the soil resulting from emissivity changes. The latter are radars that transmit a pulse of electromagnetic energy and then measure the backscattered return, which will be a function of the soil's reflectivity. This capability to sense soil moisture remotely is limited to a surface layer

3. Measurements of Surface Soil Moisture and Temperature

about 5 cm thick, and is affected by surface properties such as roughness and vegetation cover.

Dielectric Properties of Soils

When water is mixed with soil, the dielectric constant of the mixture increases dramatically from about 3 or 4 when dry to about 30 when wet. Figure 3.7 presents measurements for three soils ranging from a sandy loam to a heavy clay. These measurements were at a frequency of 1.42 GHz or a wavelength of 21 cm. Qualitatively, the behavior of the three soils is similar in that the real part of the dielectric varies from about 3 to greater than 20. However, it is clear that there are differences among the soils. All the soils have regions of slow increase and then, after a transition value, a region of

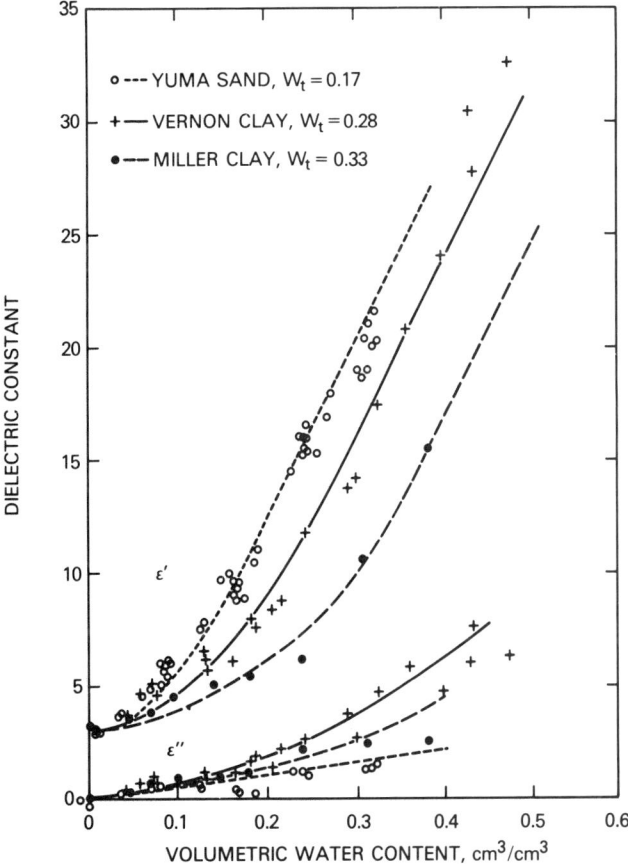

Figure 3.7. Laboratory measurements of the dielectric constant for three soils ranging from a light sand to a heavy clay at a frequency of 1.4 GHz

sharp increase. This flat region is a function of soil type, being wider for the heavier soils (Wang and Schmugge, 1980). The first water added to a soil is tightly bound to the soil surface and has dielectric properties somewhere between those of bound molecules in ice and those of the freely rotating molecules in the liquid. It is only after there are several layers of water on the particle surface that the water begins to behave like a liquid in terms of its dielectric properties and to produce a large change in the dielectric properties of wet soils.

Dobson et al. (1985) have proposed a four-component dielectric mixing model that describes the soil-water system as a host medium of dry soil solids containing randomly distributed and oriented disk-shaped inclusions of bound water, bulk water, and air. The separation of the soil water into bound and bulk components depends on the soil texture or more directly on the specific surface area of the soil. Such models allow the study of the sensitivity of a soil's dielectric properties on such things as density, texture, and salinity. For example, Figure 3.8(a) shows the dependence of the dielectric constant on density for a sandy loam soil. At zero soil moisture, the differences among the three curves are due to the different amounts of parent material present at each density. The range of densitites shown produces a 15% to 20% spread of values in the real part for wet soils. The spread in the imaginary part is almost negligible. Figure 3.8(b) shows the sensitivity to specific surface or texture at a density of 1.3 g/c^3. The soils range from a loamy sand to a clay and exhibit the type of behavior shown in Figure 3.7, that is, the flat region at low soil moistures for the clay soil.

Emissivity and Reflectivity

The reflectivity and emissivity for a soil with a smooth surface can be calculated with the Fresnel equations from electromagnetic theory. The reflectivity for the vertical polarization is:

$$R_v = |[\cos\theta - \sqrt{(k - \sin^2\theta)}]/[\cos\theta + \sqrt{(k - \sin^2\theta)}]|^2 \tag{5}$$

and for the horizontal polarization the equation is:

$$R_h = |[k\cos\theta - \sqrt{(k - \sin^2\theta)}]/[k\cos\theta + \sqrt{(k - \sin^2\theta)}]|^2 \tag{6}$$

where θ is the incidence angle and k is the dielectric constant for the soil. The horizontal polarization has the electric field of the wave parallel to the soil surface and the vertical polarization has a component of the electric field perpendicular to the soil surface. The resulting emissivities are as follows:

$$e_{h,v} = 1 - R_{h,v} \tag{7}$$

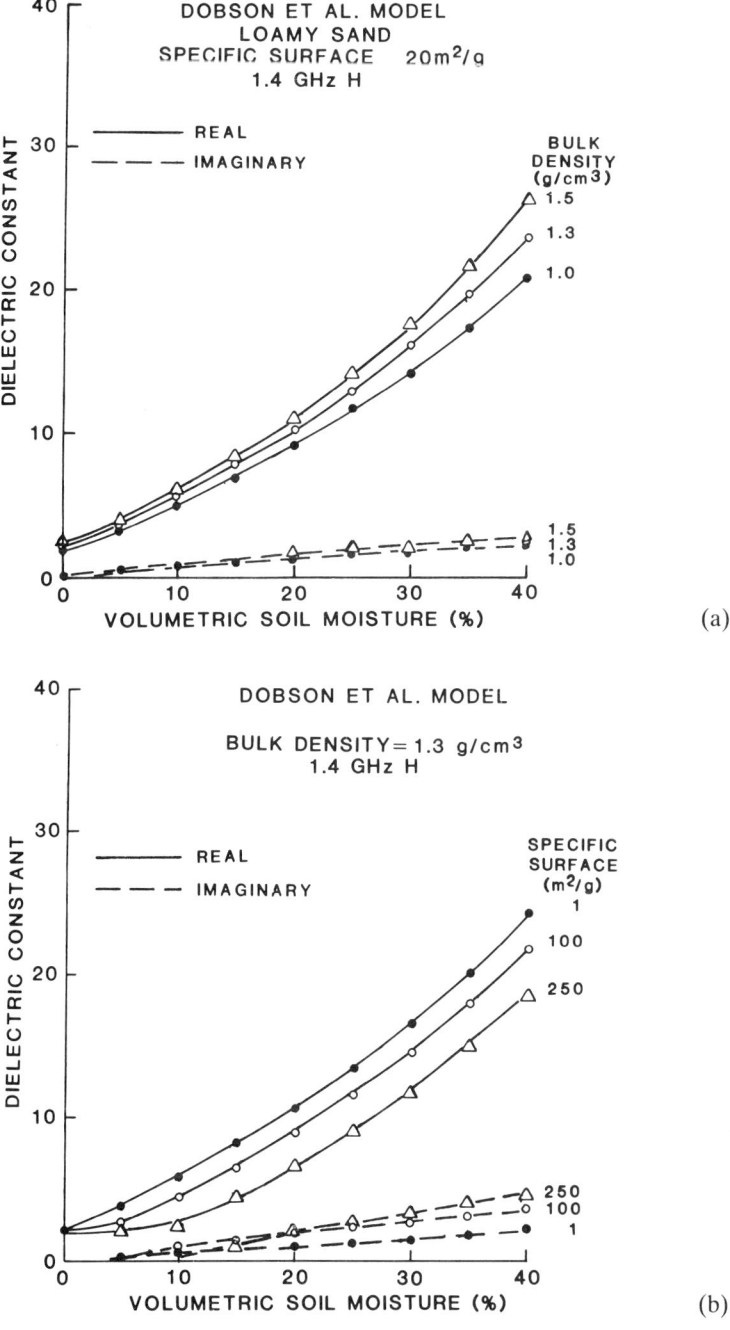

Figure 3.8. Calculations using the Dobson model of the dielectric constant showing: (a) the dependence on bulk density for a loamy sand soil, and (b) the dependence on the specific surface or texture for a soil. The soils range from a sand (1) to clay loam (250). The symbols are not measurements but just the points at which the calculations were done.

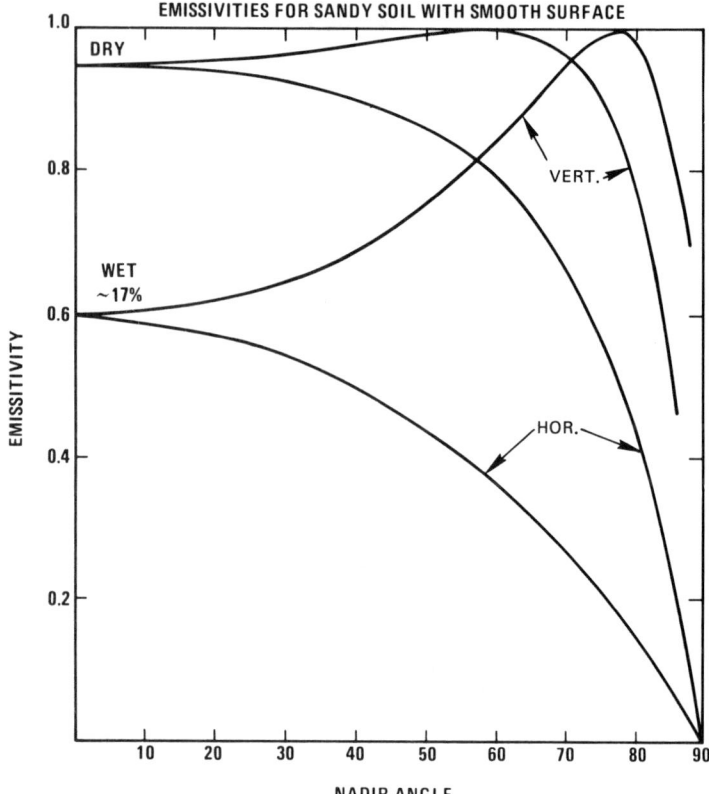

Figure 3.9. Calculations from the Fresnel equations of the emissivity for wet and dry sandy soils with smooth surfaces. The Brewster's angle for the dry soil is about 58 degrees and for the wet soil is 78 degrees.

Figure 3.9 is a plot of these equations for wet and dry soils. The plot shows that at nadir there is a decrease in the emissivity from 0.95 when dry to 0.6 when wet. While e_h decreases with θ, the difference between the wet and dry emissivities remains about the same, indicating the sensitivity of e_h to soil moisture independent of angle out to about 40 or 50°. In contrast, as a θ increases, e_v approaches 1 at what is called the Brewster's angle given by

$$\tan \theta_B = \sqrt{K} \tag{8}$$

Thus the curves for the wet and dry soils come together, indicating a loss of sensitivity. For this reason, the horizontal polarization is recommended for soil moisture sensing.

Microwave Radiometry

A microwave radiometer measures the thermal emission from the surface and at these wavelengths the intensity of the observed radiation is proportional to the product of the thermodynamic temperature of the soil and the surface emissivity (Rayleigh-Jeans approximation to the Planck radiation law, Equation 1). This product is commonly called the brightness temperature (T_B) and is given by the product of the emissivity and the soil temperature:

$$T_B = eT_{soil} \qquad (9)$$

For a radiometer at some height above the ground, atmospheric effects must be included, yielding the equation

$$T_B = \tau(RT_{sky} + (1-R)T_{soil} + T_{atm} \qquad (10)$$

where R is the surface reflectivity and τ is the atmospheric transmission. This situation is represented schematically in Figure 3.10. The first term is the reflected sky brightness, which depends on the atmospheric conditions and frequency. For the frequencies of interest to us, T_{sky} = 5 to 10 K for the normal range of atmospheric conditions, with 3 K of it being the constant cosmic background radiation. The third term is the direct atmospheric contribution and, as noted, will be about 5 K. The atmospheric transmission will typically be about 99% so we are left with the emission from the soil, that is, the second term as the main contributor to T_B. Thus Equation 10 reduces to Equation 9. This is true for an isothermal soil; however,

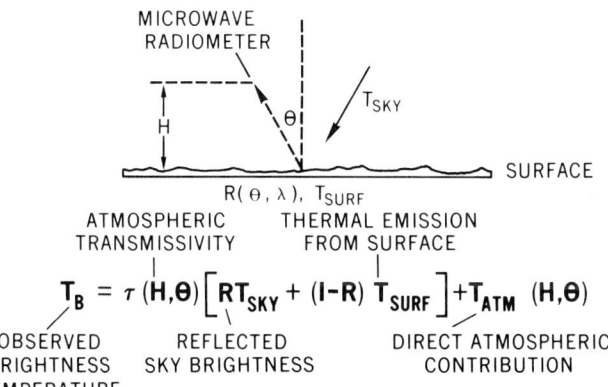

Figure 3.10. Schematic of the sources of microwave radiation observed by a radiometer at a height h and an angle θ.

when there are variations of soil temperature with depth, the factor eT_{soil} is replaced by the integral:

$$T_B = e \int_{-\infty}^{0} T(z)\alpha(z) \int_{z}^{0} \exp[-\alpha(z')dz']dz \qquad (11)$$

where $T(z)$ and $\alpha(z)$ are the temperature and absorptivity at the depth z in the soil. The latter depends on the moisture content at z. The integral essentially gives the intensity of the upwelling thermal radiation from the soil and is a weighted average of the soil temperature over the electromagnetic skin depth of the soil. The factor e, the emissivity, gives the fraction of this upwelling radiation that is transmitted into the air and it is this factor that shows the primary sensitivity to the soil moisture content. The critical factor is: What is the thickness of the layer at the surface whose dielectric properties determine the emissivity, e. It is this thickness that determines the soil moisture sampling depth for microwave sensors. An early theoretical study of the radiative transfer in a soil by Wilheit (1975) has shown that this depth is only of the order of a few tenths of a wavelength thick, or about 2 to 5 cm at the 21-cm wavelength. This result has been confirmed by the experimental work of Newton et al. (1982) and Wang (1987).

In Newton's work, the dry down of a saturated field was observed with radiometers at three frequencies (1.4, 5, and 10.7 GHz) and the temporal variation of the microwave responses was compared with those observed for the soil moisture measured in 0- to 9-, 0- to 5-, and 0- to 2-cm layers. It was found that the highest two frequencies indicate drying at a rate faster than that seen in the 0- to 2-cm layer whereas the 1.4-GHz results indicated that it is sensing a layer that is drying at a rate between those for the 0- to 2- and 0- to 5-cm layers. The conclusion was that the higher two frequencies were sampling a layer thinner than 2 cm while the 1.4 GHz was sampling a layer whose thickness was between 2 and 5 cm.

Wang's approach was to use a large number of field measurements of both the microwave emission and the soil moisture profile. Using Wilheit's model with the measured profiles, he calculated the expected emissivities and compared them with the soil moistures in several layers. By comparing these curves with measured emissivity variation with soil moisture, he concluded that the 1.4-GHz radiometer was responding to the moisture in a layer between 2 and 5 cm thick.

Field Verification

As part of a study of salinity effects on the soil's emissivity, Jackson and O'Neill (1987) made a careful series of field measurements over loamy sand soil having a smooth surface and compared the results with those expected from the dielectric constant models described above. The results

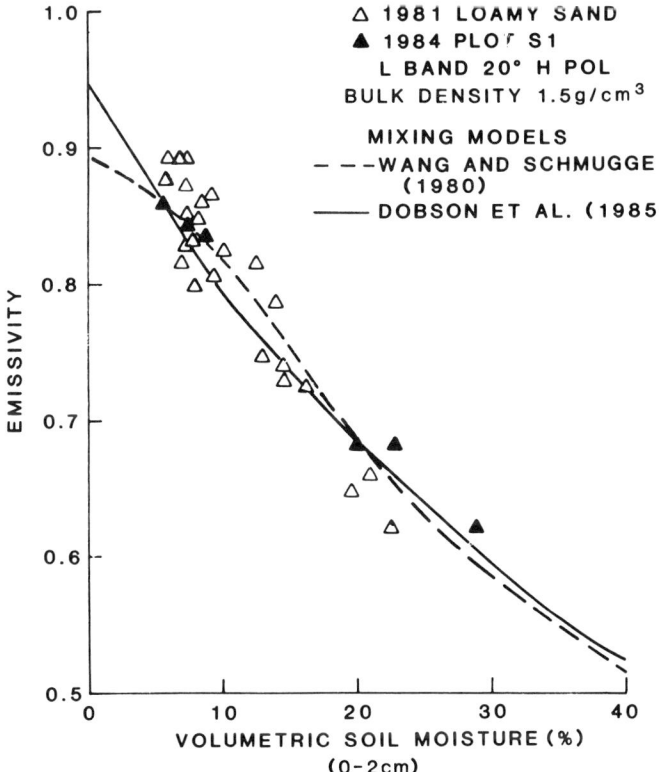

Figure 3.11. Observed and predicted relationships between emissivity and volumetric soil moisture for a bare, smooth, no-salinity loamy sand soil at 1.4 GHz (L-band). [From Jackson and O'Neill (1987).]

are shown in Figure 3.11. The emissivity ranges from 0.6 for the wet soil ($\approx$30% volumetric soil moisture) to 0.9 for the dry soil ($\approx$8%). The calculations from the two models agree reasonably well with each other and with the data. Hence the basic sensitivity of microwave emissivity to soil moisture variations is well understood and the basic theory is verified. Complications arise when such real factors as surface roughness and vegetative cover are added to the problem.

Surface Roughness

We have seen that the emissivity of a smooth surface can be calculated using the Fresnel equations (Equations 5 and 6); however, when the surface is rough, the situation is more complicated because the roughness decreases the reflectivity and thus increases the emissivity. This results

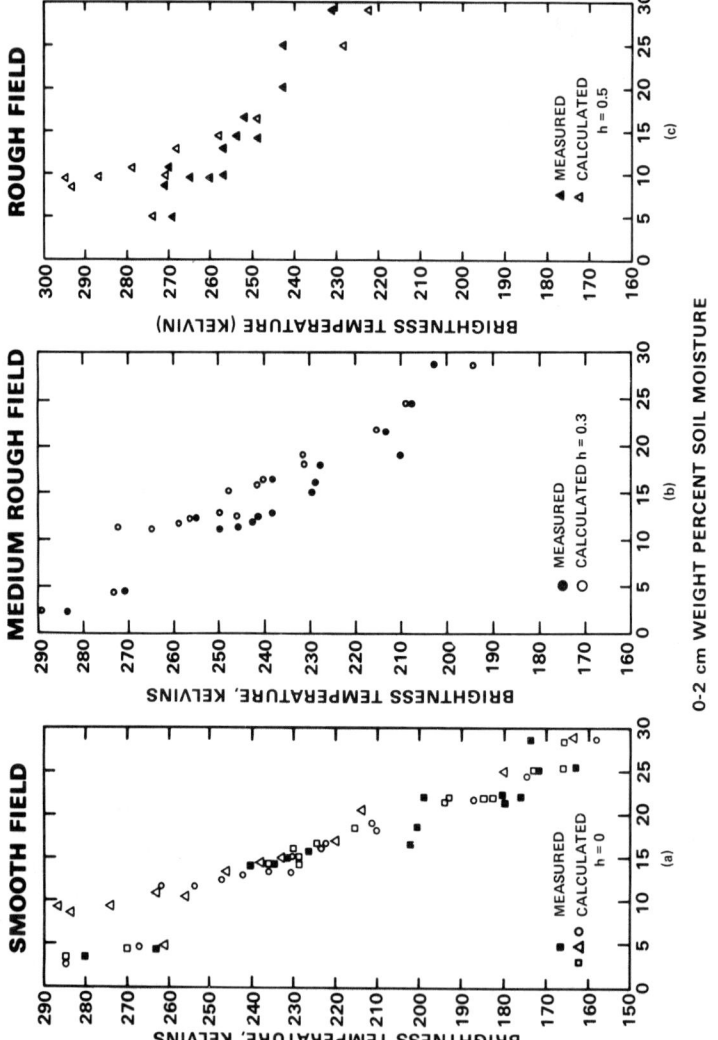

Figure 3.12. Roughness effects on the microwave emission from soils at a wavelength of 21 cm. The solid symbols are the measured values and the open symbols are calculations from Equation 13 using $h = 0$ in (a), $h = 0.3$ in (b), and $h = 0.5$ in (c). [From Choudhury et al. (1979).]

3. Measurements of Surface Soil Moisture and Temperature

from the increased surface area that can emit radiation. There have been numerous attempts to calculate quantitatively the effects of roughness, but the efforts are hampered by the difficulty of experimentally quantifying the surface roughness.

The magnitude of the effect is shown in Figure 3.12, which presents the results of field measurements at the 21-cm wavelength for three surface roughnesses classified as smooth, medium rough, and rough. These fields have surfaces with root mean square (rms) height variations of 0.92, 2.6, and 4.3 cm, respectively. The range of T_B between wet and dry conditions decreases from about 120 K for the smooth field to 40 K for the rough field. Using a simple model, Choudhury et al. (1979) have shown the increase in emissivity is given by:

$$e = R_o[1 - \exp(-h)] \qquad (12)$$

where R_o is the smooth surface reflectivity and h is an empirically derived roughness parameter that is proportional to the rms height variations with $h = 0$ for a smooth surface. For dry fields $R_o < 0.1$, the effect of roughness on the emissivity will be small, whereas for wet fields $R_o = 0.4$, the effect is larger and readily observable. Thus in Figure 3.12 there is little difference in T_B for the dry field conditions while there is about a 60 K effect for the wet field. These resuls were for fields that were specially prepared for roughness studies and represent the extremes of surface conditions, and thus it is expected that while relatively smooth surfaces do occur, the roughest condition would rarely be encountered, except, for example, immediately after deep plowing. Figure 3.13 is an example of results for more normal conditions. These were data acquired from an aircraft platform at an altitude of 300 m over a typical agricultural region in the great plains of the United States. The data were segregated according to a qualitative assessment of roughness from ground photographs and it is seen that the range of variation attributable to roughness is small for these more typical conditions. The regression slopes were not significantly altered by the roughness and there was an 8 K range in the intercepts from the smoothest to roughest. These fields covered the range of roughnesses that normally occur in an area that is mostly pastures or planted in small grains. These results point out that while roughness can have a very significant effect, its naturally occurring range may not be as great as the extremes observed in the prepared plots for truck experiments.

It is also possible to use supporting data from other sensors to correct for roughness effects. In analyzing a data set containing both 1.6-GHz scatterometer (radar) and 1.4-GHz radiometer data, Theis et al. (1986) observed that the influence of surface roughness on the radar backscatter ($\sigma°$) was substantial at large incidence angles whereas the sensitivity of $\sigma°$ to soil moisture was small at these angles. They plotted $\sigma°$ versus soil moisture at a 40-degree incidence angle and determined the intercept or 0%

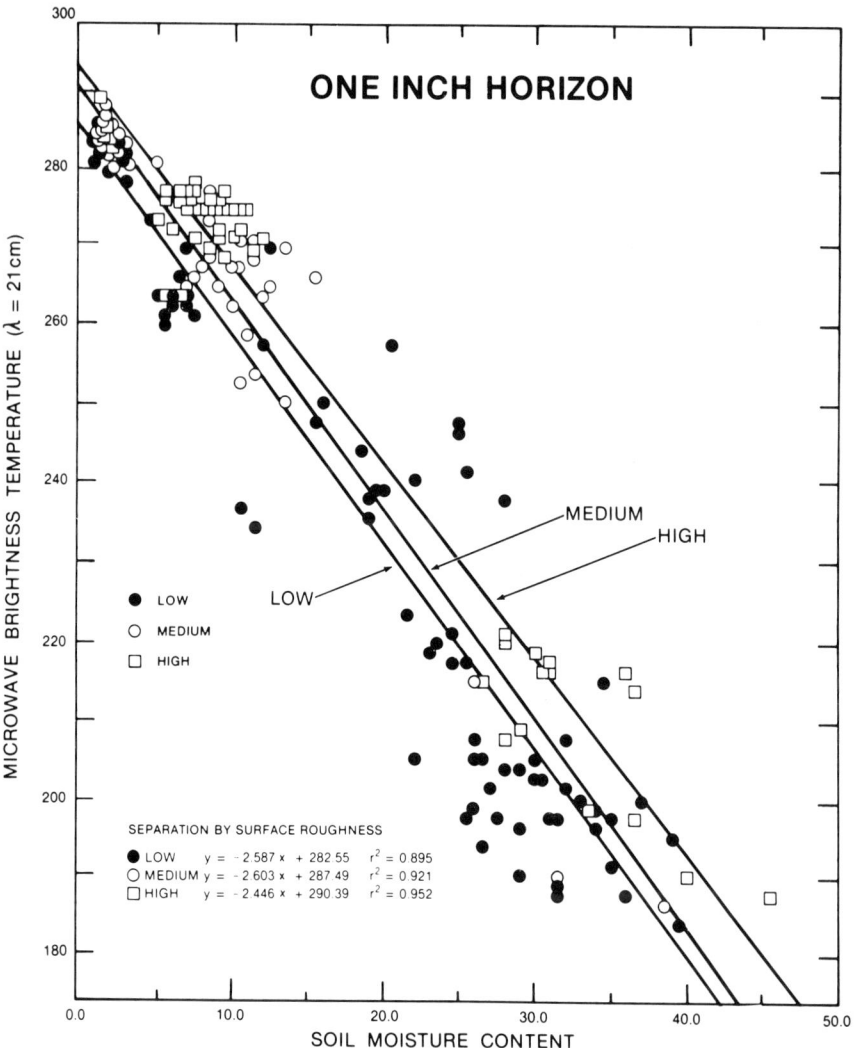

Figure 3.13. Scatter plot of T_B versus soil moisture for aircraft observations at the 21-cm wavelength over a test site in Hand County, South Dakota. The data were segregated into three roughness categories based on ground photographs. [From Owe and Schmugge (1983).]

soil moisture value of σ°, which was then used to correct for roughness in the radiometer data. Using this approach, they improved the correlation (R^2) of T_B with soil moisture from 0.69 to 0.95. Although conducted on only a limited data set, these encouraging results indicate the potential of multisensor techniques for improving microwave estimates of soil moisture.

Vegetation Effects

The simplest way to think of the effects of vegetation on the microwave emission from the soil is in terms of a cloud that can absorb and reemit radiation, that is, similar to the atmospheric situation. This assumes that scattering is negligible, which is true at the longer wavelengths. For a sufficiently thick layer of vegetation, only the radiation from the vegetation itself will be observed. The brightness temperature above the canopy is given by:

$$T_{Bp} = T_V[1 + (e_s - 1)\exp(-2\tau) + e_{sp}(T_S/T_V - 1)\exp(-\tau)] \quad (13)$$

where $\tau = l\alpha \sec \theta$ is the one-way canopy absorption factor or optical depth, l and α are the thickness and absorptivity of the vegetation, $p = h$ or v for the polarization index, and T_S and T_V are the temperatures of the soil and the vegetation (Jackson et al., 1982), respectively. If we assume that the vegetation and soil temperatures are approximately equal, this equation reduces to

$$e_p = T_B/T_S = 1 + (e_{sp} - 1)\exp(-2\tau) \quad (14)$$

Since the temperatures are expressed in degrees Kelvin, the ratio T_S/T_V typically will be within 1 or 2% of one.

Kirdiashev et al. (1979) has expressed the optical depth as:

$$\tau = \mu \sec(\theta) \alpha Q m k'' \quad (15)$$

where μ is a plant shape parameter and Q is the dry biomass in units of 100 kg/ha, k'' is the imaginary part of the dielectric constant for water, and m is the gravimetric moisture content of the vegetation. Grouping the constants together we get the following expression for τ:

$$\tau = b_2 W \quad (16)$$

where $W = Qm$ is the mass of water per unit area. Analyzing the data from a large number of field experiments, Mo et al. (1982) found that

$$\tau = 0.115 W \, (kg/m^2) \quad (17)$$

with an $r^2 = 0.95$ for conditions ranging from light grass ($W = 0.5$ kg/m^2) to mature corn ($W > 6$ kg/m^2). Using this result for τ, the expected emissivity is plotted in Figure 3.14 for conditions ranging from bare soil to a light forest cover ($W = 10$ kg/m^2). The bare soil slope is approximately that seen in Figure 3.11, whereas the slope shown in Figure 3.13 is slightly less than that shown for the grass curve. Figure 3.14 illustrates the decrease in soil

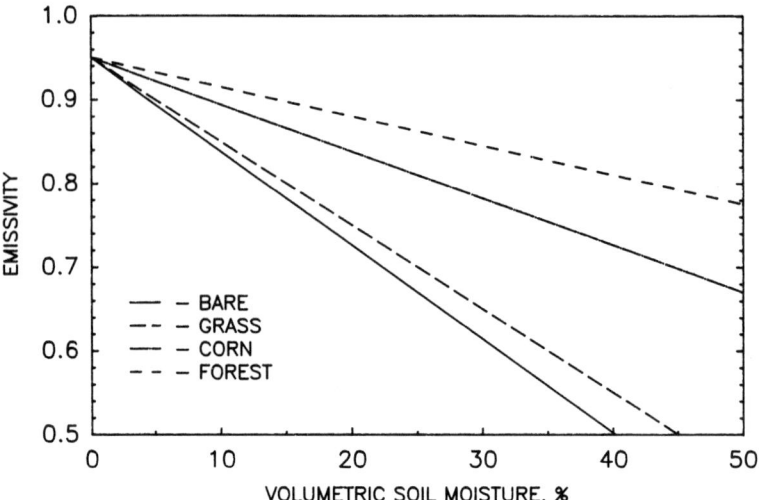

Figure 3.14. Plots of calculated emissivity versus soil moisture for a loamy sand soil with various amounts of vegetation cover.

moisture sensitivity as the vegetation cover increases. The conclusion is that microwave sensors will be of little use for soil moisture sensing when the vegetation cover exceeds that of a mature corn crop, that is, $W > 7$ kg/m². It is clear from Figure 3.14 that some estimate of the vegetation biomass covering the soil is necessary for soil moisture measurements to be made with a microwave radiometer.

A possibility for doing this lies with the use of visible and near-IR reflectance data, which are strongly dependent on vegetation conditions. In one such analysis, Theis et al. (1984) used the perpendicular vegetation index (PVI) calculated from visible and near-IR data as an estimator of plant biomass. The PVI, which is directly related to plant biomass, was used as a vegetation corrector in an algorithm to estimate soil moisture from 21-cm brightness temperature data. With this correction, the correlation between the microwave data and soil moisture improved from 0.09 to 0.75. Another example of this approach is the work done by Owe et al. (1988) in which the NDVI values calculated from the AVHRR sensor on the NOAA polar orbiting satellites were used to correct for vegetation in soil moisture estimates made using the 18-GHz ($\lambda = 1.66$ cm) data from the Scanning Multichannel Microwave Radiometer (SMMR) on the Nimbus-7 satellite. When their data were segregated according to NDVI values, the SMMR brightness temperatures accounted for about 70% of the observed soil moisture variability. They found that when NDVI > 0.4, the sensitivity of T_B to soil moisture was gone. This is an encouraging result when one considers that the wavelength used is an order of magnitude shorter than what was found to be optimum in field and aircraft experiments. These are indications of the promise that the use of multispectral data can greatly improve our ability to estimate soil moisture remotely.

An alternative approach is to consider the polarization differences at higher microwave frequencies. If we consider the difference between the vertical and horizontal polarizations in Equation 14, we get:

$$e = (e_{sV} - e_{sH}) \exp(-2\tau) \tag{18}$$

Thus the measurements of this difference can yield an estimate of the vegetation biomass. To make use of the τ measured at the higher frequencies, it is necessary to know how the dielectric properties of the vegetation change with frequency. Ulaby and El-Reyes (1987) have developed a model for the dielectric properties of vegetative material and the dominant is the dielectric constant of water whose frequency dependence is known. Recent analysis of satellite radiometer data at 37 GHz by Choudhury et al. (1987) has shown a good relation between the polarization difference and large-scale vegetation parameters.

Active Microwave Approaches

An active microwave sensor or radar sends out a pulse of microwave radiation and then measures the intensity of the radiation reflected back to it, as represented schematically in Figure 3.15. The intensity of this reflected signal is described by what is called the backscattering coefficient ($\sigma°$—sigma zero). As sketched in Figure 3.15, the intensity of the return for a smooth surface is mostly reflected or scattered into the specular direction and some scattering source, such as surface roughness, is required to obtain significant intensity in the backscatter direction. This is represented in the lower-right-hand part of the Figure, showing that the backscattered intensity is at a maximum for the 0° incidence angle for the smooth surface whereas for the rough surface the return is more isotropic.

One advantage of radar is that the energy in the received pulses can be angularly separated into the returns from different locations on the ground. Thus, if the radar is on board an aircraft, it is possible to produce a backscatter image of the ground as the aircraft flies along. If the coherent nature of radiation is utilized, it is possible to obtain very-high-resolution images from what are called synthetic aperture radars (SAR). The details of this technique are beyond the scope of this chapter and the reader is referred to the book by Ulaby et al. (1982–1986) on microwave remote sensing or to the chapter on the SAR in the *Manual of Remote Sensing* (Colwell, 1983) for more information.

Analogous to the optical reflectivity of terrain, $\sigma°$ describes the scattering properties of the surface in the direction of the illuminating source. The scattering behavior of terrain is governed by the geometrical and dielectric properties of the surface (or volume) relative to the wave properties

RADAR BACKSCATTER

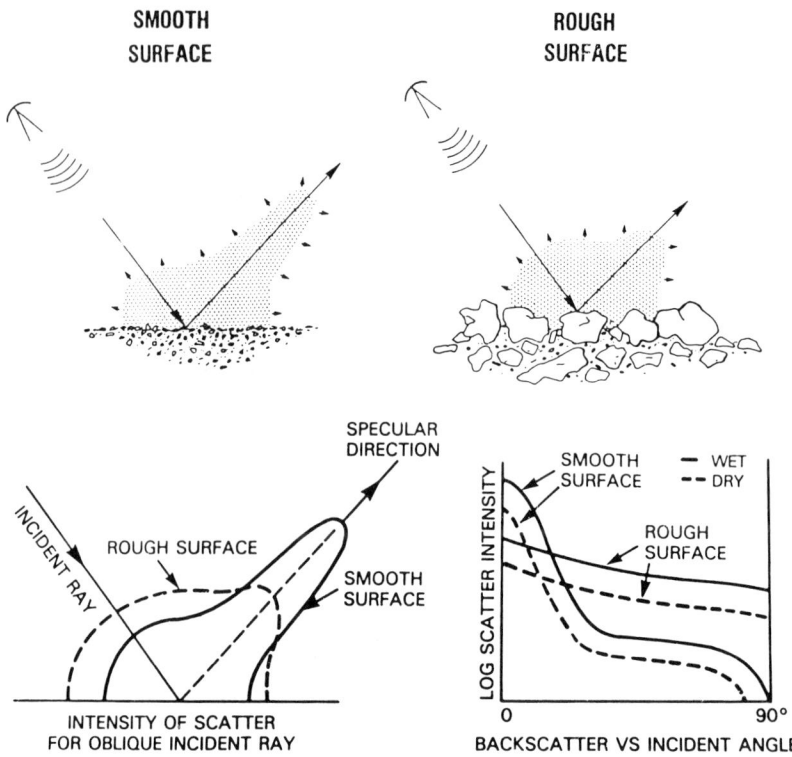

Figure 3.15. Schematic diagram of radar intensity patterns showing the different behaviors of smooth and rough surfaces. The backscatter coefficient ($\sigma°$—sigma zero) is the intensity in the direction of incidence.

(wavelength, polarization, and angle of incidence) of the incident radiation. Since the dielectric properties of the soil are strong functions of its moisture content, $\sigma°$ will be dependent on the soil moisture of a surface layer whose thickness is determined by the properties of the soil at the wavelength of observation. This thickness will be approximately the same for both active and passive microwave approaches, that is, of the order of a few tenths of a wavelength as discussed earlier. As with the passive microwave case, the dependence of $\sigma°$ on soil moisture will be a function of surface roughness and vegetation. In general, it can be said that the active microwave response is more dependent on roughness and less on vegetation than the passive microwave response. A complete description of the current status of active microwave approaches for soil moisture sensing is given in the recent review by Dobson and Ulaby (1986).

Spacecraft Results

The flights of microwave sensors on recent satellites, such as Skylab, Seasat, Nimbus 5, 6, and 7, and SIR A and B flights of the shuttle, have provided opportunities to do case studies of the remote sensing of soil moisture. The S-194 instrument on Skylab was a nonscanning 1.42-GHz (21-cm) radiometer with a 110-km field of view. With such coarse spatial resolution, it is difficult to compare the sensor response directly with ground measurements. However, there have been several indirect comparisons. In one, Wang (1985) has compared the Skylab and Nimbus 7 SMMR brightness temperatures at 6.6 and 10.7 GHz observed for several passes over the state of Texas in the United States with the Antecedent Precipitation Index (API), which is about the only moisture parameter available for comparison when retrospective analyses are performed. The data were for the same time of year, though separated by about ten years. He found that there was considerable sensitivity at all three frequencies to soil moisture/API for sparsely vegetated high plains areas. However, for more heavily vegetated east Texas, only the 21-cm data still showed any sensitivity to soil moisture variations, thus indicating the greater effectiveness of the longer wavelengths for soil moisture in the presence of vegetation.

The higher resolution of the SARs on Seasat and the shuttle afforded the opportunity to compare satellite observations directly with ground measurements. Blanchard and Chang (1983) did this for Seasat data over a test site in the Oklahoma panhandle. They compared the digitally processed Seasat $\sigma°$ data with ground measurements of soil moisture for bare, alfalfa, and milo fields. They found a strong correlation of $\sigma°$ and soil moisture with an $r^2 = 0.7$. Working with the SIR-B data from October 1984, investigators working in California (Wang et al., 1986) and in Illinois (Dobson and Ulaby, 1986) found stronger correlations ($r^2 > 0.8$) for a variety of field conditions. However, all of these investigators found that there were such strong dependencies of $\sigma°$ on such surface features as roughness and vegetation cover that it was difficult to extract soil moisture information from a single SAR observation. Most of these difficulties can be overcome by acquiring data on a number of dates, and at a number of angles or frequencies to aid in the extraction of the soil moisture information.

Discussion

The results presented here are examples of the progress that has been made in improving our fundamental understanding of the use of microwave sensors for the remote sensing of soil moisture. The next step in the process should be a demonstration of the capabilities of these sensors for determining surface soil moisture. An example of this type of result is given in Figure 3.16. Here, 1.4-GHz emissivities as measured from an air-

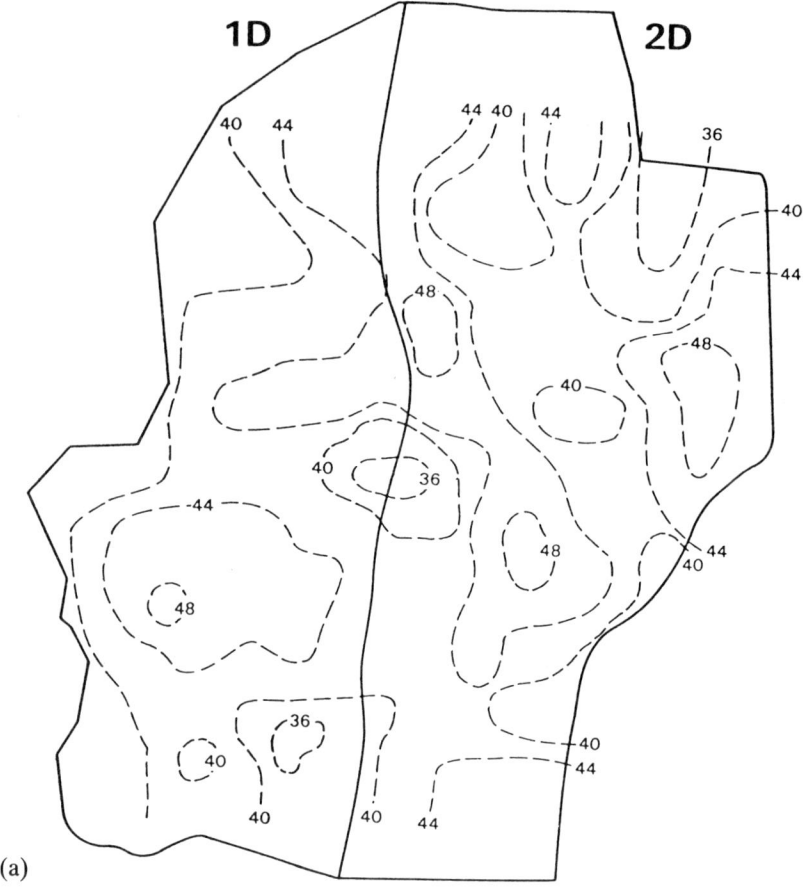

(a)

Figure 3.16. Soil moisture contours for the top 5-cm layer derived from 21-cm microwave radiometric measurements: (a) on May 28, 1987, and (b) on June 4, 1987. The data are from flights made at an altitude of 300m over two watersheds in the Konza prairie natural research area. [From Wang et al. (1989).]

craft platform were used to estimate the 0- to 5-cm soil moisture for two watersheds in the Konza prairie as part of the FIFE experiment in 1987. A linear model based on a limited number of ground samples in both 1985 and 1987 was used to convert the aircraft measurements of microwave emissivity to soil moisture. When this model was extended to a much larger area, the rms agreement with independent soil measurements was better than 3% on a volumetric basis (Wang et al., 1989). Figure 3.16 is an example of how microwave sensors can be used to map the spatial variations of soil moisture, which can be used in runoff forecast models or for the estimation of evapotranspiration.

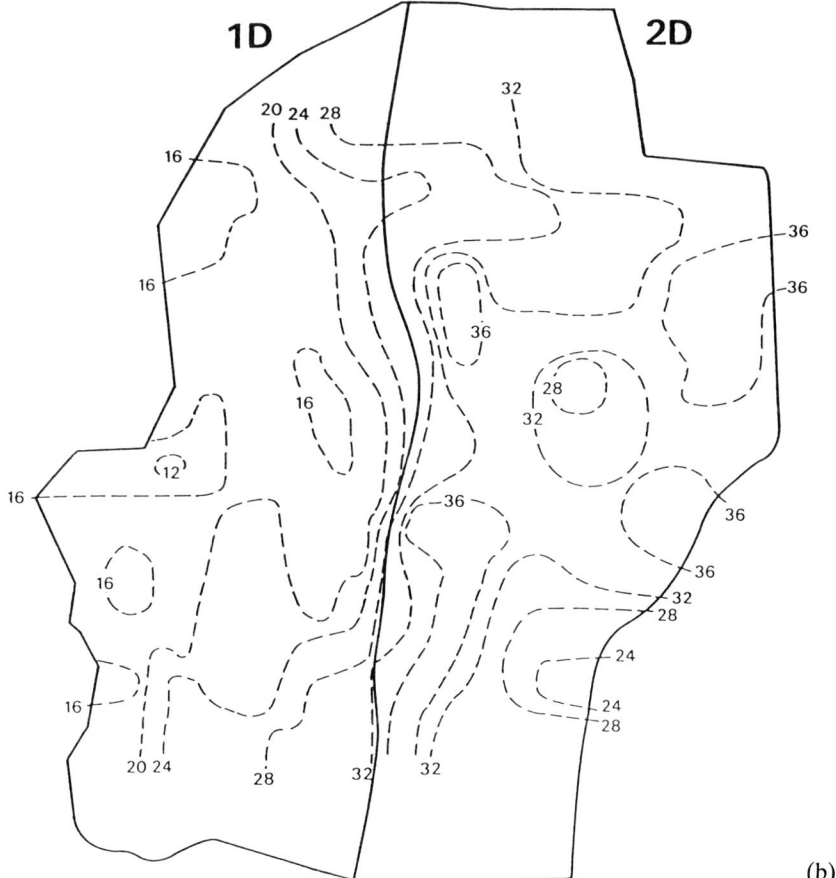

Figure 3.16. *Continued.*

As described here, microwave remote sensing techniques can provide estimates of the soil moisture content for a surface layer about 5 cm thick. This depth is shallow compared with the 1- to 2-m rooting depth of many crops. Estimating the root zone soil moisture from surface measurements has been studied using correlation techniques (Blanchard et al., 1981) and modeling studies (Jackson, 1980), which assume a moisture profile in hydraulic equilibrium. The conclusion from both approaches was that, if the water content in the top 10 cm is known, the moisture content in the top meter can be calculated to within acceptable limits, and that the lowest errors were obtained when the surface water contents were measured just before dawn—when the moisture profile is the closest to equilibrium.

Repetitive observations of the surface soil moisture can also be used to estimate soil properties. Camillo et al. (1986) coupled moisture and energy balance models with remotely sensed data of surface soil moisture to estimate the soil's hydraulic conductivity and matric potential, and a soil texture parameter. The derived conductivities agreed with values measured with an infiltration ring. This is potentially a very useful technique for estimating the soil's average hydraulic properties over large areas, as has been demonstrated using data from a helicopter-borne radar by Bernard et al. (1986).

Alternatively, knowledge of the surface layer moisture can be used to estimate moisture fluxes at the soil surface. These are of interest in themselves and can be used in moisture balance models to estimate the moisture in the profile. Barton (1978) in Australia used soil moistures as estimated from an airborne radiometer in a model to estimate the evapotranspiration from a a grassland. Bernard et al. (1981) in France used a similar approach employing simulated radar backscatter data. In a follow-up paper, they verified the technique, using field measurements with a C-band radar system (Prevot et al., 1984). Both groups reported considerable success in estimating evapotranspiration rates.

Passive microwave remote sensing of soil moisture is at a threshold at the present time. Theoretical and experimental research over the past 10 to 15 years has pretty much defined the capabilities of this approach. Briefly summarized, they are the ability to measure the moisture content of a surface layer about 5 cm thick to a relative accuracy of 10% and 15%; the fact that the measurement can be made under all weather conditions and through light to moderate vegetation conditions, and the limiting case appears to be a mature corn crop; and the fact that the factors of surface roughness and soil texture will introduce uncertainties into the soil moisture determinations that can be mostly accounted for with ancillary data. Before these systems are flown on spacecraft, it will be necessary to demonstrate the utility of such a remotely sensed soil moisture measurement with further experiments of the type described above. It is always important to realize that remote sensing measurements will not provide as accurate or as deep a measurement of soil moisture as can be obtained by conventional in-situ measurements, but they do provide a means for getting repetitive measurements over a large area of the surface soil moisture condition and thus afford unique opportunity to obtain previously unattainable information about the land surface. At the present time, there are no plans to fly microwave radiometers at wavelengths longer than 6 cm on the earth-observing system (EOS), and thus there will be no sensor for optimum observations of soil moisture.

Conclusions

In this chapter, we have described the capabilities of thermal IR sensors to observe surface temperatures and of passive microwave sensors for estimating surface soil moisture conditions. The combined measurements or estimates of these parameters can be used with coupled moisture and energy balance models to describe the fluxes from the surface. Preliminary work along this approach has been described by Camillo et al. (1986) and Soares et al. (1987). In the latter paper, remotely sensed measurements of temperature and soil moisture were used to derive regional estimates of the evapotranspiration flux.

Clearly, one of the major advantages of microwave remote sensing is its ability to observe the land surface in almost any atmospheric condition and the fact that, if we truly wish to monitor the land surface for climate change studies under all conditions, it will be necessary to make use of this capability.

References

Andre, J.C., Goutorbe, J.P., and Perrier, A. (1986). HAPEX MOBILHY: a hydrologic atmospheric experiment for the study of water budget and evaporation flux at the climatic scale. *Bull. Amer. Met. Soc.* 67:138–144.

Andre, J.C., et al. (1988). Evaporation over land-surfaces: first results from HAPEX-MOBILHY special observing period. *Ann. Geophys.* 6:477–492.

Barton, I.J. (1978). A case study comparison of microwave radiometer measurements over bare and vegetated surface. *J. Geophys. Res.* 83:3513–3517.

Becker, F. (1987). The impact of spectral emissivity on the measurement of land surface temperature from a satellite. *Int. J. Remote Sens.* 8:1509–1522.

Bernard, R., Soares, J.V., and Vidal-Madjar, D. (1986). Differential bare field drainage properties from airborne microwave observations. *Water Resources Res.* 22:869–875.

Bernard, R., Vauclin, M, and Vidal-Madjar, D. (1981). Possible use of active microwave remote sensing data for prediction of regional evaporation by numerical simulation of soil water movement in the unsaturated zone. *Water Resources Res.* 17:1603–1610.

Blanchard, B.J., and Chang, A.T.C. (1983). Estimation of soil moisture from Seasat SAR data. *Water Resources Bull.* 19:803–810.

Blanchard, B.J., McFarland, M.J., Schmugge, T.J., and Rhoades, E. (1981). Estimation of soil moisture with API algorithms and microwave emission. *Water Resources Bull.* 17:767–774.

Camillo, P.J., O'Neill, P.E., and Gurney, R.J. (1986). Estimating soil hydraulic parameters using passive microwave data. *IEEE Trans. Geosci. Remote Sens.* GE-24:930–936.

Choudhury, B.J., Schmugge, T.J., Chang, A., and Newton, R.W. (1979). Effect of surface roughness on the microwave emission from soils. *J. Geophys. Res.* 84:5699–5706.

Choudhury, B.J., Tucker, C.J., Golus, R.E., and Newcomb, W.W. (1987). Monitoring vegetation using Nimbus-7 scanning multichannel microwave radiometer's data. *Int. J. Remote Sens.* 8:533–538.

Christensen, P.R., Malin, M.C., Anderson, D.L., and Jaramillo, L.L. (1986). Thermal imaging spectroscopy in the Kelso-Baker region California. pp. 25–28. In A.B. Kahle and E. Abbott (eds.), *The TIMS Data User's Workshop, June 18 and 19, 1985*. JPL Pub 86-38, NASA Jet Propulsion Lab., California Inst. Technol., Pasadena, CA.

Colwell, R.N. (ed.) (1983). *Manual of Remote Sensing* (2nd ed.). Amer. Soc. Photogram., Falls Church, VA.

Dobson, M.C., and Ulaby, F.T. (1986a). Active microwave soil moisture research. *IEEE Trans. Geosci. Remote Sens.* GE-24:23–36.

Dobson, M.C., and Ulaby, F.T. (1986b). Preliminary evaluation of the SIR-B response to soil moisture, surface roughness, and crop canopy cover. *IEEE Trans. Geosci. Remote Sens.* GE-24:517–526.

Dobson, M.C., Ulaby, F.T., Hallikainen, M.T., and Reyes, M. (1985). Microwave dielectric behavior of wet soil Part II: Dielectric mixing models. *IEEE Trans. Geosci. Remote Sens.* GE-23:35–46.

Gillespie, A.R. (1986). Lithologic mapping of silicate rocks using TIMS. pp. 29–46. In A.B. Kahle and E. Abbott (eds.), *The TIMS Data User's Workshop, June 18 and 19, 1985.* JPL Pub. 86-38, NASA Jet Propulsion Lab., California Inst. Technol., Pasadena, CA.

Itier, B. (1982). Revision d'une methode simplifiee pour mesure du flux de chaleur sensible. *J. Rech. Atmos.* 16:85–90.

Jackson, T.J. (1980). Profile soil moisture from surface measurements. *J. Irrig. Drain.*, Div., ASCE, IR-2. pp. 81–92.

Jackson, T.J., Hawley, M.E., Shuie, J., O'Neill, P.E., Owe, M., Delnore, V., and Lawrence, R.W. (1986). Assessment of preplanting soil moisture using airborne microwave sensors. *Hydrologic Applications of Space Technology, Proc. Cocoa Beach Workshop, Florida, Aug. 1985.* IAHS Pub. 160.

Jackson, T.J., and O'Neill, P.E. (1987). Salinity effects on the microwave emission of soil. *IEEE Trans. Geosci. Remote Sens.* GE-25:214–220.

Jackson, T.J., Schmugge, T.J., Wang, J.R. (1982). Passive microwave sensing of soil moisture vegetation canopies. *Water Resources Res.* 18:1137–1142.

Kahle, A.B., and Abbott, E. (eds.) (1986). *The TIMS Data User's Workshop, June 18 and 19, 1985.* JPL Pub. 86-38, NASA Jet Propulsion Lab., California Inst. Technol., Pasadena, CA.

Kirdiashev, K.P., Chukhlantsev, A.A., and Shutko, A.M. (1979). Microwave radiation on the earth's surface in the presence of vegetation cover. *Radio Eng. Electron.* (English trans.) 24:256–264.

Kniezys, F.X., Shettle, E.P., Gallery, W.O., Chetwynd, J.H., Abreu, L.W., Selby, J.E.A., Clough, S.A., and Fenn, R.W. (1983). Atmospheric transmittance/radiance computer code LOWTRAN 6. Air Force Geophys. Lab. Rep. AFGL-TR-83-0187, Hanscom AFB, MA.

McClain, E.P., Pichel, W.G., Walton, C.C., Ahmad, Z., and Sutton, J. (1983). Multichannel improvements to satellite derived global sea surface temperatures. *Adv. Space Res.* 2:43–47.

Mo, T., Choudhury, B.J., Schmugge, T.J., Wang, J.R., and Jackson, T.J. (1982). A model for microwave emission from vegetation covered fields. *J. Geophys. Res.* 87:11229–11237.

Newton, R.W., Black, Q.R., Makanvand, S., Blanchard, A.J., and Jean, B.R. (1982). Soil moisture information and thermal microwave emission. *IEEE Trans. Geosci. Remote Sens.* GE-20:275–281.

Nieuwenhuis, G., Smit, E.H., and Thunissen, H.A.M. (1985). Estimation of regional evapotranspiration of arable crops from thermal infrared images. *Int. J. Remote Sens.* 6:1319–1334.

Owe, M., and Schmugge, T.J. (1983). Microwave radiometer response to soil moisture at the 21-cm wavelength. Presented at the Amer. Soc. Agron. Ann. Meeting, Washington, DC.

Owe, M., Chang, A., and Golus, R.E. (1988). Estimating surface soil moisture from satellite microwave measurements and satellite derived vegetation index. *Rem. Sens. Envir.* 24:331–345.

Pallucconi, F., and Meeks, G.R. (1985). *Thermal Infrared Multispectral Scanner (TIMS): An Investigator's Guide to TIMS Data*. JPL Pub. 85-32, NASA Jet Propulsion Lab., California Inst. Technol. Pasadena, CA.

Pallucconi, F.J. (1988). Land surface temperature emittance determination. pp. 98–101. In P.J. Sellers, F.G. Hall, D.E. Strebel, et al. (eds.), *First ISLSCP Field Experiment (FIFE), April 1988 Workshop Report*. Goddard Space Flight Center, Greenbelt, MD.

Prevot, L., Bernard, R., Taconet, O., and Vidal-Madjar, D. (1984). Evaporation from a bare soil evaluated from a soil water transfer model using remotely sensed surface soil moisture data. *Water Resources Res.* 20:311–316.

Price, J.C. (1984). Land surface temperature measurements from the split window channels of the NOAA-7 Advanced Very High Resolution Radiometer. *J. Geophys. Res.* 89:7231–7237.

Riou, Ch., Itier, B., and Seguin, B. (1988). The influence of surface roughness on the simplified relationship between daily evaporation and surface temperature. *Int. J. Remote Sens.* 9:1529–1533.

Schmugge, T., and Janssen, L. (1988). Aircraft remote sensing in HAPEX. *Proc. 4th Int. Colloquium of Spectral Signatures of Object in Remote Sensing*, Aussois, France, Jan. 18–22, pp. 463–467 (ESA SP-287).

Schmugge, T.J., O'Neill, P.E., and Wang, J.R. (1986). Passive microwave soil moisture research. *IEEE Trans. Geosci. Remote Sens.* GE-24:12–22.

Seguin, B., and Itier, B. (1983). Using midday surface temperature to estimate daily evaporation from satellite thermal IR data. *Int. J. Remote Sens.* 2:371–383.

Soares, J.V., Bernard, R., and Vidal-Madjar, D. (1987). Spatial and temporal behavior of a large agricultural area as observed from airborne C-band scatterometer and thermal infrared radiometer. *Int. J. Remote Sens.* 8:981–996.

Theis, S.W., Blanchard, B.J., and Blanchard, A.J. (1986). Utilization of active microwave roughness measurements to improve passive microwave soil moisture estimates over bare fields. *IEEE Trans. Geosci. Remote Sens.* GE-24:334–339.

Theis, S.W., Blanchard, B.J., and Newton, R.W. (1984). Utilization of vegetation indices to improve microwave soil moisture estimates over agricultural lands. *IEEE Trans. Geosci. Remote Sens.* GE-22:490–496.

Ulaby, F.T., and El-Reyes, M.A. (1987). Microwave dielectric spectrum of vegetation—Part II: Dual-dispersion model. *IEEE Trans. Geosci. Remote Sens.* GE-25:550–557.

Ulaby, F.T., Moore, R.K., and Fung, A.K. (1982–86). *Microwave Remote Sensing, Vols. I, II, and III*. Artech House, Dedham, MA.

Wang, J.R. (1985). The effect of vegetation on soil moisture sensing from orbiting microwave radiometers. *Remote Sens. Envir.* 11:141–151.

Wang, J.R. (1987). Passive microwave sensing of soil moisture: the frequency dependence of microwave penetration depth. *IEEE Trans. Geosci. Remote Sens.* GE-25:616–622.

Wang, J.R, Engman, E.T., Shuie, J.C., Rusek, M., and Steinmeier C. (1986). The SIR-B observations of microwave backscatter dependence on soil moisture, surface roughness, and vegetation covers. *IEEE Trans. Geosci. Remote Sens.* GE-24:510–516.

Wang, J.R., and Schmugge, T.J. (1980). An empirical model for the complex dielectric permittivity of soils as a function of water content. *IEEE Trans. Geosci. Remote Sens.* GE-18:288–295.

Wang, J.R., Schmugge, T.J., Shuie, J.C., and Engman, E.T. (1989). Mapping surface soil moisture with L-band radiometric measurements. *Remote Sens. Envir.* 27:305–312.

Wilheit, T.T. (1978). Radiative transfer in a plane stratified dielectric. *IEEE Trans. Geosci. Electron.* GE-16:138–143.

4. Estimating Terrestrial Primary Productivity by Combining Remote Sensing and Ecosystem Simulation

Steven W. Running

Beginning in 1972 with the launch of Landsat 1, estimation of terrestrial plant production has been one of the most important applications attempted of satellite remote sensing. Initial interest focused on the prediction of regional crop yields, such as wheat (Erickson, 1984). However, changing goals, hardware capabilities, and theory have produced a steady evolution in the approaches taken to calculate net primary production (NPP) of large areas. Interest has also expanded to calculating primary production of natural vegetation. The much wider array of topography, climate, canopy geometry, and life-cycle dynamics exhibited by natural vegetation make computation of primary production much more challenging than the rather controlled, organized field conditions of a crop.

The necessity to quantify more accurately global terrestrial vegetation activity was emphasized by attempts in the late 1970s to calculate a global carbon budget. Although it was clear from the Mauna Loa CO_2 concentration record that global atmospheric CO_2 was increasing, the anthropomorphic sources did not seem to balance the ocean and terrestrial sinks (Bolin, 1977; Woodwell et al., 1983; Emmanuel et al., 1984). More significantly, this problem underscored the lack of defensible measurements of terrestrial primary production at global scales (Leith and Whittaker, 1975).

More recently, attention has focused on the climate changes projected by recent Global Circulation Model (GCM) studies (Hansen et al., 1981; DOE, 1985). Predictions of global temperature increases of 4° have been challenged because the GCMs do not realistically define the land surface biological feedbacks (Dickinson, 1987; Sellers et al., 1986; Wilson et al., 1987). It has become increasingly clear that merely classifying global vegetation is inadequate; quantitative estimates of energy/mass flux activity and NPP are needed to answer critical questions in global ecology.

Beginning in 1979, a number of workshops were held by the U.S. National Aeronautics and Space Administration (NASA), the U.S. National Academy of Science (NAS), and other organizations specifically charged with developing a strategy to quantify biological activity of global terrestrial vegetation (NASA, 1983a, 1983b; NAS, 1986). The repeated conclusions of these workshops were (1) that a global biome classification was needed, as has now been done by Mathews (1983); (2) that satellite mapping of global vegetation was needed; and (3) that beyond classification, the most important carbon cycle variable for quantifying biological activity in a comparable way across biomes was net primary production (NPP). Also, for global applications, NPP needed to be monitored by satellites.

This chapter will analyze two current methodologies being developed for estimation of primary production of natural vegetation, and argue for the necessity for using both methods to deal with this complexity. One, using AVHRR/NDVI (Advanced Very High Resolution Radiometer/ Normalized Difference Vegetation Index), is based on simple principles and is already globally applicable. The second, linking satellite data with ecosystem process models, holds promise for greater accuracy, but is more complex, and has only been implemented over 1,200 km^2.

Estimates of Terrestrial NPP Using the AVHRR/NDVI

C.J. Tucker at NASA Goddard Space Flight Center first recognized the potential for using AVHRR sensor on board the TIROS meteorological satellites to get daily images of the entire earth's terrestrial vegetation. The spatial scale, or pixel size of AVHRR is rather coarse, 1.1 km, but these low-density data were precisely the reason that processing total global images was computationally possible (Yates et al., 1986). With computer-automated time compositing of the data (a procedure where data for each day of, say, one week are read and the highest value retained and mapped), a cloud-free map of global vegetation was now made possible (Holben, 1986). The vegetation of Africa (Tucker et al., 1985), North America (Goward et al., 1985), and the entire world (Justice et al., 1985) was mapped.

For this initial global work, the vegetation was "defined" as the normalized difference vegetation index (NDVI), the algebraic combination of surface radiance in the red [0.58 to 0.68 μm] and near-infrared (NIR) (0.725 to 1.1 μm), given as:

$$\frac{\text{NIR} - \text{RED}}{\text{NIR} + \text{RED}} \tag{1}$$

a dimensionless index between -1.0 and 1.0. The use of near-IR and red wavelength reflectances to characterize vegetation has had substantial development in the remote sensing disciplines (Tucker, 1979; Perry and Lautenschlager, 1984). However, with applications at continental to global scales, no direct interpretation of the NDVI against measured ground conditions was possible.

Direct validation of NDVI data is very difficult because although the AVHRR sensor has an optical resolution of 1.1 km, computer processing of the global NDVI data to date has always involved a spatial subsampling to reduce data volume (Tarpley et al., 1984). For example, during 1985, the NOAA/NESS global vegetation index (GVI, a globally generated NDVI data set) was produced by subsampling four out of five pixels in one out of every three data lines of the raw AVHRR signal, producing an aggregated 4- by 4-km pixel; then randomly subsampling again, one 4- by 4-km pixel out of each 15- by 15-km pixel array. This spatial subsampling ultimately meant that radiance data from less than 1% of the true ground area were being used for the final GVI map. Consequently, any attempts to do field measurements corresponding to specific GVI data are nearly impossible. The latest regional-scale NDVI research is using raw local area coverage (LAC) 1.1-km data to avoid these ambiguities.

Sellers (1985, 1987) derived an important interrelationship among leaf area index (LAI), absorbed photosynthetically active radiation (APAR), and NDVI that improves the utility of these biophysical variables. He found that under specified canopy properties, the APAR was linearly related to NDVI and curvilinearly related to LAI, approaching asymptotically an LAI of 6 where virtually all incident short-wave radiation is absorbed by the canopy. Sellers showed that:

$$\text{APAR} = f[\text{LAI, ISR, Canopy geometry}], \tag{2}$$

where ISR = incoming short-wave radiation, and that the AVHRR channels make:

$$\text{NDVI} = f[\text{APAR}] = f[\text{LAI}] \tag{3}$$

Consequently, given a canopy of known structure and light scattering and absorbing properties, any one measure of the canopy can be used interchangeably with the others with some algebraic manipulation of formulas. It must be recognized here that different biomes have radically different canopy structures and reflectance properties, and so can produce different NDVI while having identical LAIs. An NDVI of 0.5 may represent $\text{LAI} = 3$ in a forest but only 2.0 in a grassland. Accurate utilization of the

NDVI requires that the biome type be known so that the appropriate NDVI to LAI or APAR conversion can be made. Further, observational details such as the solar zenith angle, sensor look angle, background (soil) exposure fraction, and extent of uncorrected atmospheric interference change the NDVI-LAI-APAR relationship significantly (Sellers, 1985, 1987).

Relating NDVI, APAR, or LAI to NPP can be done using the logic developed by Monteith (1981) relating annual NPP of crops to the integrated APAR absorbed over a growing season.

$$\text{NPP} = f[\Sigma \text{APAR}] * \epsilon \tag{4}$$

where ϵ = energy-conversion efficiency, g/MJ.

This seasonally integrated APAR incorporates both meteorological conditions, primarily magnitude and seasonal duration of incoming solar radiation, and vegetation variables, the amount of leaf area and canopy radiation absorption characteristics—variables that directly determine photosynthetic production of a plant. Given the functional equivalence shown in Equation 3, this logic can be translated to satellite applications:

$$\text{NPP} = \Sigma \text{NDVI} * \epsilon \tag{5}$$

Goward et al. (1987) first related the ΣNDVI to NPP (Figure 4.1), and developed ϵ factors converting annual APAR energy in MJ to NPP in kilograms per square meter for different biome types. Important to this logic is that a static NDVI or LAI is not used; instead, the annual integration of weekly composited NDVI, analogous to the integrated APAR, is applied. Seasonal integration of the weekly composites automatically reduces cloud contamination and improves climatic sensitivity of the relationship. Although the maximum NDVIs (or LAIs) of many biomes in many climates are rather similar, higher-productivity regions have substantially longer periods of active vegetation growth, reflected in larger integrated NDVI (Goward et al., 1985; Goward, 1989). This time integration concept has proved important in accelerating the use of daily AVHRR data. If seven consecutive passes of Landsat data were composited, as is done with the weekly AVHRR data for cloud removal, only one composited image every 112 days would be obtained because of the 16-day repeat cycle of the Landsat orbit. Consequently, Landsat images are meaningful primarily as a single annual "snapshot" of the land surface. The ten times greater expense and 1,000 times greater data volume of the higher-resolution 30-m TM data also precludes use of multiple annual data sets for global studies.

A critical caveat of the Monteith (1981) logic and the derivation of Sellers (1985, 1987) is that the vegetation is not water, temperature, or nutrient limited. For annual crops that are often irrigated, fertilized, and grow during the midsummer season, these are acceptable assumptions. However, natural, perennial vegetation is routinely water stressed, usually

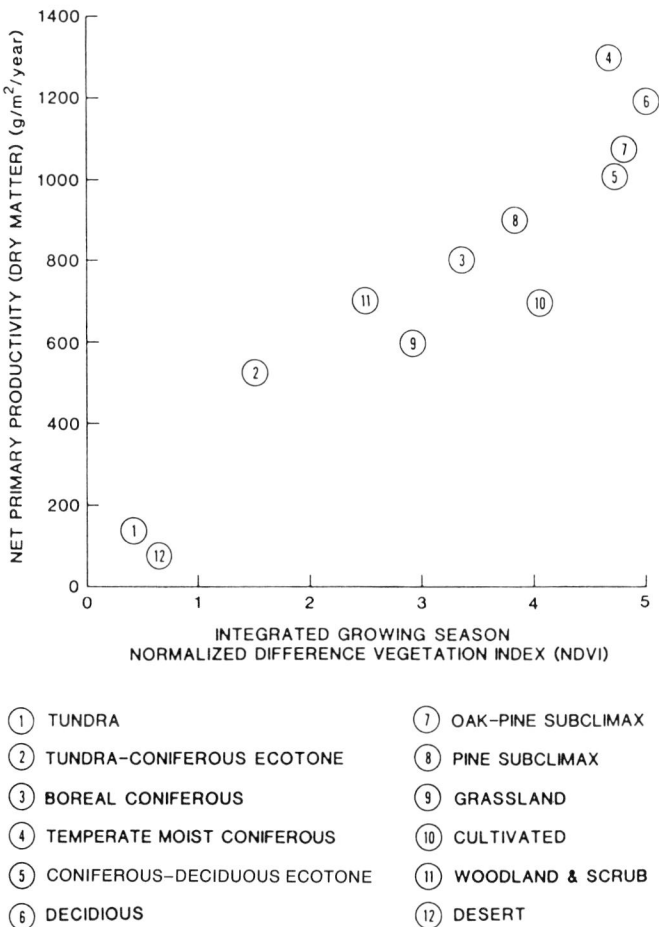

Figure 4.1. Plot of biome-average ΣNDVI versus published mean biome net primary productivity rates for North America. [From Goward et al. (1985).] NDVI data were three-week composites of at least 90 points per biome. NPP data were from the literature.

temperature limited at some point of each year, rarely has optimum nutrient availability, and often fixes substantial carbon during dormant periods while not visibly growing. Much of this variability can be subsumed in the ϵ coefficient, which then incorporates both meteorological and biochemical components. We have developed a more explict formulation that separates meteorological from biochemical components, and so should be more generally applicable under a wide variety of conditions and biome types.

$$\text{NPP} = t \Sigma\, T(\sigma * \text{NDVI}) * \epsilon \qquad (6)$$

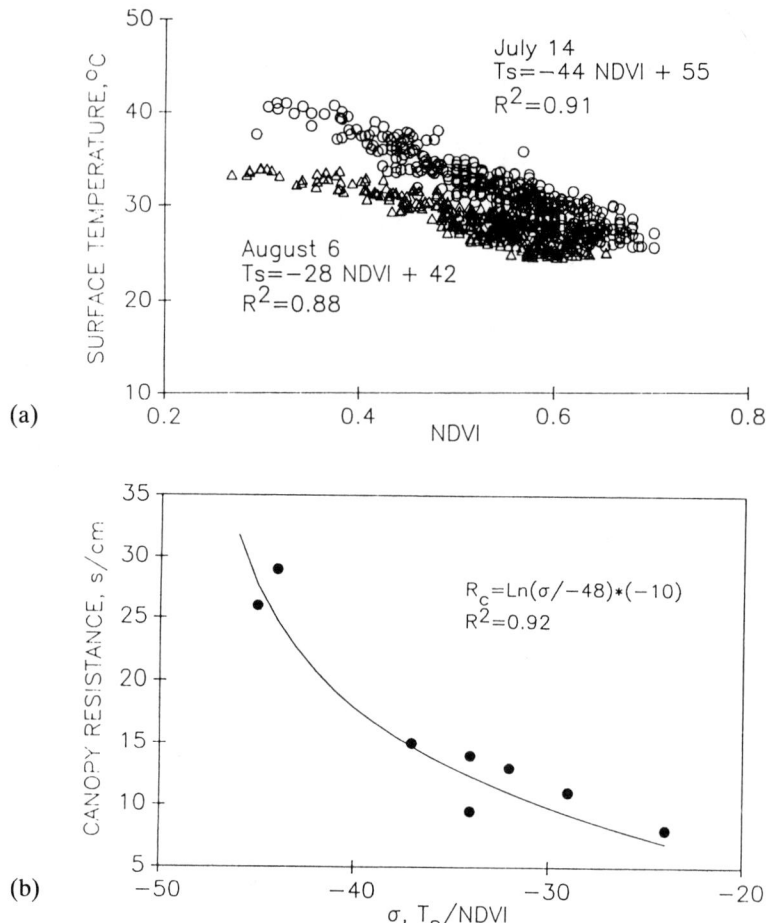

Figure 4.2. (a) Relation between surface temperature (Ts) and AVHRR/NDVI on July 14 and August 6 for a 20 by 25 pixel area of conifer forest in Montana. The change in slope from July 14 to August 6 results from 3 cm of rain on August 2 to 3. (b) The regression relationship between canopy resistance simulated by FOREST-BGC and slope of the Ts/NDVI for eight days during the summer of 1985, suggesting that the Ts/NDVI ratio from AVHRR can be used to quantify sensible/latent energy partitioning by a vegetated surface. [From Nemani and Running (1989b).]

where σ = surface resistance, Ts/NDVI; $t\Sigma T$ = temperature truncated NDVI time integration; and Ts = surface temperature, defined by AVHRR thermal channels 4 and 5.

This equation offers a number of key improvements over Equation 5. First, Nemani and Running (1989b) defined σ, a surface resistance factor, produced as Ts/NDVI, from AVHRR data over coniferous forests in Montana (Figure 4.2). We found that as water availability declined seasonally with summer drought, the surface temperature (T_s) measured by AVHRR increased when normalized to NDVI. This effectively defines a Bowen ratio energy-partitioning factor between sensible and latent heat flux, and so explicitly represents the reduction in water flux as the surface becomes water limited. Because this change in surface resistance is primarily caused by leaf stomatal closure, this σ factor should represent both H_2O and CO_2 diffusion resistance.

In order to implement this algorithm for σ, a contiguous area with a range of NDVI and surface temperature is required. However, this constraint can actually be an aid in determining appropriate spatial aggregation limits for a landscape. An area large enough to provide this needed NDVI gradient can be defined as a minimum regional cell for certain research requiring σ. Similar logic was reported by Hope (1988), relating canopy resistance to satellite-derived surface temperature ratios.

The $t\Sigma T$ notation refers to temperature-controlled time integration of seasonal NDVI, where the NDVI is not integrated over periods when the surface temperature is below 0°, and physiological activity is minimal (Figure 4.3). Typically, ΣNDVI is either done for the entire year, or is arbitrarily truncated to certain dates, because the thermal channels are not concurrently processed (Goward et al., 1985, 1987). Running and Nemani (1988) used temperature-truncated ΣNDVI but the temperature data were from an ancillary ground database.

The ϵ value, or efficiency factor, now can be used exclusively to define nutritionally related biochemical factors. For example, Mooney et al. (1987) showed that photosynthetic capacity was directly proportional to leaf N concentration. ϵ could be defined partially by leaf $N\%$, which may be remotely sensible using high-resolution imaging spectrometry (Wessman et al., 1988; Wessman, this volume). Also, separate ϵ factors can be defined for different biomes with fundamentally different physiology, such as C_3 and C_4 plants. Note that, operationally, the only addition equation 6 requires over the simple equation 5 is that surface temperatures be defined from the AVHRR channels 4 (10.5 to 11.5 μm) and 5 (11.5 to 12.5 μm). Nemani and Running (1989b) calculated these surface temperatures with the split-window technique of McClain (1980). So, ultimately, this formulation can be implemented globally just as are the current NDVI maps.

Ecosystem Modeling to Estimate NPP

Although the AVHRR/NDVI has rapidly evolved into an indispensaible tool for monitoring global NPP, it is not satisfactory for all purposes. The ΣNDVI, even with the improvements in Equation 6, still represents ecophysiological processes in only a simple empirical way, leaving it unsuited for exploring sensitive feedbacks and multifactor controls on NPP in an explicit manner. Additionally, the ΣNDVI loses dynamic sensitivity

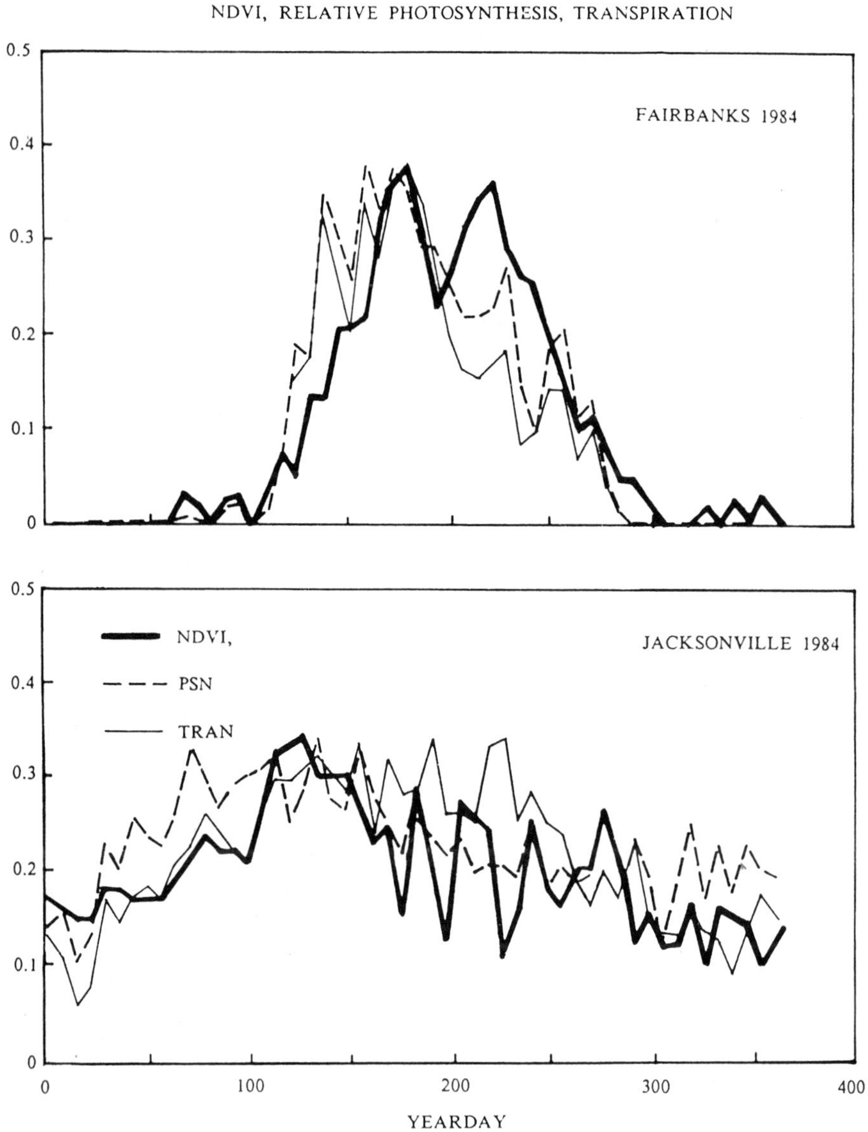

Figure 4.3. Sensitivity of the 1984 seasonal AVHRR/NDVI for defining temperature-limited growing season of forest vegetation and corresponding weekly simulations of photosynthesis and transpiration from FOREST-BGC in a boreal and subtropical climate. [From Running and Nemani (1988).]

when applied to biomes such as coniferous forests that provide a temporally inert remote sensing target. Consequently, a second approach is under development that is not constrained by these problems. This approach incorporates remote sensing of LAI with ecosystem simulation models.

Satellite Estimation of LAI

Although the NPP of grasslands, annual crops, and other seasonal biome types can be estimated by the time integration of observed developing biomass, for biome types such as forests, chapparal, and other evergreen broadleafs, permanent live biomass occupies the site continuously, causing annual NPP not to be visible from an orbiting satellite. For these biomes with continuous leaf display, a structural variable related to CO_2 exchange and comparable across biomes was required. LAI (the projected leaf area per unit ground area) provides a measure of the plant organ most directly involved in energy, H_2O and CO_2 exchange. Characterization of vegetation in terms of LAI rather than species composition was considered a critical simplification for comparison of different terrestrial ecosystems wordwide. Ecosystem analyses conducted during the International Biological Program of the 1970s had found strong correlations across biome types relating LAI to NPP (Gholz, 1982; Webb et al., 1983). A functional balance between site water availability and LAI was also found (Grier and Running, 1977), and Jarvis and McNaughton (1986) showed how evapotranspiration (ET) is directly proportional to LAI.

This logic isolated an initial specific task in global ecology—to develop a means of measuring LAI of natural vegetation by satellite. Remote sensing of LAI was first attempted for crops and grasslands, correlating spectral reflectances against direct measurement of vegetation LAI. Various combinations of near-IR and visible wavelengths have been used to estimate the LAI of wheat (Wigand et al., 1979; Asrar et al., 1984). However, for global applications, the complexities of natural, irregular canopies must be addressed. Running et al. (1986) and Peterson et al. (1987) first estimated the LAI of coniferous forests across an environmental gradient in Oregon using airborne Thematic Mapper Simulator data (Figure 4.4). A growing season site water balance ranging from +20 cm (surplus) on the Pacific coast to −80 cm water (deficit) in the interior desert produces a LAI range from 1 to 15, representing the global range of forest LAI. This work was extended to California, Montana, and Washington coniferous forests using Landsat TM data. Spanner et al. (1989) found that the strong relationships between TM NIR/RED ratios and LAI in closed-canopy, pure conifer forests of Oregon can erode in forests with mixed deciduous canopy and/or soil surface exposed.

Remote sensing of LAI was first tested with TM because the 30-m pixel size represented an area small enough to be directly measured on the ground. However, tests to AVHRR scale 1.1 km soon followed, because

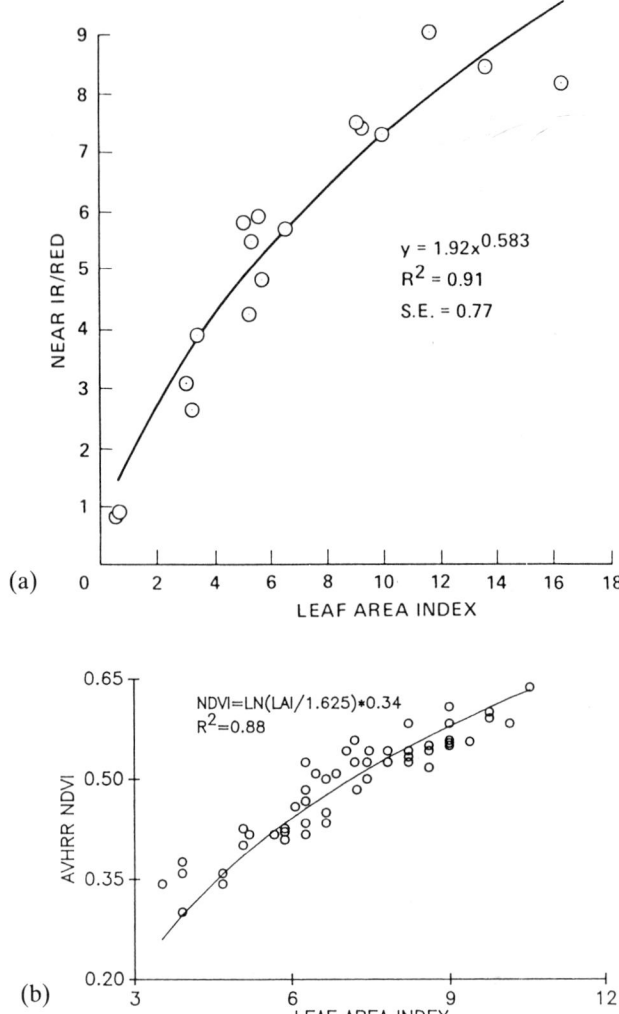

Figure 4.4. (a) Relationship between projected LAI and the Landsat TM channels 4/3 (NIR/RED) for coniferous forests across Oregon, at 30-m pixel scale. [From Peterson et al. (1987).] (b) Relationship between estimated LAI and the AVHRR/NDVI for conifer forests across Montana, at 1.1-km pixel scale. [From Nemani and Running (1989a).]

this scale is more realistic for global application. One major advance for AVHRR scale LAI validation has been the development of a portable integrating radiometer that can accurately measure forest LAI over multiple kilometer areas (Pierce and Running, 1988). Spanner et al. (1987) used the older method of measuring tree diameter or sapwood basal area and allometric equations to calculate plot LAI on conifer forest sites in

Washington, Oregon, and Montana. They found that the AVHRR NDVI correlated with LAI with a function asymptotic at LAI = 6, $R^2 = 0.76$. In a more theoretical approach, Nemani and Running (1989a) used a hydrologic equilibrium theory to estimate LAI of 52 1.1-km conifer stands in Montana, ranging from LAI = 3 to LAI = 10. AVHRR correlated with these estimated LAI highly, $R^2 = 0.88$ (Figure 4.4(b)).

A number of problems are inherent in these approaches to ground LAI measurement. Sapwood-based equations for LAI are sensitive to stand density and climatic variation, and so ideally should not be extrapolated too far from the original source (Hungerford, 1987). Optical measurements require a light extinction coefficient to be either assumed or calculated, and there is variability in this canopy property across species, and definitely biomes (Jarvis and Leverenz, 1983; Pierce and Running, 1988). The remote sensing measurements entail an equal number of potential problems, including variability in canopy bidirectional reflectance properties, soil background exposure and reflectance, atmospheric transmissivity, sun illumination angle, and sensor look angle (Curran, 1983).

Although we conclude that the LAI of conifer forests and grasslands can be measured by satellite at different spatial resolutions, this fundamental capability needs to be replicated in other biome types where canopy structure and bidirectional reflectance properties are radically different from the forests and grasslands studied to date. Also, it must be remembered that the proportionality between LAI and NPP or other canopy mass/energy flux is not constant. Biome level differences in biochemical energetics, climatic differences influencing water-use efficiency, and canopy differences in absolute light absorption efficiency are among the factors that must next be addressed to use LAI rationally as a precursor to calculation of important ecosystem processes.

Integrating LAI Into Ecosystem Models

If satellites can be used to estimate LAI regionally, the next step is to develop ecosystem process models designed to incorporate these remote sensing data. The FOREST-BGC model (Figure 4.5) is a process-level simulation that calculates the cycling of carbon, water, and nitrogen through forest ecosystems (Running and Coughlan, 1988). The model has mixed time resolutions, with hydrologic and canopy gas-exchange processes computed daily and carbon and nitrogen cycle processes computed yearly. FOREST-BGC requires daily meteorological data: maximum-minimum air temperature, dew point, incident short-wave radiation, and precipitation. The model calculates key hydrologic processes: snow melt; canopy interception and evaporation; transpiration; soil water content and outflow; carbon processes of photosynthesis, maintenence and growth respiration, carbon allocation, primary production, litterfall, and decomposition; and nitrogen processes of deposition uptake, litterfall, mineralization, and leaching.

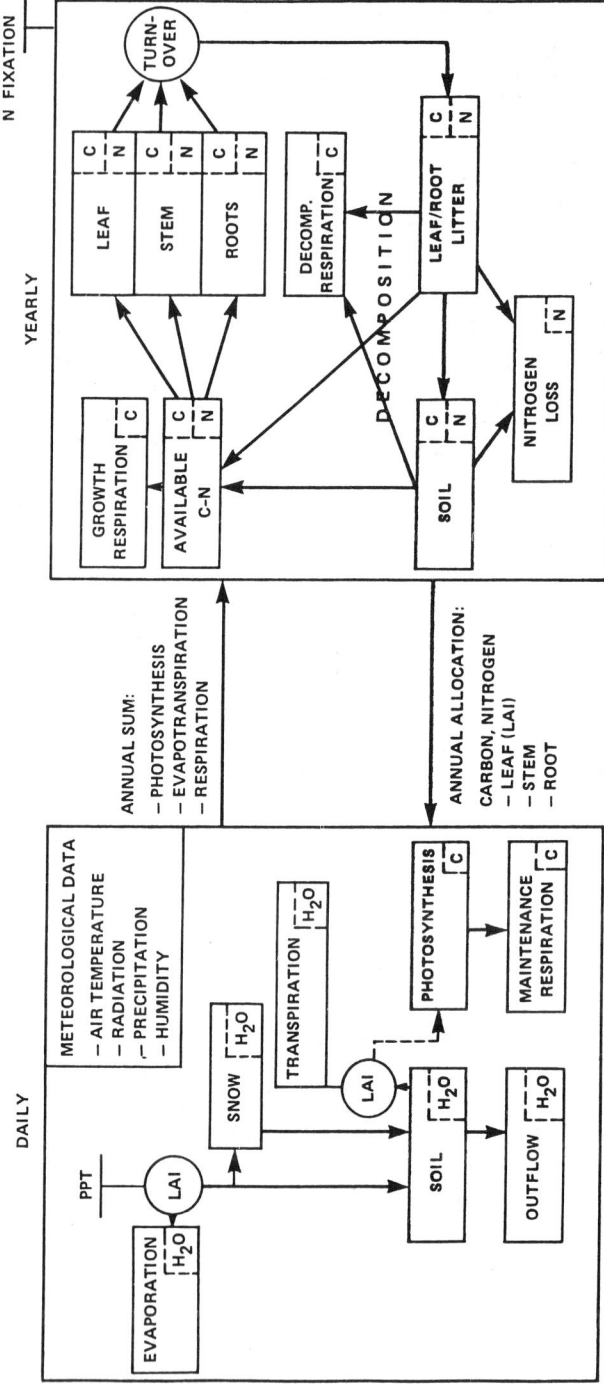

Figure 4.5. Compartment flow diagram for FOREST-BGC (Bio-Geo-Chemical cycles), a simulation of carbon, water, and nitrogen cycling processes for forest ecosystems. The two submodels integrate daily and yearly time steps, and require daily meteorological data. FOREST-BGC was designed to accept remotely sensed data, particularly of LAI as illustrated in Figure 4.4. [From Running and Coughlan (1988).]

The model was designed to be particularly sensitive to LAI because LAI can be retrieved by satellite (Figure 4.4). The following processes are controlled in part by LAI: snow melt, canopy interception and evaporation, transpiration, canopy light attenuation, photosynthesis, leaf maintenance respiration, litterfall, and leaf nitrogen turnover.

This sensitivity of FOREST-BGC to varying LAI was tested by simulating annual hydrologic balances and net photosynthesis for hypothetical forests in seven contrasting climates of North America (Running and Coughlan, 1988). These sites encompassed a temperature–moisture gradient from the cold, dry climate of Fairbanks, Alaska, with average July maximum–January minimum of 21 to −25 C and precipitation of 313 mm to Jacksonville, Florida, with temperatures of 31 to 4 C and precipitation of 1,244 mm. The hydrologic partitioning among evaporation, transpiration, and soil outflow varied significantly across this climatic range. The response to increases in LAI also varied; as the Jacksonville LAI went from 3 to 9, transpiration dramatically increased at the expense of soil outflow. However, Fairbanks was so water limited that increasing LAI produced little additional transpiration (Running and Coughlan, 1988). A similar result occurred with annual net photosynthesis, whereas in Jacksonville, with mild temperatures and ample precipitation, increasing LAI from 3 to 9 proportionally increased photosynthesis, whereas at Missoula, Montana, with a short growing season and 337-mm precipitation, little additional photosynthesis was possible (Figure 4.6).

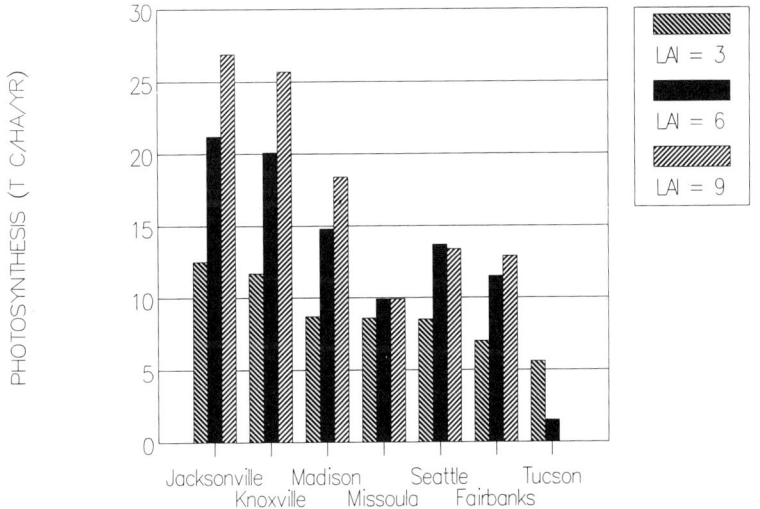

Figure 4.6. The annual forest photosynthesis at varying LAI (3, 6, 9) simulated by FOREST-BGC for seven sites of contrasting climate in North America. These results show the sensitivity to LAI designed into FOREST-BGC to optimize integration of remote sensing data from Figure 4.4. [From Running and Coughlan (1988).]

An important point here is that simulations of these plant processes can be formally validated with conventional ecological measurements. LAI is a routinely measured canopy structural variable. Photosynthesis and transpiration are measured most accurately by leaf cuvettes, leaving a challenge for whole-canopy extrapolation (Schulze et al., 1985). Whole-plant measurements are possible though, using lysimeters that follow daily transpiration mass loss by weight differences (Dunin and Aston, 1984), radioactive isotope transport methods (Waring et al., 1980), or potometers that volumetrically follow water loss (Knight et al., 1981). NPP can be measured simply as incremental increase in dry biomass over time (Gholz, 1982; Webb et al., 1983). At one step removed from the vegetation itself, aircraft flux measurements can follow diurnal changes in near-surface CO_2 and H_2O (Sellers et al., 1988; Wofsy et al., 1988). However, these atmospheric measurements must be in conjunction with canopy measurements to provide any degree of mechanistic interpretation.

Comparison of ΣNDVI with FOREST-BGC

Goward et al. (1985) related annual ΣNDVI to NPP of different biomes in North America, using NPP estimates from the literature (Figure 4.1). As an alternative means of "validating" the use of ΣNDVI as an estimate of NPP, we simulated annual NPP of a hypothetical forest on these same seven study sites, and correlated the results with the NOAA Global Vegetation Index for 1983–84 (Running and Nemani, 1988). Correlations of annual ΣNDVI across the seven sites were; $R^2 = 0.87$ with annual photosynthesis, $R^2 = 0.77$ with annual transpiration, and $R^2 = 0.72$ for net primary production (Figure 4.7). Clearly, ΣNDVI captures much of the difference between these sites in ecosystem process rates. Likewise, these results suggest that the two methodologies for estimating NPP, as derived from ΣNDVI or from ecosystem process models, are responding similarly to the critical conditions that determine NPP of natural vegetation.

Expansion of the FOREST-BGC estimates of NPP to the two-dimensional space of a landscape required an additional computational step. All critical variables—the LAI, soil-water holding capacity, and meteorological conditions—must be defined for each cell in the study area. To calculate these processes over a 28- by 55-km area of Montana required a geo-referenced data integration scheme (Figure 4.8). Data acquired from a variety of sources were used to drive a mountain microclimate simulator (MT-CLIM, Running et al., 1987) that extrapolates point-measured meteorological data into complex topography. The resulting meteorological data file was coupled with the other data files, including an AVHRR-derived LAI estimate, ultimately to run FOREST-BGC more than 1,200 times, for each 1-km cell in the area (Running et al., 1989). Ranges of estimated LAI (4 to 15), evapotranspiration (25 to 60 cm/yr) and NPP (5.7

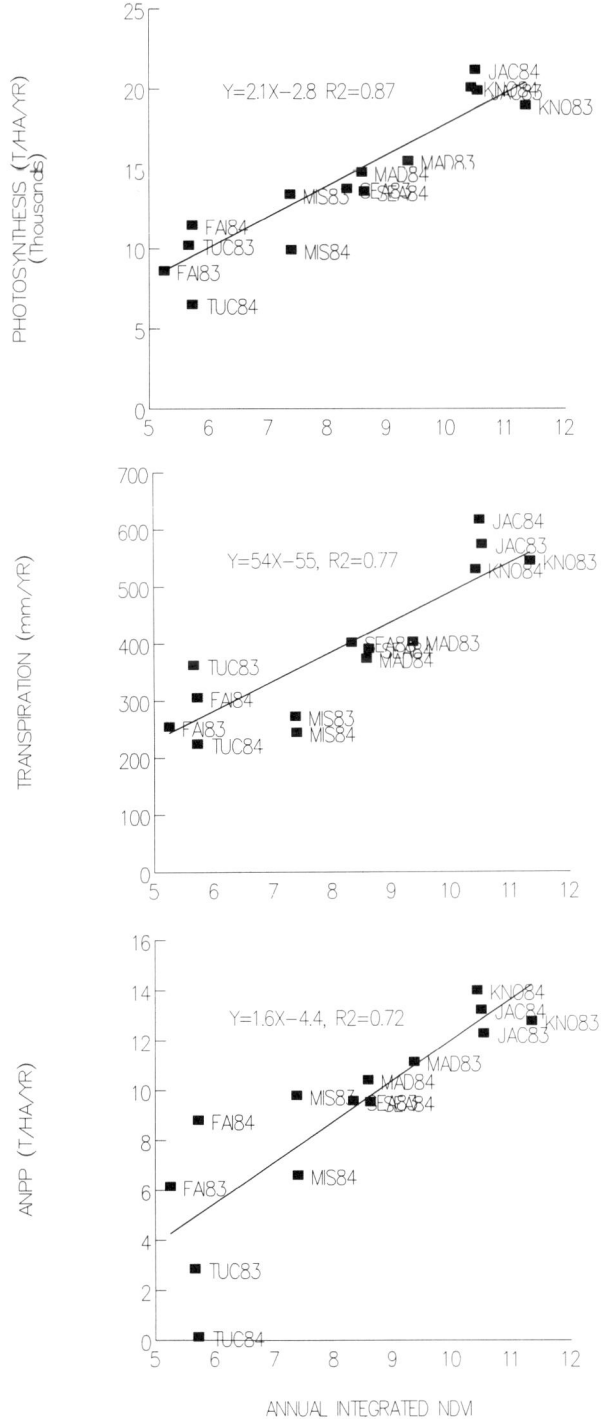

Figure 4.7. Relationship between ΣNDVI and FOREST-BGC simulations of annual photosynthesis, transpiration, and NPP for seven sites of contrasting climate in North America. These results were a test of the ability of the AVHRR ΣNDVI to infer canopy processes rates over broad regions. Abbreviations refer to the locations in Figure 4.6. [From Running and Nemani (1988).]

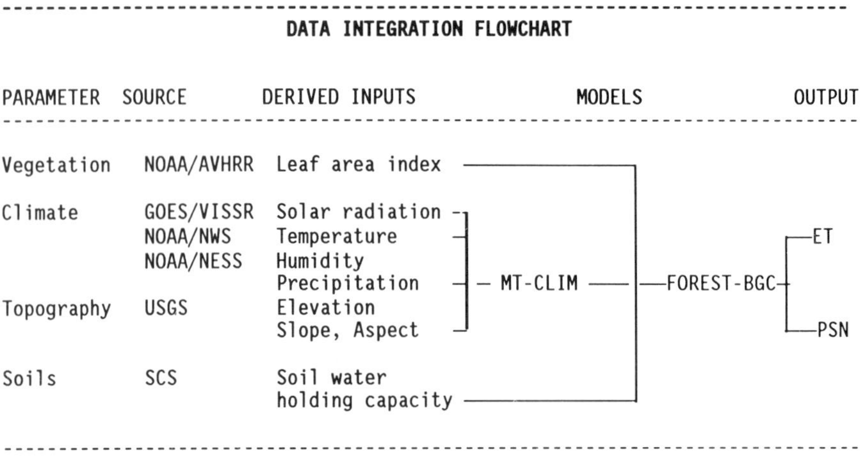

DATA INTEGRATION FLOWCHART

```
PARAMETER    SOURCE        DERIVED INPUTS              MODELS           OUTPUT

Vegetation   NOAA/AVHRR    Leaf area index ─────┐
Climate      GOES/VISSR    Solar radiation ─┐   │
             NOAA/NWS      Temperature      │   │                      ┌─ET
             NOAA/NESS     Humidity         │   │
                           Precipitation    ├─ MT-CLIM ──┤─FOREST-BGC─┤
Topography   USGS          Elevation        │   │                      └─PSN
                           Slope, Aspect ───┘   │
Soils        SCS           Soil water           │
                           holding capacity ────┘
```

NOAA = National Oceanic and Atmospheric Administration
AVHRR = Advanced Very High Resolution Radiometer
GOES = Geostationary Orbiting Environmental Satellite
VISSR = Visible and Infrared Spin Scan Radiometer
NWS = National Weather Service
NESS = National Earth Satellite Service
USGS = United States Geological Survey
SCS = Soil Conservation Service
MT-CLIM = Mountain microclimate simulator
FOREST-BGC = Forest ecosystem simulation model
ET = Evapotranspiration
PSN = Photosynthesis

Figure 4.8. An organizational diagram showing the sources of raw climatic and biophysical data, derived data, and final simulations required for generating the regional map of forest NPP in Figure 4.9. [From Running et al. (1989).]

to 14.2 T/ha/yr) across the study area follow expected trends both in magnitude and spatial pattern (Color Plate 1).

We plan next to do an ΣNDVI-generated estimate of NPP using equation 6 for the area in Color Plate 1 and compare it with the FOREST-BGC-generated NPP map. However, it is unreasonable to attempt the data-intensive computations illustrated in Figure 4.8 for 1-km cells over the entire globe. Global circulation models define the earth's surface at best to 0.5-degree grids, or about 50 by 50 km. We hypothesize, though, that ΣNDVI computations could be calibrated against high-resolution computations such as Figure 4.8 for areas representing important biome/climate types. The ΣNDVI computations for each of these biome/climate types could then be extrapolated across that region with much greater confidence.

While it is important that the ΣNDVI estimates of NPP can be implemented immediately, the benefit of FOREST-BGC simulations is that they allow the exploration of hypothetical situations that cannot be studied directly. For example, we explored the potential response of a forested region of Montana to the climate change being predicted for a doubling of atmospheric CO_2. We used a simple climatic scenario of a 4-degree increase in daily temperatures and a 10% increase in precipitation added evenly to a 1985 Missoula, Montana, data file, since current GCM results cannot be used reliably to define more exact regional scenarios (Grotch, 1988). Ecosystem responses simulated included the increase in photosynthesis rates produced by a steeper CO_2 diffusion gradient, decreased leaf conductance produced by partial stomatal closure, and re-equilibrated LAI brought about by the change in water-use efficiency and site water balance from Nemani and Running (1988a).

Results showed the higher temperatures dramatically lengthened the nonfreezing growing season in a cold climate such as in Montana from 67 to 92 days, depending on the location, accelerated snow melt dramatically, and radically changed hydrologic partitioning to evapotranspiration versus soil water outflow. Photosynthetic production increased by 5 to 30% as improved water-use efficiency and longer growing seasons combined to accelerate net photosynthesis (Color Plate 1). However, the same simulations for a hypothetical site in warm, wet Jacksonville, Florida, produced a decrease in annual photosynthesis, because the rising temperatures dramatically increased maintenence respiration losses while having proportionally less effect on photosynthesis and leaf area. Obviously, this is only speculation, but the ecosystem model provides a tool for exploring various hypotheses quantitatively and directing future research that empirical logic such as the ΣNDVI cannot provide.

Validation of Regional NPP Estimates

Although our computational ability to estimate NPP has reached continental scales, validation still requires ground measurement. The spatial scale of direct, hands-on validation measurements for most key ecosystem processes ends somewhere between the leaf and 1 km^2, depending on the process of interest. Beyond these scales, some means of defensible extrapolation must be developed. It is suggested that a well-tested ecosystem process model provides the best vehicle for doing this extrapolation. Testing of an ecosystem process model can proceed in steps of progressive complexity that, upon completion, should provide a trustworthy means of connecting ground data up to continental-scale products.

Initially, a model can be tested by individual components. Using FOREST-BGC as an example, simulations of processes like seasonal soil moisture depletion and leaf water stress have been tested in a strict predicted versus observed mode (Running, 1984; Nemani and Running,

1989a). Recognize that the successful simulations of certain key integrating variables like soil moisture depletion implicitly validate many components of the complete site hydrologic balance. Soil moisture cannot be simulated accurately if such functions as snow melt, canopy interception, or transpiration are in error. Although one can argue that internal offsetting errors could produce artifically correct results, this is not likely; if the model is tested under a range of conditions, the offsetting errors would not always be proportional.

Next, simulations can be tested in a relative, comparative way. Considering the simulation results in Figure 4.6, we know that Jacksonville, should have higher annual photosynthesis than Missoula, A model's ability to produce accurately a relative ranking of ecosystem carbon balances in contrasting sites, as partially done in Figure 4.6, provides another level of model testing and "validation."

Ultimate validation requires simulation of the variable of interest compared with direct ground measurement across a range of conditions. For example, we could define an experiment where, for the seven forest sites in Figure 4.6, we would simulate NPP under control conditions and fertilized and/or irrigated conditions. This would represent virtually a complete range of the temperature, water, nutrient, and radiation conditions known to control forest NPP. If the FOREST-BGC model could accurately simulate the magnitude of NPP under this complete range of conditions, it is suggested that the ecosystem process model would then be a trustworthy tool for extrapolating NPP estimates within the multivariate space in which it was validated. It could then provide an important intermediate link between the small-scale but ultimately accurate ground measurements and the regional to continental scales needed. This is not meant to suggest that the FOREST-BGC model is ready for this herculean test, but rather that ecosystem process models may be considered an integral component of our attempts to validate ecosystem processes at large scales.

A number of spatially integrating measurement systems are under development that may allow direct measurement of vegetation gas-exchange rates at intermediate scales. These systems include aircraft-mounted trace gas monitors, micrometeorological methods such as eddy correlation, and long-path Fourier transform IR spectroscopy.

For the ground measurements that still will be required, intelligent planning and organization may be most important. Preclassification of computed homogeneous areas may allow directed sampling, rather than a near-impossible random sample. A world network of permanent research stations routinely measuring NPP in a comparative way has been suggested as part of the International Geosphere-Biosphere Program (IGBP). The Long-Term Ecological Research (LTER) network in the United States is a good example of using existing organized data sources rather than an expensive new program.

Conclusions

Clearly, there is a necessity for a hierarchy of estimates of NPP for global ecological research. While the simplicity of the ΣNDVI logic allows immediate global implementation, mechanistic modeling, such as by FOREST-BGC is required in every major biome type to calibrate the NDVI and ϵ conversion efficiencies, and to explore hypothetical scenarios that cannot be studied otherwise. These mechanistic models must ultimately be validated against direct ground measurements. For maximum efficiency, coordinated utilization of existing ecological research stations for field NPP data will provide best results.

Acknowledgments

Much of the research in this chapter was supported by research grant #NAGW-252 from the National Aeronautics and Space Administration, Earth Sciences and Applications Division.

References

Asrar, G., Fuchs, M., Kanemasu, E.T., Hatfield, J.L. (1984). Estimating absorbed photosynthetic radiation and leaf area index from spectral reflectance in wheat. *Agron. J.* 76:300–306.

Bolin, B. (1977). Changes of land biota and their importance for the carbon cycle. *Science* 196:613–615.

Curran, P.J. (1983). Multispectral remote sensing for the estimation of green leaf area index. *Philosoph. Trans. Roy. Soc. London* 309:257–270.

Dickinson, R.E. (1987). Evapotranspiration in global climate models. *Adv. Space Res.* 7:17–26.

DOE. (1985). *Projecting the Climatic Effects of Increasing Carbon Dioxide*. In M.C. MacCracken and F.M. Luther (ed.). U.S. Dept. of Energy DOE/ER-0237 Washington, DC.

Dunin, F.R., and Aston, A.R. (1984). The development and proving of models of large scale evapotranspiration: An Australian study. *Agric. Water Manag.* 8:305–323.

Emmanuel, W.R., Killough, G.G., Post, W.M. and Shugart, H.H. (1984). Modeling terrestrial ecosystems in the global carbon cycle with shifts in carbon storage capacity by land-use change. *Ecology* 65:970–983.

Erickson, J.D. (1984). The LACIE experiment in satellite aided monitoring of global crop production. pp. 191–220. In G.M. Woodwell (ed.), *SCOPE 23: The Role of Terrestrial Vegetation in the Global Carbon Cycle*. Wiley, NY.

Gholz, H.L. (1982). Environmental limits on aboveground net primary production, leaf area, and biomass in vegetation zones of the Pacific Northwest. *Ecology* 63:469–481.

Goward, S.N. (1989). Satellite bioclimatology. *J. Clim.* 2:710–720.

Goward, S.N., Kerber, A., Dye, D.G., and Kalb, V. (1987). Comparison of North and South American biomes from AVHRR observations. *Geocarto* 2:27–40.

Goward, S.N., Tucker, C.J., and Dye, D.G. (1985). North American vegetation patterns observed with the NOAA-7 advanced very high resolution radiometer. *Vegetatio* 64:3–14.

Grier, C.C., and Running, S.W. (1977). Leaf area of mature northwestern coniferous forests: relation to site water balance. *Ecology* 58:893–899.

Grotch, S.L. (1988). *Regional Intercomparisons of General Circulation Model Predictions and Historical Climate Data*. U.S. Dept. of Energy DOE/NBB-0084, Washington, DC.

Hansen, J., Johson, D., Lacis, A., Lebedeff, S., Lee, P., Rind, D., and Russell, G. (1981). Climate impact of increasing atmospheric carbon dioxide. *Science* 213:957–966.

Holben, B.N. (1986). Characteristics of maximum value composite images from temporal AVHRR data. *Int. J. Remote Sens.* 7:1417–1434.

Hope, A.S., (1988). Estimation of wheat canopy resistance using combined remotely sensed spectral reflectance and thermal observations. *Remote Sens. Envir.* 24:369–383.

Hungerford, R.D. (1987). Estimation of foliage area in dense Montana lodgepole pine stands. *Canad. J. Forest Res.* 17:320–324.

Jarvis, P.G., and Leverenz, J.W. (1983). Productivity of temperate, deciduous and evergreen forests. In *Physiological Plant Ecology, Vol. 12D: Ecosystem Processes: Mineral Cycling, Productivity and Man's Influence*. Springer-Verlag, NY.

Jarvis, P.G., and McNaughton, K.G. (1986). Stomatal control of transpiration. Scaling up from leaf to region. *Adv. Ecolog. Res.* 15:1–49.

Justice, C., Townshend, J., Holben, B., and Tucker, C. (1985). Analysis of the phenology of global vegetation using meteorological satellite data. *Int. J. Remote Sens.* 6:1271–1318.

Knight, D.H., Fahey, T.J., Running, S.W., Harrison, A.T., and Wallance, L.L. (1981). Transpiration from 100-yr-old lodgepole pine forests estimated with whole-tree potometers. *Ecology* 62:717–726.

Leith, H., and Whittaker, R.H. (eds.) (1975). *Primary Productivity of the Biosphere*. Springer-Verlag, NY.

Matthews, E. (1983). Global vegetation and land use: new high-resolution data bases for climate studies. *J. Clim. Appl. Meteorol.* 22:474–500.

McClain, E.P. (1980). Multiple atmospheric window techniques for satellite derived sea surface temperatures. pp. 73–85. In J.F.R. Grower (ed.), *Oceanography from Space*. Plenum Press, NY.

Monteith, J.L. (1981). Climatic variation and the growth of crops. *Quart. J. Roy. Meteorol. Soc.* 107:749–774.

Mooney, H.A., Ferrar, P.J., and Slatyer, R.O. (1987). Photosynthetic capacity and carbon allocation patterns in diverse growth forms of eucalyptus. *Oecologia* 36:103–111.

NAS (Nat. Acad. Sci.). (1986). *Remote Sensing of the Biosphere. Report of the Committee on Planetary Biology*. Nat. Acad. Press, Washington, DC.

NASA (Nat. Aeronaut. and Space Admin). (1983a). *Global Biology Research Program*. MR Ramble (ed.), NASA Tech. Memo. 85629, Washington, DC.

NASA (Nat. Aeronaut. and Space Admin). (1983b). *Land-Related Global Habitability Science Issues*. NASA Tech. Memo. 85841, Washington, DC.

Nemani, R., and Running, S.W. (1989a). Testing a theoretical climate-soil-leaf area hydrologic equilibrium of forests using satellite data and ecosystem simulation. *Agric. Forest Meteorol.* 44:245–260.

Nemani, R., and Running, S.W. (1989b). Estimating regional surface resistance to evapotranspiration from NDVI and Thermal-IR AVHRR data. *J. Appl. Meteorol.* 28:276–284.

Perry, C.R., and Lautenschlager, L.F. (1984). Functional equivalence of spectral vegetation indices. *Remote Sens. Envir.* 14:169–182.

Peterson. D.L., Spanner, M.A., Running, S.W., and Teuber, K.B. (1987). Rela-

tionship of thematic mapper simulator data to leaf area index of temperate coniferous forests. *Remote Sens. Envir.* 22:323–341.
Pierce, L.L., and Running, S.W., (1988). Rapid estimation of coniferous forest leaf area index using a portable integrating radiometer. *Ecology* 69:1762–1767.
Running, S.W. (1984). Microclimate control of forest productivity: analysis by computer simulation of annual photosynthesis/transpiration balance in different environments. *Agric. Forest Meteorol.* 32:267–288.
Running, S.W., and Coughlan, J.C. (1988). A general model of forest ecosystem processes for regional applications. I. Hydrologic balance, canopy gas exchange and primary production processes. *Ecol. Model.* 42:125–154.
Running, S.W., and Nemani, R.R. (1988). Relating seasonal patterns of the AVHRR vegetation index to simulated photosynthesis and transpiration of forests in different climates. *Remote Sens. Envir.* 24:347–367.
Running, S.W., Nemani, R.R., and Hungerford, R.D. (1987). Extrapolation of synoptic meteorological data in mountainous terrain, and its use for simulating forest evapotranspiration and photosynthesis. *Canad. J. Forest Res.* 17:472–483.
Running, S.W., Nemani, R.R., Peterson, D.L., Band, L.E., Potts, D.F., Pierce, L.L., and Spanner, M.A. (1989). Mapping regional forest evapotranspiration and photosynthesis by coupling satellite data with ecosystem simulation. *Ecology* 70:1090–1101.
Running, S.W., Peterson, D.L., Spanner, M.A., and Teuber, K.B. (1986). Remote sensing of coniferous forest leaf area. *Ecology* 67:273–276.
Schulze, E.D., Cermak, J., Matyssek, R., Penka, M., Zimmerman, R., Vasicek, F., Gries, W., and Kucera, J. (1985). Canopy transpiration and water fluxes in the xylem of the trunk of Larix and Picea trees—a comparison of xylem flow, porometer and cuvette measurements. *Oecologia* (Berlin) 66:475–483.
Sellers, P.J. (1985). Canopy reflectance, photosynthesis and transpiration. *Int. J. Remote Sens.* 6:1335–1372.
Sellers, P.J. (1987). Canopy reflectance, photosynthesis and transpiration. II. The role of biophysics in the linearity of their interdependence. *Remote Sens. Envir.* 21:143–183.
Sellers, P.J., Hall, F.G., Asrar, G., Strebel, D.E., and Murphy, R.E. (1988). The first ISLSCP field experiment (FIFE). *Bull. Amer. Meteorol. Soc.* 69:22–28.
Sellers, P.J., Mintz, Y., Sud, Y.C., and Dalcher, A. (1986). A simple biosphere model (SiB) for use with general circulation models. *J. Atmos. Sci.* 43:505–531.
Spanner, M.A., Peterson, D.L., Running, S.W., and Pierce, L. (1987). The relationship of AVHRR data to the leaf area index of western coniferous forests. pp. 358–360. In *Space Life Sciences Symposium: Three Decades of Life Sciences Research in Space*. Nat. Aeron. and Space Admin., Washington, DC, June 21–26.
Spanner, M.A., Pierce, L.L., Peterson, D.L., and Running, S.W. (1989). Remote sensing of temperate coniferous forest leaf area index: The influence of canopy closure, understory vegetation and background reflectance. *Int. J. Remote Sens.* (in press)
Tarpley, J.D., Schneider, S.R., and Money, R.L. (1984). Global vegetation indices from the NOAA-7 meteorological satellite. *J. Clim. Appl. Meteorol.* 23:491–494.
Tucker, C.J. (1979). Red and photographic infrared linear combinations for monitoring vegetation. *Remote Sens. Envir.* 8:127–150.
Tucker, C.J., Townshend, J.R.G., and Goff, T.E. (1985). African land cover classification using satellite data. *Science* 227:369–374.
Tucker, C.J., Vanpraet, C., Boerwinkle, E., and Gaston, A. (1984). Satellite re-

mote sensing of total dry matter accumulation in the Senegalese Sahel. *Remote Sens. Envir* 13:461–474.

Waring, R.H., Whitehead, D., and Jarvis, P.G. (1980). Comparison of an isotopic method and the Penman-Monteith equation for estimating transpiration of Scots pine. *Canad. J. Forest Res.* 10:555–558.

Webb, W.L., Lauenroth, W.K., Szareck, S.R., and Kinerson, R.S. (1983). Primary production and abiotic controls in forests, grasslands, and desert ecosystems in the United States. *Ecology* 64:134–151.

Wessman, C.A., Aber, J.D., Peterson, D.L., and Melillo, J.M. (1988). Remote sensing of canopy chemistry and nitrogen cycling in temperate forest ecosystems. *Nature* 335:154–156.

Wiegand, C.L., Richardson, A.J., and Kanemasu, E.T. (1979). Leaf area index estimates for wheat from Landsat and their implications for evapotranspiration and crop modeling. *Agron. J.* 71:336–342.

Wilson, M.F., Henderson-Sellers, A., and Dickinson, R.E. (1987). Sensitivity of Biosphere-Atmosphere-Transfer Scheme (BATS) to the inclusion of variable soil characteristics. *J. Clim. Appl. Meteorol.* 26:341–363.

Wofsy, S.C., Harriss, R.C., and Kaplan, W.A. (1988). Carbon dioxide in the atmosphere over the Amazon Basin. *J. Geophys. Res.* 93:1377–1388.

Woodwell, G.M., Hobbie, J.E., Houghton, R.A., Melillo, J.M., Moore, B., Peterson, B.J., and Shaver, G.R. (1983). Global deforestation: Contribution to atmospheric carbon dioxide. *Science* 222:1081–1086.

Yates, H.W., Strong, A., McGinnis, D., and Tarpley, D. (1986). Terrestrial observations from NOAA operational satellites. *Science* 231:463–470.

5. Remote Sensing of Litter and Soil Organic Matter Decomposition in Forest Ecosystems

John D. Aber, Carol A. Wessman, David L. Peterson, Jerry M. Melillo, and James H. Fownes

Remote sensing is increasingly recognized as an important tool for landscape or regional estimation of ecosystem function, and for determination of biosphere-atmosphere interactions. Existing remote sensing systems have been used to monitor the seasonal phenology of standing green biomass and its production on a continental scale (Tucker et al., 1985); to measure changes in forest canopy leaf area index over large environmental gradients (Spanner et al., 1984, Running et al., 1986, Peterson et al., 1987); to track deforestation in tropical regions (Woodwell et al., 1986), and for the detection of ecosystem stress and forest decline (Rock et al., 1986). These approaches have relied on the detection of large structural changes in canopy properties that relate directly to processes controlling net primary productivity.

Decomposition of litter and soil organic matter is the complementary process to primary production, and can play a major role in controlling long-term ecosystem function through regulation of nutrient availability. Estimation of decomposition rates by remote sensing could provide information on the spatial variability of this process at scales relevant to regional-scale research. Can the decomposition process be measured in any meaningful way by remote sensing, and by parameters that do not also relate to rates of net primary productivity?

The purpose of this chapter is to present a brief review of the factors controlling decomposition in forest ecosystems, and then to discuss the potential for estimating rates of decay by remote sensing. We will present, as an example, a recent attempt to measure key forest canopy chemistry parameters, and to predict from them rates of nitrogen cycling. Results will be used to suggest the nature of ultimate controls of nutrient cycling through decomposition in forest ecosystems, and how those rates might be predicted by remote sensing.

Decomposition of Fresh Litter

Few processes have received more study in forest ecosystems than the decay of fresh litter, particularly foliar litter. Combinations of a few key parameters, usually nitrogen and lignin content, have been found to predict both weight loss and nutrient dynamics for a very wide range of litter types.

For example, Melillo et al. (1982) showed that the ratio of lignin to nitrogen in fresh foliar litter predicted initial rates of weight loss for six northern hardwood species. Similar analyses applied to a study of different foliar litters in North Carolina (Cromack, 1973) yielded similar results (Figure 5.1). Aber et al. (1980) showed that the relationship between carbon and nitrogen dynamics in most litter decay studies could be expressed as an inverse linear relationship between percent weight remaining and the nitrogen concentration in that remaining material (Figure 5.2). Melillo et al. (1982) then showed that the slope of this relationship was a function of the lignin content of the foliage. Further analysis (Aber and Melillo, 1982) produced a generalized relationship for total nitrogen immobilized in the first year of decomposition as a function of both initial lignin and initial nitrogen concentration (controlling both rate of decay and carbon-nitrogen interactions).

Decay rates of fresh litter also show a strong relationship with climate. Meentemeyer (1978) and Meentemeyer and Berg (1986) have used data sets from several sites to develop multiple regression equations that predict early litter decay rates as a function of litter quality (lignin content) and climate (summarized as actual evaportranspiration, AET).

These results suggest that estimates of foliar litter decay rates could be made from remote sensing data if lignin and nitrogen content of produced litter could be predicted with some accuracy. A confounding process here is retranslocation: the movement of nutrients, particularly nitrogen, out of foliage and into twigs during senescence (Ryan and Bormann, 1982; Flanagan and Van Cleve, 1983). This causes litter nitrogen content to be very different from that of green leaves, and requires, in turn, either that remote sensing data be acquired very near the end of the senescence process, or that litter nitrogen be predictable from some characteristic of green foliage that can be estimated by remote sensing.

Figure 5.1. Decomposition rates for several different species of leaf litter as a function of their initial lignin to nitrogen ratio. Differences in rates between New Hampshire and North Carolina litters reflects differences in climate between sites. [Melillo et al. (1982).]

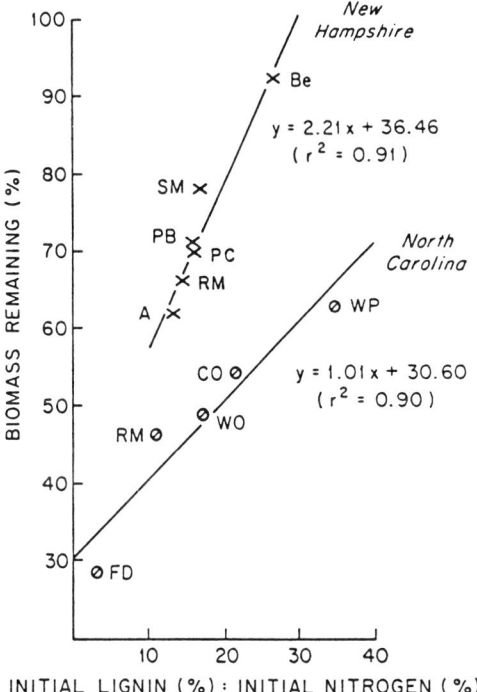

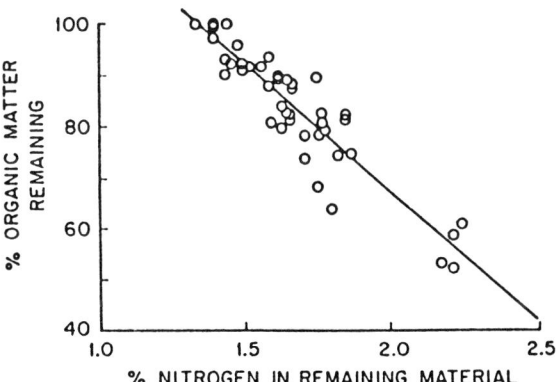

Figure 5.2. Example of inverse linear relationship between percent original weight remaining and nitrogen concentration in remaining material commonly found in decomposing foliar litter. These data are for red pine needle litter placed in forests in Alaska, Massachusetts, New Mexico, Wisconsin, Indiana, and Georgia.

Soil Organic Matter Decomposition

In stark contrast with litter decay studies, our ability to predict the decay rate and nutrient release from soil organic matter is very poor. Soil organic matter can be chemically fractionated in several ways related to molecular size, nitrogen content, chemical structure, and particle density (e.g., Schnitzer, 1978; Stevenson, 1985; Spycher et al., 1983), and yet no algorithm has been presented for predicting carbon or nitrogen release from organic matter as a function of these fractions. All estimates of soil organic matter decay involve some form of direct measurement under laboratory or field conditions. Computer models of soil organic matter dynamics generally treat this compartment as having one to three conceptually separated components that vary only in estimated turnover rate (Parton et al., 1988; Aber et al., 1982; Pastor and Post, 1986). The relative amount of material entering each compartment may be a function of the initial chemical quality of litter.

While litter and humus decay are clearly similar processes, there has been very little work that links the two. Litter decay work has concentrated on biochemical characteristics controlling microbial processing, whereas soil organic matter research has focused on characterizing the type of compounds present and on theories of humus formation.

Recent efforts have used long-term litter decomposition to attempt to bridge the gap between litter and humus decay research. These studies have looked at continuous changes in organic fractions through time or through soil horizons (e.g., Berg and Agren, 1984; Berg et al., 1982). At least two characteristics of decaying organic matter show continuous changes from the litter to humus stages.

The first is the inverse-linear relationship between weight loss and nitrogen concentration, which holds through 75 to 80% of initial weight lost (e.g., Figure 5.2). A cross-site experiment has also demonstrated that the slope of this relationship is not altered by climate, although the rate of decay is (Melillo et al., 1988).

A second continuous change is in the ratio of lignin to cellulose, expressed as the ligno-cellulose index (LCI, lignin/[lignin + cellulose]). This increases continuously during decay as cellulose decays more rapidly than lignin. Long-term studies suggest that the decay rate of organic matter slows dramatically once the LCI reaches approximately 0.7, and that the inverse linear relationship between carbon and nitrogen dynamics breaks down at this point as well (Melillo et al., 1988; McClaugherty and Berg, 1987; Melillo, unpublished).

Long-term litter decay studies provide insight into the dynamics of the most rapidly decaying fraction of soil organic matter. That these dynamics are an extension of early litter decay processes suggests that long-term litter decay dynamics may be predictable from litter chemistry. Yet the fraction of total nutrient release in forest ecosystems that comes from this "old litter" and the fraction that comes from "humus" (older material with an LCI greater than about 0.7) remain unclear. If chemical fractionation and prediction of humus dynamics are not well known in the field, what are the chances that they can be predicted by remote sensing?

Prediction of Nitrogen Mineralization by Remote Sensing—a Case Study

Nitrogen mineralization from soil organic matter is a crucial by-product of the decay process. Nitrogen is generally the most limiting nutrient in temperate forest ecosystems, and continued primary production depends on a continuous supply of mineral nitrogen from decayed soil organic matter. Remote sensing of the mineralization process itself is not a realistic possibility. The estimation of nitrogen mineralization by remote sensing will, then, require that some strong, causal relationship exist between characteristics of forest ecosystems that are accessible to remote sensing technology and the factors controlling the decay process. We have explored this possibility using experimental high-spectral-resolution imaging spectrometers currently under development through the National Aeronautics and Space Administration (NASA).

The goal of this study was to determine whether differences in the lignin and nitrogen content of whole canopies from a diverse set of well-studied forest ecosystems could be detected and predicted using spectral reflectance data from imaging spectrometers, and whether those differences related to field measured rates of nitrogen mineralization. Results were also compared with data on soil texture and net primary productivity (above ground) to identify other ecosystem characteristics that might be related to canopy chemistry and nitrogen mineralization.

This study was made possible by the availability of 18 previously studied stands in the University of Wisconsin Arboretum in Madison and on Blackhawk Island near Wisconsin Dells (Wessman et al., 1988; Nadelhoffer et al., 1983; Pastor et al., 1984). The Blackhawk Island sites represent a set of mature, undisturbed stands that show a remarkable correlation between soil type and vegetation (Pastor et al., 1982). In contrast, most of the Arboretum sites are restored ecosystems in which vegetation has been manipulated to attempt reconstruction of particular forest types (Jordan et al., 1987). Dominant species in these stands included red and sugar maple; red, white and black oak; red and white pine; black cherry, and basswood. The use of both mature and disturbed sites allowed testing of the generality of the relationships derived.

All of the stands used had been previously characterized for annual nitrogen mineralization rate using the on-site incubation technique for total above-ground productivity and for soil characteristics, including soil texture (Pastor et al., 1982, 1984; Nadelhoffer et al., 1983, 1985; Lennon et al., 1985). Nitrogen mineralization rates for Blackhawk Island sites used here are somewhat higher than reported by Pastor et al. (1984) because of the inclusion of measured rates for the 0- to 10-cm layer in mineral soil.

Total canopy chemistry was determined in each stand by combining analyses of green foliar lignin and nitrogen content with measured rates of foliage litter fall, multiplied by a measured foliar retention time for the pine species.

Near-infrared (IR) reflectance of whole canopies for 18 of our sites was measured in August, 1985 using the Airborne Imaging Spectrometer (AIS, Goetz et al., 1985) mounted on a NASA C-130 aircraft. Developed at the Jet Propulsion Laboratory for NASA, the AIS is an experimental instrument using 32- by 32-element mercury cadmium telluride arrays to measure reflected solar radiance in 32 wavelength bands for 32 picture elements (pixels) simultaneously. As the aircraft advances the equivalent of one pixel in ground distance, four grating positions are used to record reflectance in 128 wavelength bands. Data used in this paper were collected for 128 wavelength bands between 1.2 and 2.4 μm (micrometers). It was discovered during data quality analysis that values obtained above 1.6 μm were subject to second-order contamination from shorter wavelengths, restricting this analysis to the region between 1.2 and 1.6 μm. For a more detailed discussion of the remote sensing techniques, see Wessman et al. (1987) and Peterson et al. (1988).

Spectra were averaged from a three-by-three pixel area covering each of the study sites sampled in 1985 (pixel size was roughly 11 by 11 m, study site plots were 30 by 30 m).

Variation in Canopy Characteristics and Ecosystem Function

The several canopy and ecosystem parameters measured in all stands show quite different degrees of variability (Figure 5.3). A fivefold range in nitrogen mineralization rates is translated into less than a twofold range in total canopy nitrogen content. This small range results from higher foliar biomass in the pine stands being offset by lower nitrogen concentration compared with hardwood stands (Appendix 1). In contrast, pine foliage is high in lignin content relative to most hardwood species, such that total lignin content of canopies varies nearly ninefold. Total lignin content within the hardwood stands alone is also more variable than total nitrogen content, because of large differences in lignin concentration between oaks and maples (Appendix 1). Above-ground NPP varies only threefold across the sites, also less than nitrogen mineralization, suggesting a decline in nitrogen-use efficiency (Vitousek, 1982) with increasing nitrogen mineralization.

Remote Sensing of Total Canopy Chemistry

Results of the remote sensing analyses have been presented in detail elsewhere (Wessman et al., 1987, 1988a). Using a three-wavelength model

5. Remote Sensing of Litter and Organic Matter Decomposition   93

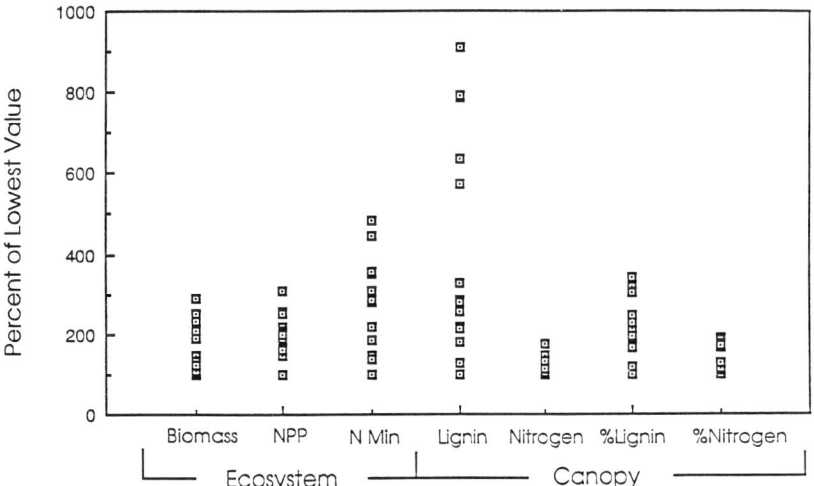

Figure 5.3. Variation in canopy and ecosystem parameters for stands used in this step. All data are expressed as a percentage of the lowest value for that parameter. Data on net annual nitrogen mineralization, NNP, and above-ground live biomass are from Nadelhoffer et al. (1983, 1985), Pastor et al. (1984), and Lennon (1985).

based on first-order derivative spectra, a strong correlation was obtained between measured lignin concentration in whole canopies and near-IR reflectance as measured by the AIS (Figure 5.4). This analysis includes both disturbed and undisturbed stands.

Prediction of nitrogen content or concentration in whole canopies was

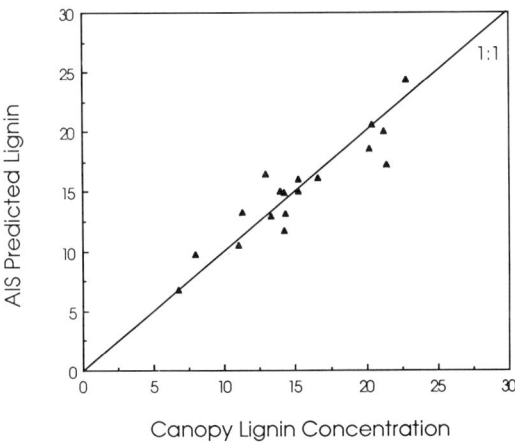

Figure 5.4. Measured lignin concentration in whole canopies compared with concentration as predicted by multiple regression using first derivative of reflectance at three wavelengths (1256, 1555, 1311 nm) extracted from spectra acquired with the Airborne Imaging Spectrometer (AIS) (n = 18, R^2 = 0.85). [Wessman et al. (1988a).]

less successful (Wessman et al., 1987), mainly because of the narrow range of values measured in the field. Because of this, further discussion in this chapter will be limited to canopy lignin concentrations.

Canopy lignin concentrations were available for three stands measured in 1984 that were not used in the above analysis. The standard error of prediction for these three stands using 1984 field data and 1985 AIS data was 1.6% lignin (11% of the mean value of 15.0% lignin for this three stands, Wessman et al., 1988a).

Relationships Between Canopy Lignin Concentration and Ecosystem Parameters

There is a very strong inverse relationship between field-measured, whole-canopy lignin concentration and measured rates of net annual nitrogen mineralization (Figure 5.5). Using only undisturbed stands on Blackhawk Island, the relationship is particularly strong ($R^2 = 0.99$, $n = 7$, $p < 0.0001$, SEE = 3.19). Adding the Arboretum stands, which are in different stages of recovery from human use, the correlation is less strong ($R^2 = 0.21$, $n = 14$, $p = 0.0347$, SEE = 30.1), largely because of a single outlier. Removing this one point yields again a strong correlation ($R^2 = 0.78$, $n = 13$, $p < 0.0001$, SEE = 15.3).

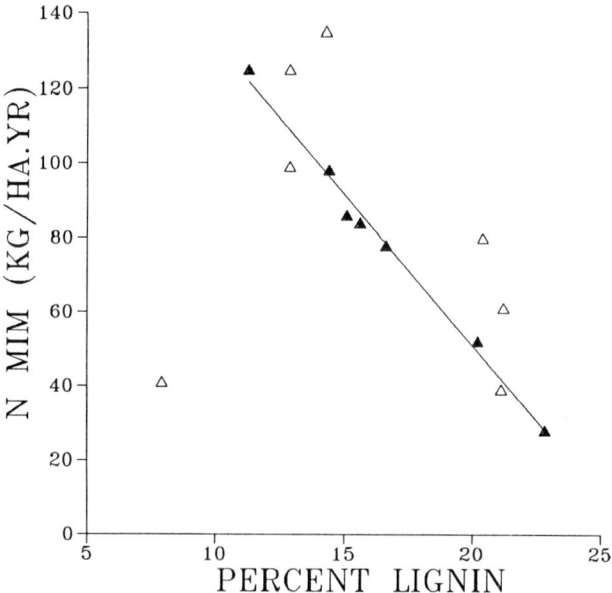

Figure 5.5. Relationship between field-measured whole canopy lignin concentration and field-measured rates of net annual nitrogen mineralization (filled triangles = undisturbed sites, Blackhawk Island; open triangles = disturbed and restored sites, Arboretum. Line is best fit to Blackhawk Island data.).

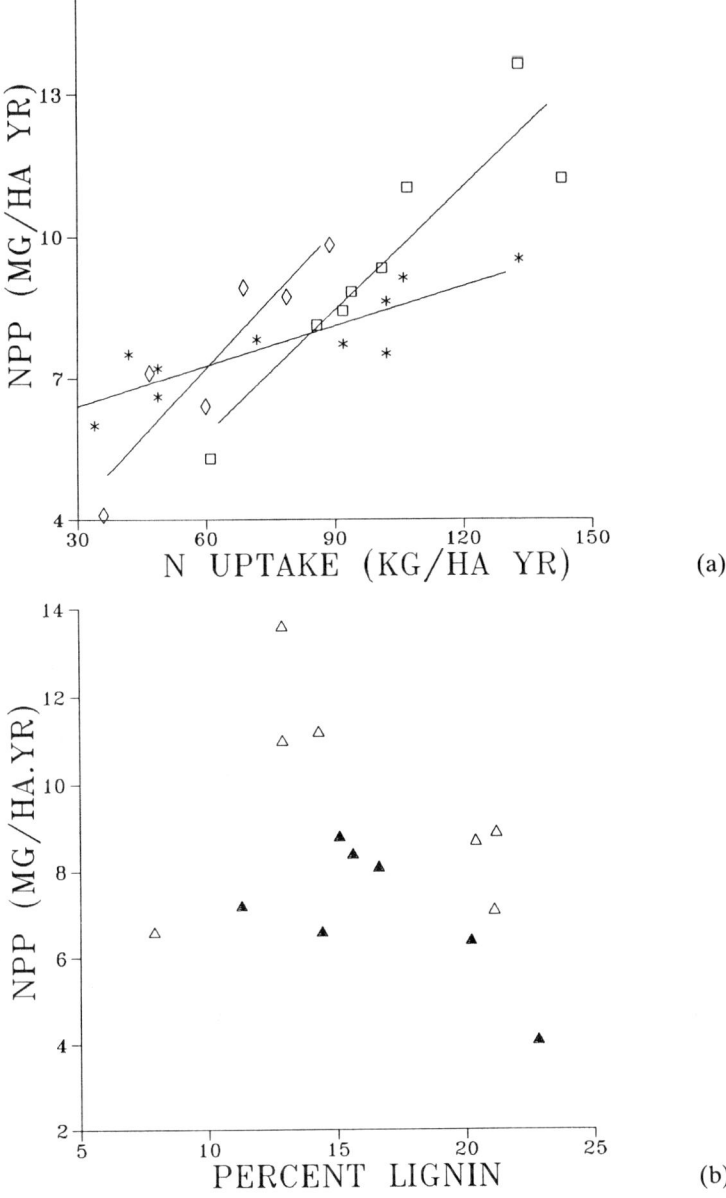

Figure 5.6. (a) Genera-specific relationships between measured rates of net annual nitrogen mineralization and measured NPP above-ground (data compiled from Pastor et al., 1984; Nadelhoffer et al., 1985; Lennon et al., 1985; Aber et al., 1983). Nitrogen mineralization data from Blackhawk Island are slightly higher than reported by Pastor et al. (1984) due to inclusion of measured rates in the mineral soil. (∗ = *Acer* [$R^2 = 0.76$, $n = 10$, $p < 0.01$], □ = *Quercus* [$R^2 = 0.84$, $n = 8$, $p < .01$], ◇ = *Pinus* [$R^2 = 0.84$, $n = 6$, $p < 0.5$]). (b) Field-measured above-ground NPP versus field-measured whole-canopy lignin concentration (filled triangles = Blackhawk Island, open triangles = Arboretum).

95

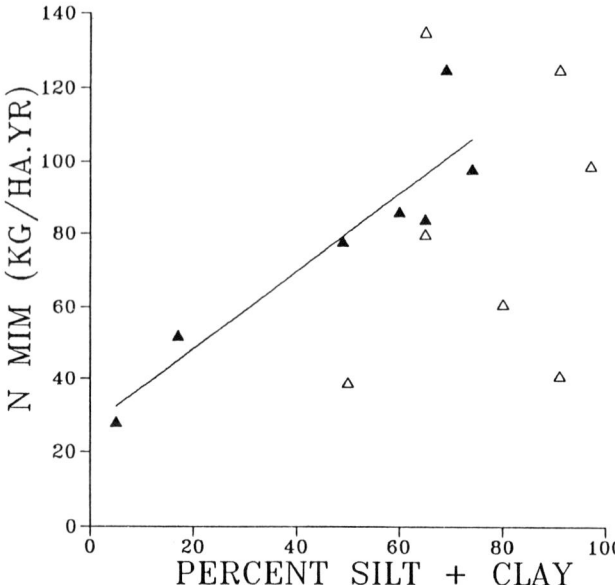

Figure 5.7. Measured relationship between soil texture (percent silt + clay) and net annual nitrogen mineralization. (Filled triangles = Blackhawk Island, open triangles = Arboretum. Line represents best fit to Blackhawk Island values $R^2 = 0.82, p = 0.002$).

We have used a two-step model to predict net annual nitrogen mineralization for all of Blackhawk Island from AIS data (Wessman et al, 1988a). First, AIS data are used to estimate the percent foliar lignin for each pixel within the 0.75-km^2 study area. Patterns in this scene correspond very well with known patterns of species distribution (Pastor et al., 1982) and measured species foliar lignin values (Appendix 1). Second, net annual nitrogen mineralization is estimated from canopy lignin concentration using the regression above (Figure 5.5) for undisturbed stands only. Validation of these estimates using four stands on Blackhawk Island not included in the regression analysis, but for which net nitrogen mineralization has been measured, indicates a standard error of prediction of 7.5 kg N·ha^{-1}·yr^{-1} ($n = 4$, mean = 74.25; Wessman et al., 1988a).

In previous papers, we have shown that a strong relationship exists between nitrogen mineralization and total above-ground NPP for different subsets of sites (Nadelhoffer et al., 1985; Pastor et al., 1984; Lennon et al., 1985). Combining data from all studies, the relationship becomes even stronger, but genus specific (Figure 5.6(a)). Thus the relationship of the remotely sensed variable, lignin concentration, with NPP, is weaker than with nitrogen mineralization (Figure 5.6 (b)).

There is also a strong relationship between nitrogen mineralization and soil texture (a surrogate for water-holding capacity in the deep, stone-free soils of the study sites) for the undisturbed sites only. It does not apply to the disturbed sites (Figure 5.7; Nadelhoffer et al., 1983; Pastor et al., 1982, 1984). This results in no relationship between canopy lignin and soil texture for the full set of stand.

Implications of These Results for General Predictability of Organic Matter Decay Rates and Ecosystem Function

Correlational analyses of this type always raise the question of causation versus chance correlation. To what extent can the results be extended to other systems? How generally can rates of nitrogen mineralization (and to some extent soil organic matter decay) be predicted from foliar lignin concentration? These questions can be posed, to some extent, in two ways: (1) Statistically—what part of the potential universe of forests are included in the analysis? (2) Theoretically—is there a theoretical underpinning that suggests these relationships should occur?

Statistically, the accurate prediction of lignin concentration for canopies of very different species composition and foliar morphology and chemisty suggests that the remote sensing results are generalizable, at least across temperate forest ecosystems. However, theoretical support for this relationship is lacking, as the near-IR absorption spectrum for lignin, a large, amorphous aromatic polymer with variable structure, is not yet well defined. However, we have recently succeeded in developing an algorithm for predicting lignin content of dried, ground foliage using near-IR reflectance characteristics (Wessman et al., 1988b). Still, it is possible that some other canopy component or characteristic both is strongly related to total lignin content and is responsible for the differences in derivative reflectance spectra found in this analysis.

The diversity of forest types used in this analysis also suggests that the relationship between canopy lignin concentration and nitrogen mineralization rates is generalizable. In this case, there is a theoretical basis for the relationship in that leaf litter is a major component of total biomass and nutrient inputs to forest soils, and the lignin content of litter is an important determinant of rates of litter decay and nitrogen release (Berg and Agren, 1984; Berg et al., 1982; Melillo et al., 1982; Aber and Melillo, 1982).

However, the single major outlier, a sugar maple plantation established on an old field in the Arboretum, indicates that the relationship between canopy lignin and nitrogen mineralization is not one of immediate cause and effect. Chemical analyses of foliage collected across the full range of nitrogen mineralization rates showed no detectable changes in lignin content for leaves of a particular species (Fownes,1985). Changes in nitrogen concentration were also small and nonsignificant for individual species. Thus foliar chemistry within a species is not drastically altered in response to nitrogen mineralization (compare the results of fertilization experiments, e.g., Turner, 1977; Safford et al., 1977; Miller et al., 1981; Weetman and Fournier, 1984). This suggests that species replacement will have to occur for canopy chemistry to adjust to altered nitrogen mineralization rates. This also suggests an interesting question: Does the lignin content of the array of species that comes to dominate a disturbed site reflect the nitrogen mineralization rate on that site?

Similarly, nitrogen mineralization rates do not change immediately in response to foliar lignin concentrations. It appears that 50 years of occupation of an old field site by sugar maple has not altered the nitrogen mineralization rate significantly, in spite of a precipitation input rate of 8 kg inorganic $N \cdot ha^{-1} \cdot yr^{-1}$ (Nadelhoffer et al., 1985) and rapid decay and incorporation of foliar litter into the soil. This may be due to long-term replenishment of soil humus lost during years of cultivation. The question of rates of change in nitrogen mineralization within forest soils is one that deserves further attention.

The tight four-way relationship among foliar lignin, soil texture, species occurrence, and NPP for the undisturbed sites supports both theoretical-physiological (Mooney and Gulmon, 1982; Coley et al., 1985) and modeling (Pastor and Post, 1986) research that suggests that water availability is a prime determinant of the inherent productive capacity of a site, and hence controls realizable NPP, species composition, and concentrations of secondary compounds, such as lignin, in foliage. It further supports the idea that the equilibrium rate of nitrogen cycling achieved is controlled proximally by the nitrogen content and decay rate of litter, and ultimately by the fundamental productive capacity of the site, in these stands determined by water-retention capacity. And finally, it suggests that the adjustment between site and species in terms of bringing foliar chemistry in line with nitrogen availability does occur within the successional lifetime of a forest.

Both the NPP and soil-texture analyses imply that relationships between remote sensing data and ecosystem processes will become less precise the further removed the predicted process is from direct interactions with canopy chemistry. Canopy lignin content and nitrogen mineralization are coupled through effects of lignin on decomposition. Relationships with NPP and soil texture depend on more complex, long-term interactions and are therefore "noisier" and less likely to apply to disturbed sites.

We conclude that remote sensing of important canopy chemical features is feasible. Predicting rates of important ecosystem processes from remotely sensed data is less certain, and depends on strong correlations between canopy variables and the process of interest. Our results suggest that those correlations will become less predictive for processes that are not functionally related to the measured canopy variables. With regard to rates of organic matter decay, it appears that the tight coupling of canopy chemistry with nitrogen mineralization may allow a generalized predictive ability at least for this close surrogate of decomposition.

Acknowledgments

This research was supported through grants from NSF (BSR-8317531), the NASA-Ames Research Center (NCA 2-28), and the College of Agriculture and Life Sciences, University of Wisconsin. Steve McNulty, Sue Kaat, Rich Hallett, Mary Martin, and Mike Spanner assisted with sample collection, handling, and analysis. Andy Turner performed the lignin analyses.

References

Aber, J.D., and Melillo, J.M. (1980). Litter decomposition: measuring relative contribution of organic matter and nitrogen to forest soils. *Canad. J. Bot.* 58:416–421.

Aber, J.D., and Melillo, J.M. (1982). Nitrogen immobilization in decaying hardwood leaf litter as a function of initial nitrogen and lignin content. *Canad. J. Bot.* 60:2263–2269.

Aber, J.D., Melillo, J.M., and Federer, C.A. (1982). Predicting the effects of rotation length, harvest intensity and fertilization on fiber yield from northern hardwood forests in New England. *Forest Sci.* 28:31–45.

Berg, B., and Agren, G. (1984). Decomposition of needle litter and its organic-chemical components-theory and field experiments. Long-term decomposition in a scots pine forest III. *Canad. J. Bot.* 62:2880–2888.

Berg, B., Hannus, K., Popoff, T., and Theander, O. (1982). Changes in organic chemical components of needle litter during decomposition. Long term decomposition in a scots pine forest I. *Canad. J. Bot.* 60:1310–1319.

Coley, P.D., Bryant, J.P., and Chapin, F.S. (1985). Resource availability and plant antiherbivore defense. *Science* 230:895–899.

Cromack, K. (1973). Litter production and litter decomposition in a mixed hardwood watershed and in a white pine watershed at Coweeta Hydrologic Station, North Carolina. Dissertation, Univ. of Georgia, Athens.

Flanagan, P.W., and Van Cleve, K. (1983). Nutrient cycling in relation to decomposition and organic matter quality in taiga ecosystems. *Canad. J. Forest Res.* 13:795–817.

Fownes, J.H. (1985). Water use and primary production of Wisconsin hardwood forests. Dissertation, Univ. of Wisconsin, Madison.

Goetz, A.F.H., Vane, G., Solomon, J., and Rock, B.N. (1985). Imaging spectrometry for earth remote sensing. *Science* 228:1147–1153.

Jordan, W.R., Gilpin, M.E., and Aber, J.D. (1987). Restoration ecology: Ecological restoration as a technique for basic research. pp. 3–22. In W.R. Jordan, M.E. Gilpin, and J.D. Aber (eds.), *Restoration Ecology: A Synthetic Approach to Ecological Research.* Cambridge Univ. Press, Cambridge, England.

Lennon, J.M., Aber, J.D., and Melillo, J.M. (1985). Primary production and nitrogen allocation of field grown sugar maple in relation to nitrogen availability. *Biogeochem.* 1:135–154.

McClaugherty, C.A., and Berg, B. (1987). Cellulose, lignin and nitrogen concentrations as rate regulating factors in late stages of forest litter decomposition. *Pedobiologia* 30:101–112.

Meentemeyer, V. (1978). Macroclimate and lignin control of litter decomposition rates. *Ecology* 59:465–472.

Meentemeyer, V., and Berg, B. (1986). Regional variation in rate of mass loss of *Pinus sylvestris* needle litter in Swedish pine forests as influenced by climate and litter quality. *Scand. J. Forest Res.* 1:167–180.

Melillo, J.M., and Aber, J.D. (1984). Nutrient immobilization in decaying litter: an example of carbon-nutrient interactions. In: H. Cooley and F. Goley (eds.), *Trends in Ecological Research*. Plenum, NY.

Melillo, J.M., Aber, J.D., Linkins, A.E., Ricca, A., Fry, B., and Nadelhoffer, K.J. (1989). Carbon and nitrogen dynamics along the decay continuum: Plant litter to soil organic matter. *Plant and Soil* (in press).

Melillo, J.M., Aber, J.D., and Muratore, J.M. (1982). Nitrogen and lignin control of hardwood leaf litter decomposition dynamics. *Ecology* 63:621–626.

Miller, A.G., Miller, J.D., and Cooper, J.M. (1981). Optimum foliage nitrogen concentration in pine and its change with stand age. *Canad. J. Forest Res.* 11:563–572.

Mooney, H.A., and Gulmon, S.L. (1982). Constraints on leaf structure and function in relation to herbivory. *BioScience* 32:198–206.

Nadelhoffer, K.J., Aber, J.D., and Melillo, J.M. (1983). Leaf-litter production and soil organic matter dynamics along a nitrogen mineralization gradient in southern Wisconsin (USA). *Canad. J. Forest Res.* 13:12–21.

Nadelhoffer, K.J., Aber, J.D., and Melillo. J.M. (1985). Fine root production in relation to total net primary production along a nitrogen availability gradient in temperate forests: A new hypothesis. *Ecology* 66:1377–1390.

Parton, W.J., Stewart, J.W.B., and Cole, C.V. (1988). Dynamics of C, N, P and S in grassland soils: a model. *Biogeochem.* 5:109–132.

Pastor, J., Aber, J.D., McClaugherty, C.A., and Melillo, J.M. (1982). Geology, soils and vegetation of Blackhawk Island, Wisconsin. *Amer. Midland Naturalist* 108:266–277.

Pastor, J., Aber, J.D., McClaugherty, C.A., and Melillo, J.M. (1984). Aboveground production and N and P cycling along a nitrogen mineralization gradient on Blackhawk Island, Wisconsin. *Ecology* 65:256–268.

Pastor, J., and Post, W.M. (1986). Influence of climate, soil moisture and succession on forest carbon and nitrogen cycles. *Biogeochem.* 2:3–28.

Peterson, D.L., Aber, J.D., Matson, P.A., Card, D.H., Swanberg, N., Wessman, C., and Spanner, M. (1988). Remote sensing of forest canopy and leaf biochemical contents. *Remote Sens. Envir.* 24:85–108.

Rock, B.N., Vogelmann, J.E., Williams, D.L., Vogelmann, A.F., and Hoshizaki, T. (1986). Remote detection of forest damage. *BioScience* 36:439–445.

Running, S.W., Peterson, D.L., Spanner, M.A., and Teuber, K.B. (1986). Remote sensing of coniferous forest leaf area. *Ecology* 67:273–276.

Ryan, D.F., and Bormann, F.H. (1982). Nutrient resorption in northern hardwood forests. *BioScience* 32:29–32.

Safford, L.O., Young, H.E., and Knight, T.W. (1977). Effect of soil and urea fertilization on foliar nutrients and basal area growth of red spruce. *Univ. Maine Life Sci. Agric. Exp. Sta. Tech. Bull.* 740.

Schnitzer, M. (1978). Humic substances: Chemistry and reactions. pp. 1–64. In M. Schnitzer and S.U. Khan (eds.). *Soil Organic Matter*. Elsevier, Oxford, England.

Spanner, M.A., Peterson, D.L., Hall, M.H., Wrigley, R.C. Card, D.H., and Running, S.W. (1984). Atmospheric effects on the remote sensing estimation of forest leaf area index. pp. 1295–1308. In *Proc. Eighth International Symposium on Remote sensing of Environment*. Univ. of Michigan, Ann Arbor.

Spycher, G., Sollins, P., and Rose, S.L. (1983). Carbon and nitrogen in the light fraction of a forest soil: Vertical distribution and seasonal patterns. *Soil Sci.* 135:79–87.

Stevenson, F.J. (1985). Geochemistry of soil humic substances. In D.M. McKnight (ed.), *Humic Substances in Soil, Sediment and Water: Geochemistry, Isolation and Characterization*. Wiley, NY.

Tucker, C.J., Townshend, J.R.G., and Goff, T.E. (1985). African land-cover classification using satellite data. *Science* 227:369–375.
Turner, J. (1977). Effects of nitrogen availability on nitrogen cycling in a Douglas-fir stand. *Forest Sci.* 23:307–316.
Vitousek, P.M. (1982). Nutrient cycling and nutrient use efficiency. *Amer. Naturalist* 119:553–572.
Weetman, G.F., and Fournier, R.M. (1984). Ten-year growth and nutrition effects of a straw treatment and of repeated fertilization on jack pine. *Canad. J. Forest Res.* 14:416–423.
Wessman, C.A, Aber, J.D., and Peterson, D.L. (1987). Estimation of forest canopy characteristics and nitrogen cycling using imaging spectrometery. pp. 114–18. In G. Vane (ed.), *Imaging Spectroscopy II, Proceedings of the International Society for Optical Engineering, Vol. 834*. San Diego, CA.
Wessman, C.A., Aber, J.D., Peterson, D.L., and Melillo, J.M. (1988a). Foliar analysis using near infrared spectroscopy. *Canad. J. Forest Res.* 18:6–11.
Wessman, C.A., Aber, J.D., Peterson, D.L., and Melillo, J.M. (1988b). Remote sensing of canopy chemistry and nitrogen cycling in temperate forest ecosystems. *Nature* 335:154–156.

Appendix 1

Sampling of green foliage chemistry as part of this project yielded a good deal of information regarding the variation in nitrogen and lignin content, and also weight-to area ratios by species with height. The following table summarizes those results. In general, lignin and nitrogen concentrations did not change significantly or consistently with height, although weight-to-area ratios did.

Table 5.1. Weight-to-Area Ratios and Percent Lignin and Nitrogen Content for Foliage for Different Species Collected From Four Levels Within the Canopy (1 = Dominant, 2 = Codominant, 3 = Intermediate, 4 = Overtopped). Data From 1984 Collection. Pine Foliage Area Values Are One-Sided Projection

							Canopy Position								
		1			2			3			4			All	
Species	N	Mean	SD	N	Mean	SD	N	Mean	SD	N	Mean	SD	N	Mean	SD
Nitrogen (percent dry weight):															
Red pine	2	1.18	0.05	6	1.23	0.13	1	1.26	—	—	—	—	9	1.22	0.11
White pine	2	1.34	0.06	7	1.44	0.17	6	1.55	0.20	1	1.50	—	17	1.47	0.17
Red maple	3	2.04	0.07	1	1.99	—	3	1.81	0.36	4	1.89	0.15	11	1.92	0.23
Sugar maple	2	1.97	0.11	3	2.16	0.27	4	2.14	0.22	6	2.15	0.19	15	2.13	0.21
Black oak	3	2.31	0.08	—	—	—	2	2.42	0.00	—	—	—	5	2.35	0.08
Red oak	7	2.47	0.17	5	2.45	0.11	2	2.68	0.10	—	—	—	14	2.49	0.17
Black cherry	1	2.23	—	4	2.61	0.26	2	2.41	0.23	4	2.70	0.33	11	2.57	0.31
White oak	9	2.64	0.20	5	2.64	0.13	5	2.52	0.19	3	2.53	0.12	22	2.60	0.18
Basswood	—	—	—	2	2.80	0.18	2	2.83	0.07	4	2.94	0.23	8	2.88	0.20
Lignin (percent dry weight):															
Basswood	—	—	—	2	7.7	1.63	—	—	—	2	7.1	1.35	4	7.4	1.53
Sugar maple	1	9.6	—	3	7.2	0.45	3	11.8	2.27	3	6.6	1.29	10	8.6	2.68
Black cherry	1	9.4	—	4	12.2	2.29	—	—	—	2	7.1	1.38	7	10.3	2.94
Red maple	2	14.6	0.47	—	—	—	2	9.7	0.22	4	12.7	2.10	8	12.4	2.33
White oak	8	13.7	2.29	4	12.0	1.03	5	13.5	2.00	2	11.4	5.00	19	13.1	2.63
Black oak	2	18.2	2.40	—	—	—	2	18.3	0.24	—	—	—	4	18.2	1.71
Red oak	6	20.3	2.57	4	18.8	2.58	1	14.7	—	—	—	—	11	19.2	2.93
White pine	2	18.7	0.01	7	21.2	1.23	6	22.0	0.93	1	22.9	—	17	21.3	1.40
Red pine	2	21.3	1.80	6	21.8	1.40	1	20.4	—	—	—	—	9	21.5	1.47

5. Remote Sensing of Litter and Organic Matter Decomposition   103

Dry weight per unit area (mg/cm²):

Species														
Basswood	—	—	2	5.4	0.28	2	5.5	2.21	4	4.0	1.43	8	4.7	1.67
Sugar maple	9.5	0.03	3	6.0	1.14	4	4.4	0.40	6	3.4	0.41	15	5.0	2.11
Red maple	6.2	0.62	1	6.3	—	3	5.5	1.70	4	5.2	1.43	11	5.7	1.36
Black cherry	5.2	—	4	7.6	1.09	2	6.3	0.38	4	4.2	0.45	11	5.9	1.62
White oak	10.9	1.95	5	8.1	2.04	5	7.1	1.25	3	4.9	0.08	22	8.6	2.73
Red oak	10.8	0.85	5	9.6	0.82	2	6.6	0.78	—	—	—	14	9.8	1.63
Black oak	10.3	0.91	—	—	—	2	9.5	0.55	—	—	—	5	10.0	0.89
White pine	39.5	18.3	64	41.6	29.3	48	33.5	6.2	18	42.2	8.0	146	38.8	20.8
Red pine	60.6	4.1	45	56.3	6.0	8	49.7	2.2	—	—	—	68	56.5	5.3

6. Water and Energy Exchange

Robert E. Dickinson

The goal of obtaining water and energy exchange on continental scales from remote sensing is fundamental to the questions not only of ecosystem functioning but also of land climate processes and regional hydrology. A general conceptual framework is described here for carrying out this work. Progress up to now has been limited for several reasons: the current sensing systems are probably inadequate for the task, the information content of potential future systems has not been adequately characterized by modeling sensitivity studies, and the linked remote sensing and modeling infrastructure has not yet been developed that is needed to carry out this activity.

An optimum approach to water and energy exchange at the land surface involves combining several kinds of observations with an appropriate modeling framework. Observations known to contain information about water and energy fluxes are radiative skin temperatures over the diurnal cycle; rainfall (with as much spatial and temporal detail as possible over continental scale areas); divergence of moisture flux in an atmospheric column; descriptions of the surface vegetation cover in terms of parameters that affect evapotranspiration and surface albedo, and likewise for soils and terrain in terms of their effects on surface hydrology; any direct measures of soil moisture that are possible such as can be inferred from microwave emissivities; all the observations needed for an atmospheric model to provide adequately surface air temperatures, winds, and relative humidity;

and adequate information on atmospheric cloud, aerosol, and humidity structure to estimate surface incident solar and longwave radiation.

Partial approaches to estimating surface water and energy fluxes are possible by using subsets of the above information and that is what has been done up to now, as reviewed in the following sections. An integrated approach using General Circulation Models (GCMs) linked to remote sensing would be expedited by use of the same physical process algorithms for both model and remote sensing. For pedagogical reasons, this chapter introduces simplified versions of algorithms suitable for this purpose.

The Skin-Temperature Method

Skin temperature is the surface temperature inferred from thermal emission. It will generally be some average of canopy and soil temperature, according to the source of emission. The simplest version of using skin temperature to infer surface energy fluxes is the application of a hand-held radiometer and air thermometer to measure the difference in temperature between a canopy and surface air. As a canopy becomes more water stressed, it becomes warmer as a result of the decrease in evaporative cooling. Thus, although the canopy air temperature difference is most directly dependent on evapotranspiration, it can be correlated with crop water stress, leaf water potential, or soil moisture, according to what related local site measurements have been made (e.g., Idso et al., 1975; Idso and Ehler, 1976, 1977; Jackson and Reginato, 1976; Jackson, 1985; Ehler et al., 1978). This empirical approach has been developed for use with satellite data (e.g., Seguin and Itier, 1983).

The bare-soil version of the use of skin temperature in remote sensing has been known as "the thermal inertia method." In the absence of air and water, only radiative heating is present and so the diurnal variation of temperature depends only on the thermal conductivity and specific heat of the soil or rock (e.g., as applied to Mars by Sinton and Strong, 1960). This approach has also caught the attention of terrestrial geologists (e.g., Kahle, 1977; Price, 1977; Watson and Hummer-Miller, 1981). We simply show what might be inferred from the thermal inertia approach using the force restore approximation for surface soil temperature (e.g., Deardorff, 1978; Dickinson, 1988), which is now commonly used in GCMs. This approximation is based on replacing the temperature conduction term at the surface, depending on the vertical temperature gradient, by surface temperature and its time derivative with factors chosen so that the latter two terms are equivalent to the conductive term for sinusoidal forcing. For this purpose, T_g = surface soil temperature, $\nu = 2\pi/86{,}400$ is the diurnal frequency in s^{-1}, λ = thermal conductivity of the top 0.1 m or so of the soil, and $D = (2\lambda/\nu c)^{1/2}$, the penetration depth, where c = volumetric specific heat. Then according to this approximation:

6. Water and Energy Exchange

$$\frac{\lambda}{D}\left[\frac{1}{\nu}\frac{\partial T_g}{\partial t} + T_g\right] = Q(t) \tag{1}$$

where $Q(t)$ = surface heat input, assumed to be largely diurnal. The amplitude of the diurnal wave is evidently proportional to the amplitude of Q and is inversely proportional to $(\lambda c)^{1/2}$. The principle of the thermal initial approach is to measure T_g, so that for given Q, the product λc is determined. The latter depend on the surface composition and on soil moisture. Generally, $D \approx 0.1$ m, and λ lies between 1 and 2 W m^{-1} K^{-1}, so a diurnal range of T_g of 20 K requires a range of Q of about 300 Wm^{-2}.

Equation 1 is readily generalized to a more accurate treatment of the soil thermal diffusion. However, its usefulness without further information is limited by the neglect of vegetation, and even worse, by the dependence of Q on sensible heat and evaporative fluxes, denoted H and E, respectively, which depend strongly on the difference between T_g and the overlying air temperature T_a, and that between the mixing ratio of water at the soil surface q_g and in the overlying air q_g. Using ρ_a = air density, C_p = specific heat of air, these fluxes are written:

$$H = \frac{\rho_a C_p (T_g - T_a)}{r_H} \tag{2}$$

$$E = \frac{\rho_a (q_g - q_a)}{r_E} \tag{3}$$

The terms r_H and r_E are resistances, inversely proportional to the strength of the surface wind. These are determined from boundary layer theory as a function of the height of the atmospheric variables, surface roughness, and atmospheric stability. For conditions of neutral atmospheric stability and a wet surface:

$$r_H = r_E = \frac{C}{U} \tag{4a}$$

where U is the atmospheric wind and

$$C = \frac{(\ln z/z_o)^2}{k^2} \tag{4b}$$

is the inverse "drag coefficient." The term z is the level at which the atmospheric variables are provided, measured from the "zero-displacement" level of the canopy, $k \approx 0.4$ is the Von Karman constant, and z_o is the roughness parameter. Typically, z is between 1 and 10 m and z_o ranges from 0.01 m to 1 m, going from smooth to very rough land surfaces. Hence C ranges from about 30 for rough surfaces to 300 for smooth land surfaces (and up to 700 for very smooth surfaces such as water.).

Methods to include the sensible and latent heat flux given surface atmospheric observations, have been developed, for example, by Soer (1980), Hechinger et al. (1982), Camillo et al. (1983), and Van de Griend et al. (1975). The latter two papers emphasize the need to model the variation of soil moisture as well as surface temperature and fluxes.

The q_g in Equation 3 is usually calculated as the saturated mixing ratio at T_g, so that if the surface is not saturated, the resistance $r_E = r_H$ plus an additional term that depends on the rate of moisture diffusion to the surface and hence on soil moisture (e.g., Choudhury and Monteith, 1988). Carlson et al., (1981) (see also, Carlson, 1986; Flores and Carlson, 1987) eliminate the need for surface observations by introducing a planetary boundary layer model that is integrated forward in time from radiosonde soundings. They set

$$r_E = \frac{r_s}{M} \tag{5}$$

where r_s is r_E for a saturated surface and M is called the "moisture availability" or the "wetness factor." The term Q in Equation 1 is defined as:

$$Q = R_N - H - LE \tag{6}$$

where R_N = net radiation (absorbed solar − net thermal infrared [IR] loss) and L = latent heat of vaporization. With a thermal diffusion version of Equation 1, Carlson et al. used the diurnal variation of surface skin temperature (i.e., the temperature inferred from satellite radiances) to infer moisture availability M, assuming Equation 5.

This moisture availability approach has provided a useful observationally based representation of evapotranspiration for use in mesoscale models. However, because it does not necessarily mimic the physics of either soil evaporation or canopy transpiration very well, its extrapolation of evapotranspiration to times when observational skin temperature data are not available may be inaccurate. Furthermore, the soil thermal diffusion and inferred thermal inertia become irrelevant for surfaces largely shaded by vegetation. Wetzel et al. (1984) and Wetzel and Woodward (1987) have developed a variation on the moisture availability approach using the morning rise of skin temperature obtained from the GOES thermal channel. They argue for an empirical relationship equivalent to $M = W^2/E$, where W is a measure of soil moisture content. They validate their estimates of day-to-day variation of W with an antecedent rainfall index (described in next section). Diak et al. (1986) compared surface temperatures for a mesoscale model and using various approaches with GOES inferred skin temperatures. They note that Bowen ratios are insensitive to M except for low values, 0.1 or less.

For terrestrial applications, the thermal inertia method depends on the fluxes of latent and sensible heat to the atmosphere, as expressed by Equations 1 and 5. Indeed, for most situations, the variation of surface temperature, depending on E and H, is relatively insensitive to the likely variation of thermal inertia compared with that of surface roughness, surface winds, and soil moisture. Many GCMs, in particular the standard models of the Geophysical Fluid Dynamics Laboratory (GFDL) and the National Center for Atmospheric Research (NCAR), that average over the diurnal cycle, neglect soil thermal inertia and assume instantaneous energy balance; that is, $Q = 0$. If we make these assumptions, we can get a simple expression for the difference between air and skin temperature. Assuming a wet surface, we take

$$q_g - q_a = q_g - q_{as} + q_{as} - q_a \approx \frac{\partial q_s}{\partial T}(T_g - T_a) + q_{as} - q_a \tag{7}$$

where q_{as} is atmospheric water vapor mixing ratio at saturation. Then, from Equations 2 and 3 and $H + LE = R_N$, we have

$$(T_g - T_a) = \frac{(R_N - LE_o)}{\alpha} \tag{8a}$$

taking

$$\alpha = \rho_a \left(\frac{C_p}{r_H} + \frac{L \partial q_s / \partial T}{r_E} \right) \tag{8b}$$

$$E_o = \frac{\rho_a}{r_E}(q_{as} - q_a) \tag{8c}$$

The term E_o corresponds to the component of evaporation driven by dryness of the air (vapor-pressure deficit). For nonwet surfaces, the models multiply the wet surface $q_g - q_a$ of Equation 7 by a wetness factor M, which can be included in the definition of r_E by Equation 5. In Equation 8, r_H (given the surface roughness) and the vapor pressure deficit depend only on atmospheric conditions. With these and R_N as given, the values of $T_g - T_a$ depend only on the water flux resistance, r_E, or conversely, a measurement of $T_g - T_a$ will determine r_E. After this inversion, Equations 2 and 3 are used to obtain the surface fluxes of sensible and latent heat. The term $L \partial q_s / \partial T$ ranges from the value of C_p at 7 C to $5 \times C_p$ at 35 C. For cooler temperatures, and for greater roughness (i.e., smaller r_H), it is evident from Equations 8a and 8b that the first term in α may be comparable to, or even larger, than the second, making it more difficult to infer r_E from the surface–air temperature difference.

A canopy model approach using satellite data to infer surface water and energy exchange has been developed by Taconet et al. (1986a, 1986b) (see

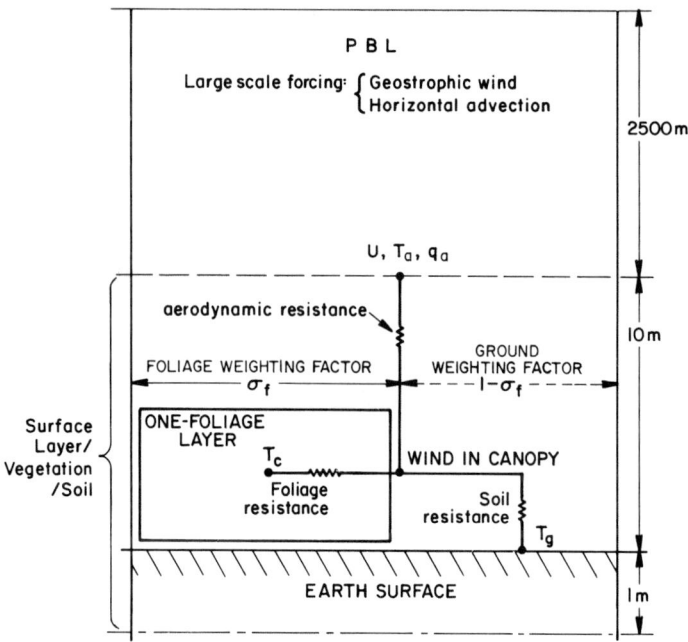

Figure 6.1. Schematic redrafted from Taconet et al. (1986a) of application of satellite-derived skin temperature to infer evapotranspiration and soil moisture. [Reproduced with permission of the American Meteorological Society.]

also Serafini, 1987). It corresponds to the energy balance approach just described except that r_E in Equation 8b must include, in addition to the aerodynamic resistance (4a), a canopy resistance r_c representing the integrated effect of the stomatal resistance of individual leaves. Taconet et al. worked with 1400 LT thermal channel data from the Advanced Very High Resolution Radiometer (AVHRR) of the NOAA-7 polar orbiter, but in a more recent study have also considered diurnal cycle data from the METEOSAT geostationary satellite. The basis of their procedure as illustrated in Figure 6.1 is as follows: The satellite data provide the midday peak (or diurnal variation) of surface skin temperature. A joint canopy/soil model is integrated in time, driven by measured net radiation, atmospheric temperatures, water vapor deficit, and winds. This model calculates soil moisture, evapotranspiration, and canopy temperature. The satellite-measured skin temperature is used in such an approach to infer the canopy resistance on the basis of surface energy balance, as described above. This procedure thus also yields surface evapotranspiration and sensible heat fluxes, and soil moisture from an assumed relationship between stomatal resistance and soil moisture.

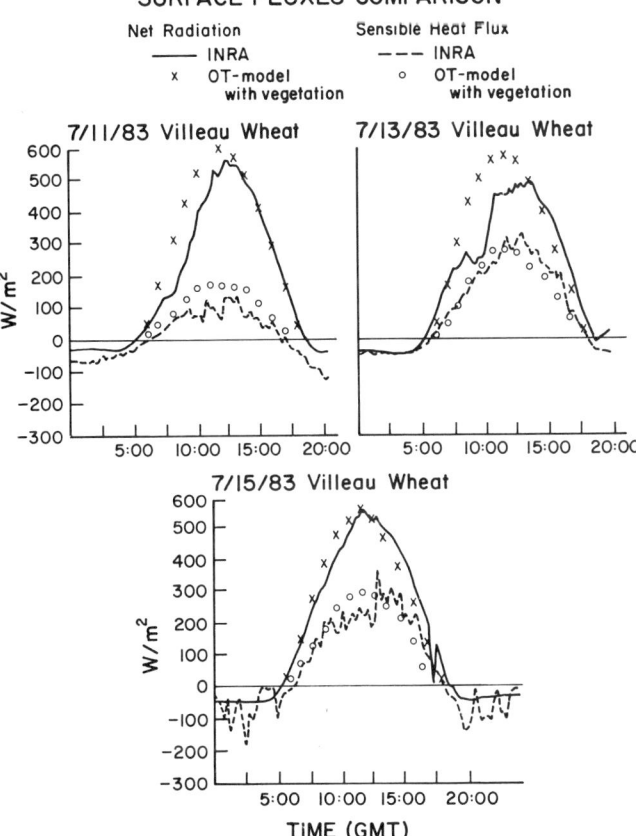

Figure 6.2. Comparison by Taconet et al. (1986b) of satellite-derived estimates of and sensible energy fluxes with surface measurements. [Reproduced with permission of the American Meteorological Society.]

Taconet's model requires data on surface winds, atmospheric temperature, and humidity over a diurnal cycle. These were obtained using one of two approaches. First, averages of local surface measurements were used. Second, a planetary boundary layer model was initialized by measurements from the nearest radiosonde stations and integrated over the diurnal cycle to provide the needed surface atmospheric terms. The latter was found to give more satisfactory results. For validation, she compared her results with local measurements of sensible heat flux (as shown in Figure 6.2) and of soil moisture.

That skin temperature procedure is also sensitive to the prescribed atmospheric winds and temperatures is seen from Equations 4 and 8. Tarpley (1988) shows for sites in Kansas that variations in wind have an effect comparable to that of variations in soil moisture on the morning increase of skin temperature inferred from the GOES thermal channel. Under ideal conditions—that is $r_c \gg r_H$ but $r_c \ll (L\partial q_s/\partial T)r_H/C_p$—we can neglect aerodynamic resistance terms as well as the sensible heat flux. The solution for r_c from Equation 8, with T_g replaced by canopy temperature T_c, then simplifies to

$$r_c \simeq r_E \simeq \rho_a L \frac{[(T_c - T_a)\partial q_s/\partial T + (q_{as} - q_a)]}{R_N} \quad (9)$$

which shows that, at best, the relative error in r_c will be proportional to the relative error in estimating the canopy air-temperature difference and air relative humidity and that of the inverse of net radiation. Midday conditions in a warm climate would seem optimum for estimating r_c according to Equation 9. For example, with $T_c = 35$ C, $T_c - T_a = 8$ C, $R_N = 800$ W m^{-2}, and relative humidity of 0.5, Equation 9 would give $r_c = 100$ s m^{-1}, typical of a well-watered canopy, with the temperature and humidity differences contributing approximately equally to r_c.

To the extent that $H \ll LE$ and that heat storage is negligible, as assumed for Equation 9, LE can be obtained directly from R_N; see Equation 6. The generally close correspondence between LE and R_N under well-watered conditions has suggested direct use of R_N to define potential evapotranspiration. This has long been done by Budyko in the Soviet Union. The most popular such approach has been that of Priestly and Taylor (1972). Specification of incident solar radiation over large areas requires the application of satellite remote sensing. See Diak and Gautier (1983), Pinker and Ewing (1986), Darnell et al. (1988), and Sellers et al. (1988) for currently available algorithms.

The temperature accuracies required to estimate r_c adequately may be difficult to achieve over large areas because of variations in surface emissivities, the need to correct satellite-inferred surface temperatures for atmospheric water vapor emission, and the likely presence of thin or sparse cloud cover and continental aerosol even on days ascertained to be cloud-free. Also, the effective canopy temperature determining sensible and latent heat fluxes will in general not be the same as the effective radiative temperature. A more fundamental limitation of the approach, as applied, is its restriction to completely clear days, which may be as little as 10% of the total time. A lack of prescription of vegetation cover properties, such as leaf area index (LAI) and aerodynamic roughness, over large areas may also limit the application of these procedures. Taller vegetation will normally have a smaller air canopy temperature difference, since Equation 8 will usually not simplify to Equation 9 and r_H, which becomes smaller with taller vegetation, will increase the magnitude of α in Equation 8. Hence the method may be more difficult to apply in forests.

The Rainfall Approach

A very important factor in determining evapotranspiration not only instantaneously but especially over weeks or longer is the incident rainfall. An early approach to estimating evapotranspiration based on rainfall is that of Thornthwaite and Mather (1955), most recently used in modified form by Willmott et al. (1985) to generate a global climatology of evapotranspiration (shown in Figure 6.3), and by Walsh et al. (1985) to examine how much of the variability of monthly surface temperatures that is uncorrelated with atmospheric 700-mb (millibar) heights is correlated with soil moisture and snow cover. The basic principle of using rainfall to estimate evapotranspiration is that of water balance. All implementations have used crude models for estimating evapotranspiration and runoff, usually on a monthly basis because of the greater accessibility of monthly rainfall data. According to the Thornthwaite method, runoff occurs only when monthly precipitation P exceeds "potential evapotranspiration," E_o, a quantity estimated by correlating monthly surface air temperature with lysimeter data. Under excess P, it is assumed that

$$E = E_o$$
$$R = P - [E + (w^* - w)] \geq 0 \tag{10}$$

whereas when $P < E_o$

$$E = P + M(E_o - P)$$

where the wetness factor M depends on the soil moisture w relative to the bucket capacity w^* and where R is runoff. In principle, the rainfall approach can take much smaller time steps than monthly, and use remotely sensed, as well as local, rain-gauge measurements. Essentially the same approach is currently used but with short (approximately one-half hour) time steps in most GCMs (e.g., Manabe, 1969) and referred to as a "bucket model." Pinker and Corio (1987) and Corio and Pinker (1987) have applied a variation of this method developed by Lettau that uses measured solar incident radiation and rainfall to estimate the monthly evapotranspiration over Kansas. The main limitation of the method is its crude empirically based description of evapotranspiration and runoff, which become increasingly unrealistic on shorter time scales and cannot be safely applied to climates differing from those for which they were derived. Thus the Thornthwaite bucket model approach is probably especially suspect for tall vegetation and tropical conditions and for describing the diurnal cycle of evapotranspiration. Use of more physically based models for evapotranspiration and runoff that could be calculated on a shorter time step (daily or over diurnal cycle) would make this approach more robust and reliable.

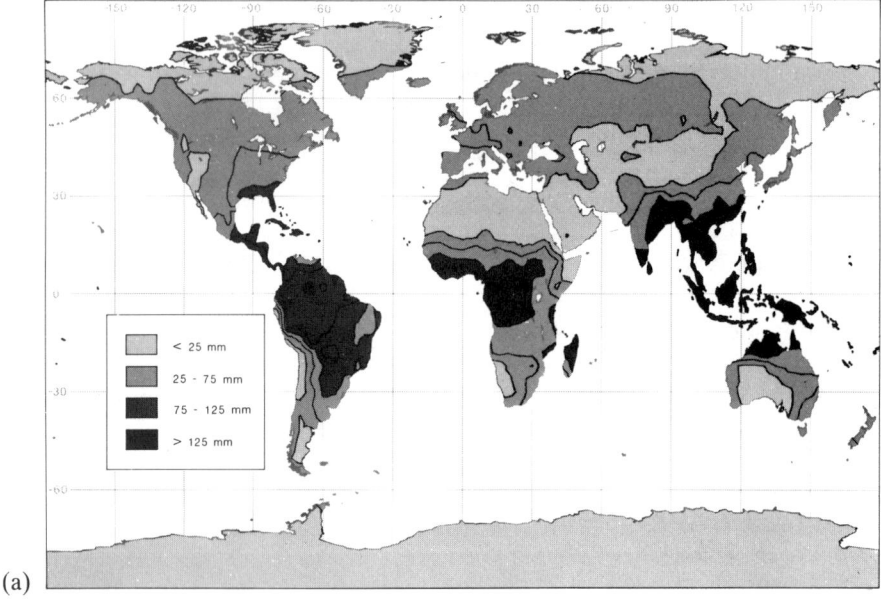

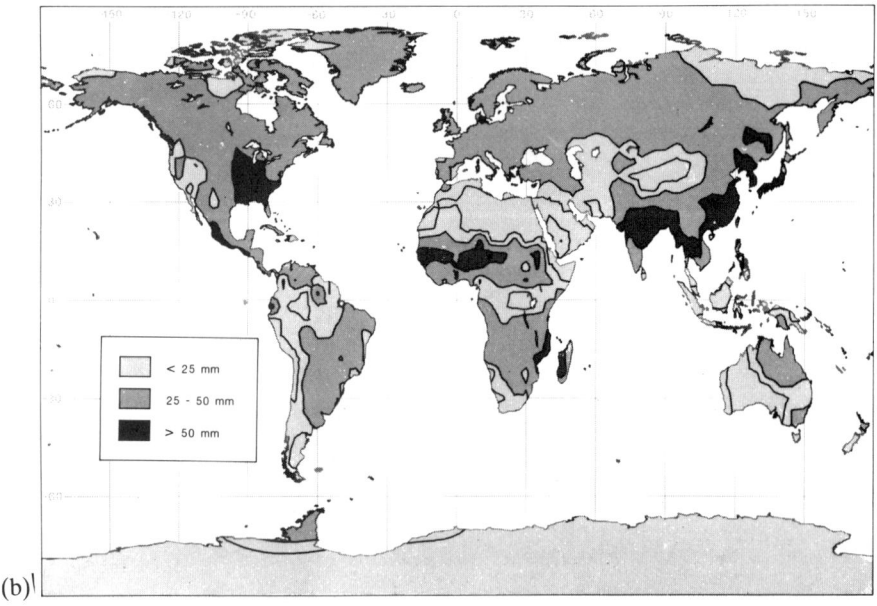

Figure 6.3. Annual mean and interannual variability of global evapotranspiration as inferred by Willmott et al. (1985). [Reproduced with permission of the Royal Meteorological Society.]

An even simpler and more empirical approach to the use of rainfall to infer soil moisture is the antecedent precipitation method (Blanchard, 1981; Choudhury and Blanchard, 1983). This method essentially assumes that the rainfall is a random forcing defining a first-order Markov process, that is,

$$API_j = K(API_{j-1} + P_{j-1}) \tag{11}$$

where the subscripts refer to a particular day, API is the antecedent precipitation index, and K is a regression coefficient. Equation 11 is mathematically equivalent to a finite difference version of a very simple bucket model, that is,

$$\frac{\partial w}{\partial t} + \frac{w}{\tau} = P \tag{12}$$

where

$$\tau = \frac{\Delta t}{(1 - K)}$$

with Δt the interval of time over which Equation 11 is applied (usually one day). Equation 12 neglects runoff and assumes that evapotranspiration depends only on the soil water W and the time scale τ. Choudhury and Blanchard have used the Manabe (1969) bucket model to estimate values of K lying between 0.9 and 0.99 or, equivalently, τ lying between 10 and 100 days. More recently, Delworth and Manabe (1988) fitted soil moisture decay time scales to a 50-year GCM simulation, as illustrated in Figure 6.4.

One of the practical objectives of determining soil moisture is to establish the presence and severity of drought. The most widely used index for this purpose is the Palmer Drought Severity Index (PDSI), as reviewed by Alley (1984). Conceptually, it uses potential evapotranspiration from the Thornthwaite approach and other considerations to compute the difference between actual precipitation and that "climatologically appropriate for existing conditions." A weighted measure of this precipitation is used to force an equation equivalent to Equation 11 with $K = 0.897$, and to keep track of three separate indexes, "wet spell becoming established," "drought becoming established," and "wet spell or drought that has become established," which are reinitialized to zero under appropriate conditions. The large number of rules governing these procedures cannot be summarized here, except to quote Alley (1984): "The methodology used to normalize the values of PDSI is based on very limited comparisons and is only weakly justified on a physical or statistical basis." Karl (1986) has found statistically significant correlations between the PDSI and subsequent seasonal and monthly mean temperatures in the United States and calls for improved

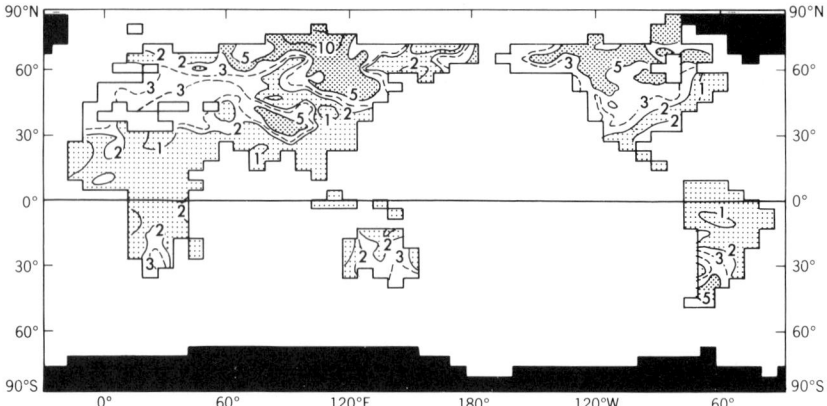

Figure 6.4. Time scale of soil moisture loss from Delworth and Manabe (1988). Units are months. [Reproduced with permission of the American Meteorological Society.]

estimates of evapotranspiration for seasonal and monthly weather forecasting.

One key issue in application of a rainfall-based method to evapotranspiration is the quality of the measurement of rainfall. Because of the large spatial variability of rainfall, the conventional rain gauges are inherently inaccurate except in high-density networks. Thus the need for quantitative measurement of rainfall suggests the development of spatially averaging remote sensing techniques. Weather radars have been developed as a tool to do this from the ground (e.g., Austin, 1987). Up to now, satellite remote sensing of rainfall (e.g., Griffith, 1987; Simpson et al., 1988a) has largely used correlations of precipitation with cloud-covered area and cloud-top emission temperature, with some limited attempts to use passive microwave instruments. The TRMM project (Simpson et al., 1988b) is intended to provide large improvements in satellite capabilities, including radar to measure rainfall in the tropics.

The Atmospheric Water Vapor Divergence Method

Large-scale atmospheric circulation patterns act to redistribute horizontally water vapor and thermal energy carried by the air. On the global scale, the large radiative imbalance observed by satellites at the top of the atmosphere, that is, net absorption in equatorial latitudes and net radiative loss at high latitudes, is maintained by atmospheric transport of latent and thermal energy, and to a lesser extent by ocean circulations. The atmospheric energy transport, dominated by water vapor outside of the winter

6. Water and Energy Exchange

polar regions, is the source of energy for atmospheric winds. Since the horizontal transport of water is such an important feature of the general circulation, its detailed patterns have long been inferred from observed winds and water vapor mixing ratios. The basic atmospheric conservation law constraining surface fluxes of water vapor is that the difference between surface evaporation E and precipitation P is

$$E - P = \int_0^\infty \left[\mathrm{div}\left(\rho \mathbf{C} q\right) + \frac{\partial \rho q}{\partial t} \right] dz \tag{13}$$

where ρ is atmospheric density, $\mathbf{C}$ the horizontal wind, q the mixing ratio of atmospheric water vapor (i.e., ρq is the water vapor density), and z is altitude. Since the atmosphere can carry, at most, 0.05 m of water, the second term representing change of atmospheric storage is negligible for averages of a month or longer.

Equation 13 in principle applies at a point, but in practice it is not practical to apply except as an average over a region that is large enough to contain a reasonable network of wind and water vapor observations from the surface up to at least 5 km. These observations have traditionally been supplied by meteorological radiosondes, but other surface and remote sensing approaches from the surface and space to measure atmospheric water vapor and winds are becoming available.

Rasmusson (1968) showed that reliable time means, monthly or longer, of $E - P$ over North America could be obtained from Equation 13 using wind and water vapor data from radiosonde stations, provided an area 2 by 10^6 km^2 or larger was considered. Figure 6.5, from that study, illustrates the inferred seasonal change of soil moisture over the eastern and central United States and southern Canada. Kellogg and Zhao (1988) have used the analysis of $E - P$ by Rasmusson as a test of the reality of the seasonal

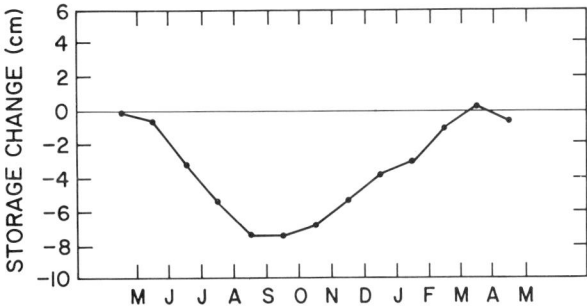

Figure 6.5. Mean monthly surface water storage (i.e., $E - P$ with sign reversed) computed by Rasmusson (1968) from the divergence of atmospheric water vapor flux. [Reproduced with permission of the American Meteorological Society.]

cycle of soil moisture over North America in the various GCMs that have projected climate change from doubling of carbon dioxide.

Bryan and Oort (1984) have applied the atmospheric flux approach globally with ten years of data. The computed $E - P$ averaged over all continents was apparently of low accuracy (e.g., of positive sign over midlatitudes). However, application of the 1979 FGGE data, produced by four-dimensional data assimilation techniques, gave reasonable agreement with Baumgartner and Reichel's (1975) global data based on the Thornthwaite approach.

Other recent studies (Peixoto and Oort, 1983; Savijärvi, 1988) have used the radiosonde station data approach but have concentrated on the global aspects of the atmospheric hydrological cycle rather than on the implications for continental soil moisture. The advantages of four-dimensional data assimilation could be achieved by use of the data sets for atmospheric moisture and winds generated at numerical weather prediction centers (in particular the U.S. National Meteorological Center (NMC) and the European Centre for Medium-range Weather Forecasts (ECMWF). These data are generated by assimilating the observed station data along with other correlated sources of information, such as satellite radiances and global model predictions, to obtain an optimum estimate (e.g., DiMego, 1988). Such an approach forms the foundation for initializing winds and temperatures in the forecast models. However, the usefulness of their moisture analyses is still questionable (Nogues-Paegle and Daley, 1988). Trenberth et al. (1987) report that the analyzed relative humidity from ECMWF dropped on the average 22% on May 1, 1985, as a result of changes in the prediction model, the database being unchanged.

The Vegetation Index Approach

Vegetation, because of its chloroplasts, absorbs much more solar radiation at visible (i.e., less than 0.7 micrometer) than at near-IR wavelengths. Vegetation indexes constructed on the basis of this property (e.g., the NDVI parameter of the AVHRR sensor, c.f., Running, this volume, Chapter 4) can indicate the amount of green leaf material present. Microwave at 37 GHz also appears to be sensitive to the water, and perhaps structure, of canopy foliage (Choudhury, et al., 1987). Tucker and Choudhury (1987) have compared the two methods as means of detecting drought and Becker and Choudhury (1988) have theoretically examined both and have shown how they are related. They suggested that, at low values of vegetation cover, the microwave will be more sensitive than NDVI to changing vegetation cover. Running and Nemani (1988) have shown a good correlation between modeled annual transpiration with only meteorological inputs allowed to vary and annual integrated NDVI. This correlation suggests the plausible conclusion that where there is less soil moisture to transpire, vegetation will be more sparse (e.g., Eagleson, 1982; Woodward, 1987) but could be determined in part by the dependence of both measured quantities on the length of the growing season. Nemani and Running (1989) show a close correlation for Montana conifers between LAI and simulated seasonal transpiration, the latter largely controlled for their study sites by the amount of water retained in the soil at spring snow melt. Their observations show that excluding stands of larch (a unique deciduous conifer), the NDVI obtained from Landsat and from AVHRR is related to the LAI of the individual sites.

Sellers (1985) has provided an analytic framework for relating vegetation index measurements to canopy water flux. The crux of his approach is to apply a reasonably accurate two-stream model for radiative transfer in a canopy to obtaining analytic solutions for canopy albedo, canopy resistance, and photosynthesis. These properties are related through the radiative transfer model to fundamental leaf light level properties, in particular, a leaf scattering model, and models for leaf stomatal resistance and photosynthesis. Sellers constructs a vegetation index from the difference between broad-band visible and near-IR canopy albedos. This index differs from those based on satellite data in that the latter refer to narrow wavelength bands and bidirectional reflectances, that is, reflected radiation viewed from a single direction.

Sellers shows theoretically that a remotely sensed vegetation index can be used to specify canopy resistance under the following conditions:

1. The leaf stomatal properties are known.
2. There is complete canopy cover, or alternatively, the optical properties of the underlying soil and the fraction of exposed soil are known.
3. The foliage is all green, or alternatively, the fraction of nongreen foliage and its optical properties are known.
4. The vegetation is not water or temperature stressed.

These conditions in total are too restrictive for the approach to be readily applicable for relating measured NDVI to evapotranspiration. However, the basic principle involved is clear. As NDVI increases, and hence also the canopies' leaf area index, the canopies' resistance to transpiration decreases. At low values of leaf area index, the stomatal resistance of individual leaves will be maximum and both NDVI and canopy conductance (inverse of resistance) will vary linearly with leaf area index. At large values of leaf area index, both NDVI and canopy conductance become insensitive to leaf area index as heavily shaded leaves contribute little to either (Sellers, 1985; Choudhury, 1987). Hence NDVI and canopy conductance under well-watered conditions maintain a near-linear proportionality to each other even at large leaf area index (Sellers, 1985; Hope et al., 1986; Hope, 1988).

The assumption of no water stress (water stress would increase r_c) is contrasted with the skin temperature approach, which should work best when the canopy is water stressed, the only condition for which the skin temperature approach offers the possibility of a quantitative estimate of soil moisture through its control of r_c.

In summary, the NDVI is now fairly well validated as a measure of seasonal greenness of grasslands in tropical conditions. It should be capable of establishing springtime development of canopies and autumn leaf drop for mid-latitude crops and deciduous forests, and, at least under some conditions, the LAI during the growing season, but probably not short-term water stress in the absence of leaf drop or senesence. The NDVI helps constrain the specification of the maximum canopy capacity to transpire water, but considerable additional information is required before actual transpiration can be calculated. Application of NDVI or some other such vegetation index to specify seasonal variations of vegetation green leaf area (or other such parameter) could significantly improve global modeling of evapotranspiration, provided modeling of evapotranspiration adequately included the role of vegetation canopies.

It is still difficult to interpret in detail the NDVI obtained from satellites because of several difficulties. Because of resolution that is very coarse compared with the scale of individual plants, there can be considerable nonlinearities from spatial heterogeneities. Contributing factors include variability of vegetation cover, leaf optical properties, stems, and branches. Sun and viewing angles and atmospheric structure must be considered for quantitative consideration of land characteristics.

Direct Measurements of Soil Moisture

Passive and active microwave signals contain considerable information on the moisture contained in the near-surface soil layers (Schmugge, 1987). Corrections are needed for soil roughness and vegetation cover and little or no signal is transmitted through thick canopies (Schmugge, this volume, Chapter 3). Eagleman and Lin (1976) reported a correlation of 0.96 between soil moisture estimated from a version of the Thornthwaite approach at 118 stations in Texas, New Mexico, and Kansas and soil moisture inferred from the 1.4-GHz radiometer on Skylab. Blanchard et al. (1981) have reported good correlations between data from the 19-GHz ESMR microwave radiometer on Nimbus 5 and an antecedent precipitation index and Choudhury and Golus (1988) have reported good correlations for the 6-GHz band on the Nimbus-7 SMMR instrument. Schmugge (this volume, Chapter 4) reviews other spacecraft results. Jackson and O'Neill (1987) discuss some conditions under which quantitative application of microwave data to infer soil moisture may be difficult. Little or no information can be gained from forested areas. Large amounts of thatch from unburned and ungrazed glassland may obscure the soil moisture signal (Schmugge et al., 1988). On the other hand, for regions of sparse and/or short vegetation, microwave emissions can provide a powerful approach for obtaining the moisture of the upper soil layers (0.01 to 0.05 m). This information, obtained from remote sensing and used to adjust the soil moisture in a detailed surface model, could give large improvements in estimates of evapotranspiration globally over much of the land area. Local application of this approach with a bare soil model has been demonstrated by Prevot et al. (1984).

Use of Soil and Terrain Information and Spatial Variability

Versions of the standard rainfall water budget approach (reviewed earlier) have frequently been employed to provide useful models for river basin runoff through tuning of model parameters to bring predicted runoff into agreement with observed hydrographs (i.e., the time series of river or stream flow). Gleick (1986) argues for the use of such models in evaluating the hydrological impacts of climate change. However, the empirical basis of such models makes them inapplicable on a global basis.

In principle, a knowledge of rainfall and runoff completely determines evapotranspiration. Thus physically based models of runoff are needed as part of a general approach to evapotranspiration. A key issue is how to represent the role of terrain, that is, the distribution of slopes and, to a lesser extent, the soil cover, in promoting the basin runoff. The development of physically based models of runoff emphasizing terrain effects is currently a research priority of the hydrological community (e.g., Band, 1986; Sivapalan et al., 1987; Beven et al., 1988; Wood et al., 1988).

Besides soil and terrain, the spatial and temporal patterns of rainfall are important determinants of runoff and, by inference, evapotranspiration. Much of the spatial variability is on scales small compared with the resolution of atmospheric models and needs to be represented statistically (e.g., Rodriguez-Iturbe, 1986; Bell, 1987) and parameterized in the models as a subgrid scale process. Spatial variability of rainfall imposes spatial variability of soil moisture that can have major effects on large-scale evapotranspiration (Wetzel and Chang, 1988).

The Combined Method

The proposed combined method is an extension of the systems that are used numerically for numerical weather forecasting (e.g., Bengtsson et al., 1980; Smith et al., 1988; Bengtsson and Shukla, 1988). It consists of integrating in time a three-dimensional atmospheric model coupled to a model for land surface processes. The atmospheric model optimally combines predicted fields of wind, temperature, and moisture with observations of these fields as provided by radiosonde, satellite soundings (reviewed by Isaacs et al., 1986), and other instrumental systems using principles summarized below. Currently, such systems are least successful with the moisture fields (e.g., see Nogues-Paegle and Daley, 1988). Further observations of cloud cover, evapotranspiration, and rainfall could presumably improve this aspect of the models, as would better model parameterizations of the hydrological cycle and higher vertical resolution of the moisture analyses in the planetary boundary layer. The variables from the atmospheric models to be combined with observations for calculating surface evapotranspiration are solar and long-wave radiation incident at the surface, precipitation and surface winds, temperature, and humidity. Besides the precipitation and humidity fields, the modeled surface solar radiation may be questionable at present because of its dependence on possibly inadequately simulated cloud and aerosol properties.

In addition to the atmospheric model, the proposed method requires a realistic model of land surface processes, treating the role of soils and plant canopies in surface evapotranspiration and runoff with adequate realism. Perhaps future improved versions of the SiB model (Sellers et al., 1986) or the BATs model (Dickinson, 1984, Dickinson et al., 1986; e.g., Figure 6.6) or other such approaches involving both vegetation and soil processes (e.g., Choudhury and Monteith, 1988) will serve this function. Significant defects of current treatments are the neglect of slope effects on runoff and of the subgrid scale distribution of rainfall in model hydrology.

A complete global system as visualized here for determining evapotranspiration from remote sensing could not be provided by small individual research groups, because of the intensive data transfers and model computing required. Rather, it most logically would eventually be implemented by extension of the current operational weather services (e.g., NMC or ECMWF), which, in any case, will require such a system to exploit, for their weather-forecasting applications, the improved scientific understanding and observational capabilities of the hydrological cycle expected to evolve over the next decade. Besides providing the short-term surface interactions needed to improve forecasting, a natural by-product would be a means to afford a physically based definition of drought much superior to the current Palmer index.

Development of the optimum procedures for combining model and observational data will require considerable further efforts. All that can be presented here are some of the elementary concepts behind such an approach. To combine two estimates of a given parameter, for example, P_1 and P_2, to estimate, for example, precipitation P, one would use

$$P = \frac{(W_1 P_1 + W_2 P_2)}{(W_1 + W_2)} \tag{14}$$

where W_1 and W_2 are weights:

$$W_1 \sim \frac{1}{\sigma_1^2}; \quad W_2 \sim \frac{1}{\sigma_2^2} \tag{15}$$

6. Water and Energy Exchange 123

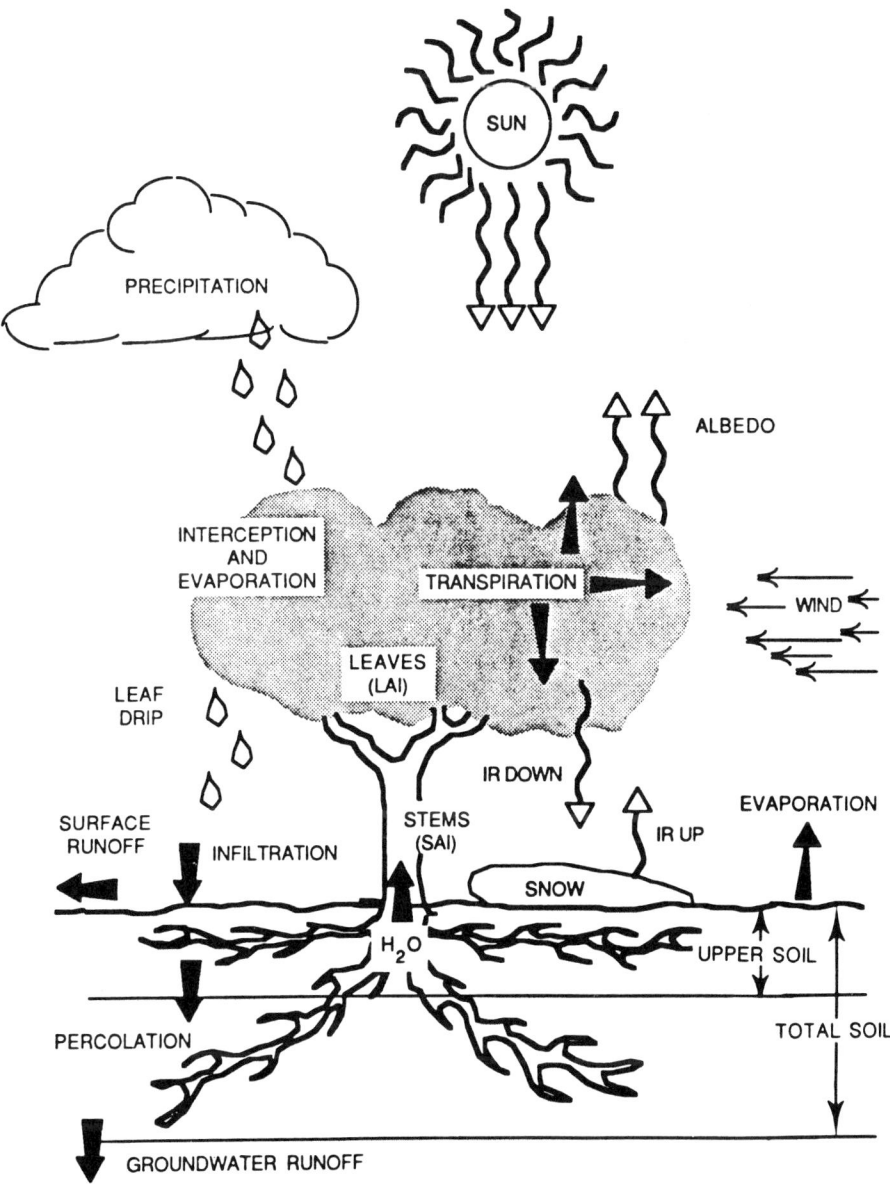

Figure 6.6. Conceptualization of the BATs model [Dickinson et al. (1986)].

where σ_1^2 and σ_2^2 are measures of the uncertainty of the estimates, for example, variances if some statistical means of estimating a variance is available. Thus, in the absence of observations, the best estimates are those provided by the models. Where and when observations are available, they are combined with the model fields according to the relative confidence assigned the observational and model results and constrained by the appropriate model equations. Sources of error in model and observational results need to be adequately characterized to carry out this synthesis. Also, attention must be given to time and space continuity; that is, the correlation of a variable at a given point with itself over some domain of influence in time and space is used to extrapolate point observations to provide information over this domain.

Figures 6.7 through 6.10 give examples of some of the information now furnished by the BATs surface model coupled to the NCAR Community Climate Model that might become part of such a global system. These figures all represent a blowup over North America of one-month July aver-

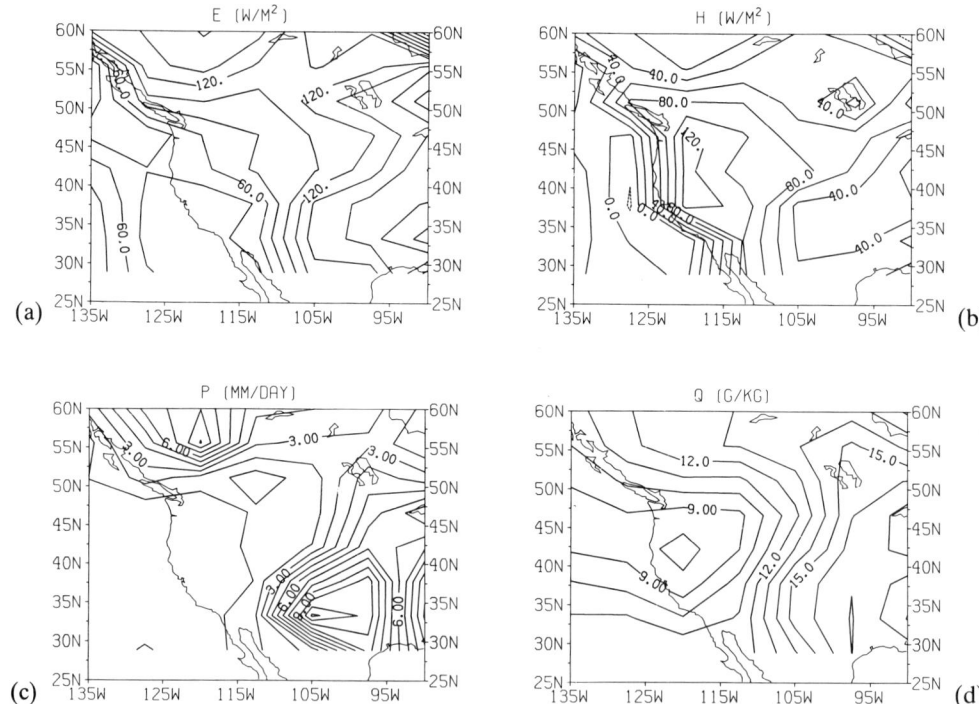

Figure 6.7. Surface hydrological and energy fluxes for a July using the BATs surface package (Dickinson et al., 1986) coupled to the NCAR Community Climate Model: (a) evapotranspiration, (b) sensible heat fluxes in W m^{-2}, (c) rainfall in mm d^{-1}, and (d) water vapor mixing ratio at a pressure of 0.991 of surface, in units of g kg^{-1}.

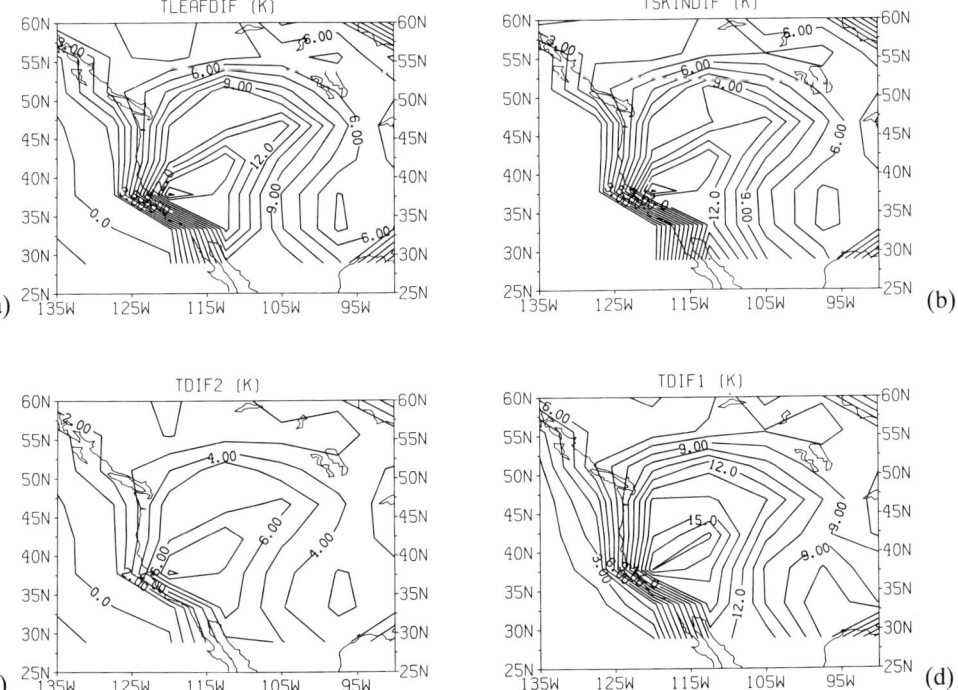

Figure 6.8. As in Figure 7, but four measures of day–night temperature difference: (a) the difference between average daytime and average nighttime canopy temperature; (b) the difference between average daytime and average nighttime skin temperature (combining canopy and ground temperatures according to their exposure to space); (c) the difference between average daytime and average nighttime air temperature, corresponding to that measured under Stevenson screen conditions; and (d) the same as (c) except determined by selecting the maximum daytime values and minimum nighttime values.

ages taken from a recent three-year annual cycle simulation. Figure 6.7 illustrates the model simulation of surface (1) water flux, (2) sensible heat energy fluxes, (3) the corresponding monthly rainfall, and (4) the water vapor mixing ratio over the lowest model level (approximately 70 m above the surface). Figure 6.8 illustrates four different possible measures of day–night surface temperature variation. Which is best to consider from observational or physical grounds is yet to be established. Figure 6.9 shows surface and top of the atmosphere radiative fluxes. Figure 6.10 shows a "normalized vegetation index," as defined in Sellers (1985). All the figures present a pattern consistent with a dry, nearly cloud-free southwestern United States and a secondary region of seasonal aridity centered at 60N, 105W in central Canada. The latter is exaggerated compared with reality; that is, there is a defect in the model. The wettest parts of the model over

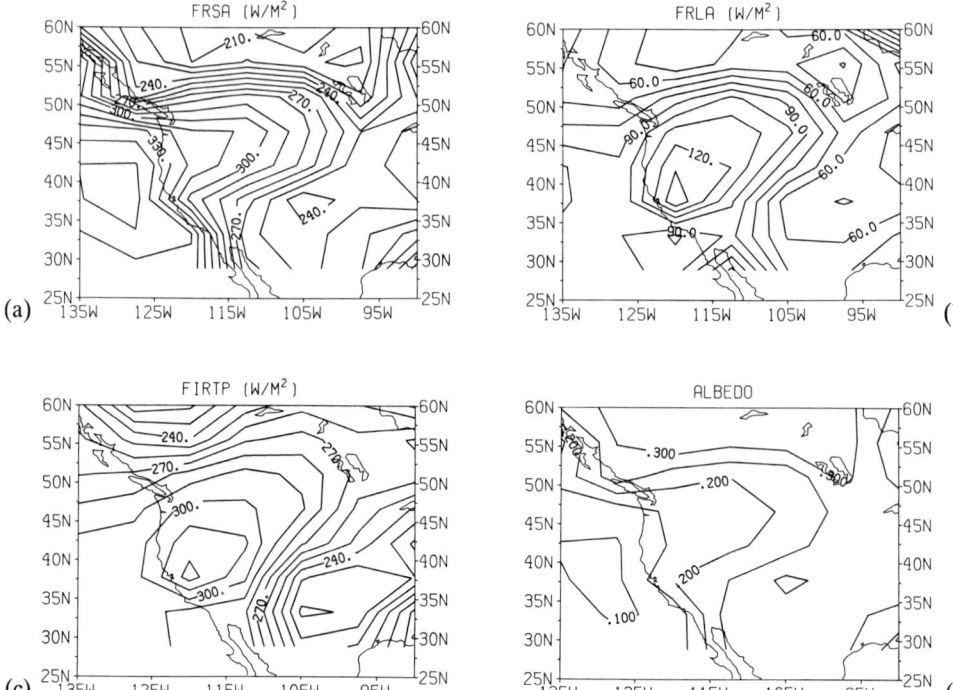

Figure 6.9. Net radiation quantities: (a) net solar radiation absorbed at the surface, (b) net thermal IR emitted from the surface, (c) thermal IR emitted from the top of the atmosphere, and (d) integrated albedo at the top of the atmosphere. All quantitites are in units of W m^{-2} except (d), which is dimensionless.

this domain are seen to be the southeastern United States and northwestern United States/Canada. Unfortunately, the broad-band albedo definitions of visible and near-IR fluxes do not distinguish very well in the model between vegetation and soils so that the resulting pattern is determined primarily by the atmospheric water vapor removal of the near infrared fluxes.

The system for a model-based approach to remote sensing of evapotranspiration is sketched in Figure 6.11. Especially important are model variables that are integrated forward in time, for example, soil moisture and soil temperature, which are referred to as "prognostic" variables. The combination of a time and space stream of observational and model data to adjust the values of model prognostic variables is known as "four-dimensional data assimilation."

In such a system, rainfall would be provided to the land surface models on an instantaneous basis using such an optimum combination of the measured rainfall and that provided by the atmospheric model. The rainfall is

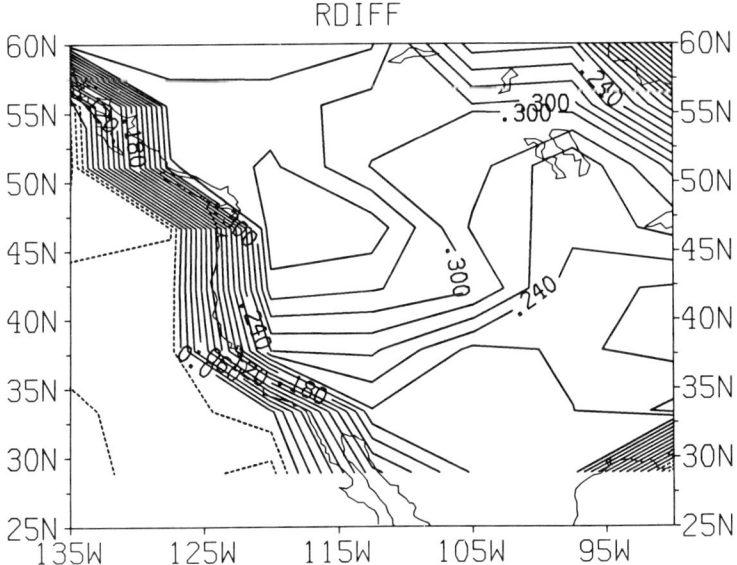

Figure 6.10. "Vegetation index" defined as the difference between reflected broadband near-IR and visible radiation at the surface divided by their sum.

partitioned by the land model into soil moisture, interception, and surface runoff, and the soil moisture is removed by evapotranspiration or infiltration to below the root zone.

The interception and evapotranspiration would depend on atmospheric conditions and on net surface radiation provided by the atmospheric model, with its could cover parameters adjusted on the basis of observed cloud cover information, as inferred from satellite-measured long-wave and reflected solar radiation. One important parameter of the surface model would be day–night skin temperature differences, which may be available only during clear-sky conditions. Model estimates of soil moisture would be adjusted as described above to force the model values of skin temperature when they can be observed to agree more closely with satellite data. Additional adjustment of surface layer soil moisture to agree better with microwave measurements could be especially important during cloudy conditions.

A major difficulty is the linking of the various models and data flows so that they work without degradation by any and all sources of error that creep into the operation of such complex systems. Much research and development work is needed to determine how best to combine the highly redundant but individually "noisy" model inputs.

Some versions of the approach just described are essential to provide global estimates of evapotranspiration and sensible heat fluxes with adequate accuracy. They would provide these estimates on a daily to inter-

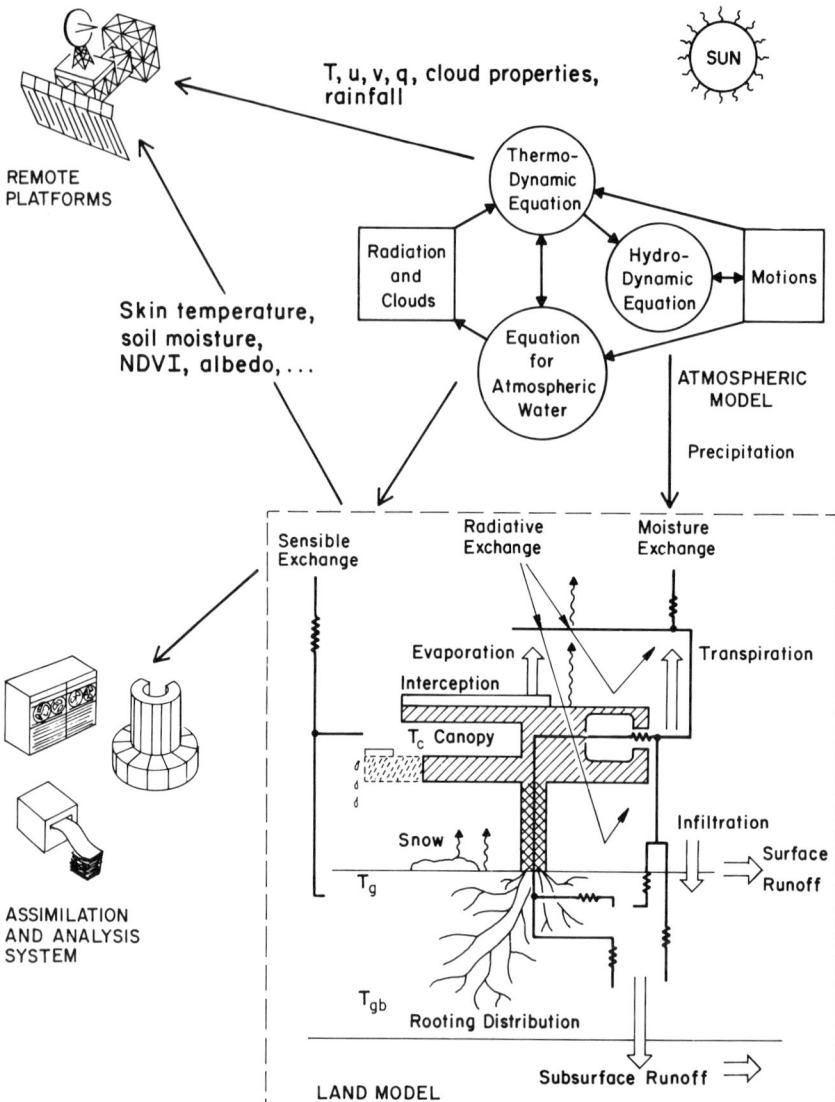

Figure 6.11. Schematic of the proposed combined method using models and remote sensing data to infer energy and water fluxes over land.

annual basis, and in doing so not only would quantify the hydrological functioning of land ecosystems, but also would improve our understanding of the information content of the models and observing systems.

Acknowledgments

The author wishes to thank R. Daley, R. Gurney, P. Morel, G. McBean, J. Shukla, and E. Wood for past discussions of the theme of this chapter —that an integrated approach is needed for a successful global system for remote sensing of surface hydrological processes; B. Choudhury, S. Running, T. Schmugge, P. Sellers, and an anonymous reviewer for helpful comments on the initial manuscript; C. Borchardt for quickly and accurately helping to produce the manuscript; and P. Kennedy for providing help with the computer graphics.

References

Alley, W.M. (1984). The Palmer Drought Severity Index: limitations and assumptions. *JCAM* 23:1100–1109.
Austin, P.M. (1987). Relation between measured radar reflectivity and surface rainfall. *Mon. Wea. Rev.* 115:1053–1070.
Band, L.E. (1986). Topographic partition of watersheds with digital elevation models. *Water Resources Res.* 22(1):15–24.
Baumgartner, A., and Reichel, E. (1975). *The World Water Balance.* Elsevier, NY.
Becker, F., and Choudhury, B.J. (1988). Relative sensitivity of normalized difference vegetation index (NDVI) and microwave polarization difference index (MPDI) for vegetation and desertification monitoring. *Remote Sens. Envir.* 24:287–311.
Bell, T.L. (1987). A space-time stochastic model of rainfall for satellite remote-sensing studies. *J. Geophys. Res.* 92(D8):9631–9643.
Bengtsson, L., Ghil, M., and Källén, E. (1980). *Dynamic Meteorology: Data Assimilation Methods.* Springer-Verlag, NY.
Bengtsson, L., and Shukla, J. (1988). Integration of space and in situ observations to study global climate change. *Bull. Amer. Meteorol. Soc.* 69(10):1130–1143.
Beven, K.J., Wood, E.F., and Sivapalan, M. (1988). On hydrological heterogeneity-catchment morphology and catchment response. *J. Hydrology* 100:353–375.
Blanchard, B.J., McFarland, M.J., Schmugge, T.J., and Rhoades, E. (1981). Estimation of soil moisture with API algorithms and microwave emission. *Water Resources Bull* . 17(5):767–774.
Bryan, F., and Oort, A. (1984). Seasonal variation of the global water balance based on aerological data. *J. Geophys. Res.* 89(D7):11, 717–11, 730.
Camillo, P.J., Gurney, R.J., and Schmugge, T.J. (1983). A soil and atmospheric boundary layer model for evapotranspiration and soil moisture studies. *Water Resources Res.* 19(2):371–380.
Carlson, T.N. (1986). Regional-scale estimates of surface moisture availability and thermal measurements. *Remote Sens. Rev.* 1:197–247.
Carlson, T.N., Dodd, J.K., Benjamin, S.G., and Cooper, J.N. (1981). Satellite estimation of the surface energy balance, moisture availability and thermal inertia. *J. Appl. Meteorol.* 20:67–87.
Choudhury, B.J. (1987). Relationships between vegetation indices, radiation

absorption, and net photosynthesis evaluated by a sensitivity analysis. *Remote Sens. Envir.* 22:209–233.

Choudhury, B.J., and Blanchard, B.J. (1983). Simulating soil water recession coefficients for agricultural watersheds. *Water Resources Bull.* 19(2):241–247.

Choudhury, B.J., and Golus, R. E. (1988). Estimating soil wetness using satellite data. *Int. J. Remote. Sens.* 9(7):1251–1257.

Choudhury, B.J., and Monteith, J.L. (1988). A four-layer model for the heat budget of homogeneous land surfaces. *Quart. J. Meteorol. Soc.* 114:373–398.

Choudhury, B.J., Tucker, C.J., Golus, R.E., and Newcomb, W.W. (1987). Monitoring vegetation using Nimbus-7 scanning multichannel microwave radiometer data. *Int. J. Remote Sens.* 8:533–538.

Corio, L.A., and Pinker, R.T. (1987). Estimating monthly mean water and energy budgets over the central U.S. Great Plains. Part II: Evapoclimatonomy experiments. *Mon. Wea. Rev.* 115:1153–1160.

Darnell, W.L., Staylor, W.F., Gupta, S.K., and Denn, F.M. (1988). Estimation of surface insolation using sun-synchronous satellite data 1. *J. Climate* 8:820–835.

Deardorff, J. (1978). Efficient prediction of ground temperature and moisture with inclusion of a layer of vegetation. *J. Geophys. Res.* 83:1889–1903.

Delworth, T.L., and Manabe, S. (1988). The influence of potential evaporation on the variabilities of simulated soil wetness and climate. *J. Climate* 1:523–547.

Diak, G., and Gautier, C. (1983). Improvements to a simple physical model for estimating insolation from GOES data. *JCAM* 22:505–508.

Diak, G., Heikkinen, S., and Bates, J. (1986). The influence of variations in surface treatment on 24-hour forecasts with a limited area model, including a comparison of modeled and satellite-measured surface temperatures. *Mon. Wea. Rev.* 114:215–232.

Dickinson, R.E. (1984). Modeling evapotranspiration for three-dimensional global climate models. pp. 58–72. In J.E. Hansen and T.Takahashi (eds.), *Climate Processes and Climate Sensitivity*. Geophys. Mono. 29. Amer. Geophys. Union, Washington, DC.

Dickinson, R.E. (1988). The force-restore model for surface temperatures and its generalizations, *J. Climate* 1:1086–1097.

Dickinson, R.E., Henderson-Sellers, A., Kennedy, P.J., and Wilson, M.F. (1986). *Biosphere Atmosphere Transfer Scheme (BATS) for the NCAR Community Climate Model*. Nat. Center for Atmos. Res. Tech Note/TN275 + STR.

DiMego, G.J. (1988). The national meteorological center regional analysis system. *Mon. Wea. Rev.* 116:977–1000.

Eagleman, J.R., and Lin, W.C. (1976). Remote sensing of soil moisture by a 21-cm passive radiometer. *J. Geophys. Res.* 81(21):3660–3666.

Eagleson, P.S. (1982). Ecological optimality in water limited natural soil-vegetation systems. 1. Theory and hypothesis. *Water Resources Res.* 18:325–340.

Ehler, W.L., Idso, S.B., Jackson, R.D., and Reginato, R.J. (1978). Diurnal changes in plant water potential and canopy temperature of wheat as affected by drought. *Agron. J.* 70:999–1004.

Flores, A.L., and Carlson, T.N. (1987). Estimation of surface moisture availability from remote temperature measurements. *J. Geophys. Res.* 92(D8):9581–9585.

Gleick, P.H. (1986). Methods for evaluating the regional hydrologic impacts of global climatic changes. *J. Hydrology* 88:97–116.

Griffith, C.G. (1987). Comparisons of gauge and satellite rain estimates for the central United States during August 1979. *J. Geophys. Res.* 92(D8):9551–9566.

Hechinger, E., Becker, F., and Raffy, N. (1982). Comparison between the accuracies of a new discretization method and an improved Fourier method to evalu-

ate heat transfers between soil and atmosphere. *J. Geophys. Res.* 87:7325–7339.
Hope, A.S. (1988). Estimation of wheat canopy resistance using combined remotely sensed spectral reflectance and thermal observations. *Remote Sens. Envir.* 24:369–383.
Hope, A.S., Petzold, D.E., Goward, S.N., and Ragan, R.M. (1986). Simulated relationships between spectral reflectance, thermal emissions, and evapotranspiration of a soybean canopy. *Water Resources Bull.* 22(6):1011–1019.
Idso, S.B., and Ehler, W.L. (1976). Estimating soil moisture in the root zone of crops. A technique adaptable to remote sensing. *Geophys. Res. Lett.* 3:23–25.
Idso, S.B., Jackson, R.D., and Reginato, R.T. (1975). Estimating evaporation: A technique adaptable to remote sensing. *Science* 189:991–992.
Idso, S.B., Jackson, R.D., and Reginato, R.J. (1977). An equation for potential evaporation from soil, water and crop surfaces adaptable to use by remote sensing. *Geophys. Res. Lett.* 4:187–188.
Isaacs, R.G., Hoffman, R.N., and Kaplan, L.D. (1986). Satellite remote sensing of meteorological parameters for global numerical weather prediction. *Rev. Geophys.* 24(4):701–743.
Jackson, R.D. (1985). Evaluating evapotranspiration at local and regional scales. *Proc. IEEE* 73(6):1086–1096.
Jackson, R.D., and Reginato, R.J. (1976). Compensating for environmental variability in the thermal inertia approach to remote sensing of soil moisture. *J. Appl. Meteorol.* 15:811–817.
Jackson, T.J., and O'Neil, P. (1987). Temporal observations of surface soil moisture using a passive microwave sensor. *Remote Sens. Envir.* 21:281–296.
Kahle, A.B. (1977). A simple thermal model of the earth's surface for geologic mapping by remote sensing. *J. Geophys. Res.* 82:1673–1680.
Karl, T.R. (1986). The relationship of soil moisture parameterizations to subsequent seasonal and monthly mean temperature in the United States. *Mon. Wea. Rev.* 114:675–686.
Kellogg, W.W., and Zhao, A. (1988). Sensitivity of soil moisture to doubling of carbon dioxide in climate model experiments. Part 1: North America. *J. Climate* 1:348–366.
Manabe, S. (1969). Climate and ocean circulation:I. The atmospheric circulation and the hydrology of the earth's surface. *Mon. Wea. Rev.* 97:739–774.
Nemani, R.R., and Running, S.W. (1989). Testing a theoretical climate-soil-leaf area hydrologic equilibrium of forests using satellite data and ecosystem simulation. *Agric. Forest Meteorol.* 44:245–260.
Nogues-Paegle, J., and Daley, R. (1988). Summary of the global weather experiment workshops on the hydrological cycle and data assimilation. *Bull. Amer. Meteorol. Soc.* 69(4):377–382.
Peixoto, J.P., and Oort, A.H. (1983). *The Atmospheric Branch of the Hydrological Cycle and Climate. Variations in the Global Water Budget.* Reidel, Norwell, MA, pp. 5–65.
Pinker, R.T., and Corio, L.A. (1987). Estimating monthly mean water and energy budgets over the central U.S. Great Plains. Part I: Evapoclimatonomy model formulation. *Mon. Wea. Rev.* 115(6):1140–1152.
Pinker, R.T., and Ewing, J.A. (1986). Effect of surface properties on the narrow to broadband spectral relationship in clear sky satellite observations. *Remote Sens. Envir.* 20:267–282.
Prevot, L., Bernard, R., Taconet, O., Vidal-Madjar, D., and Thony, J.L. (1984). Evaporation from a bare soil evaluated using a soil water transfer model and remotely sensed surface soil moisture data. *Water Resources Res.* 20(2):311–316.

Price, J.C. (1977). Thermal inertia mapping: A new view of the earth. *J. Geophys. Res.* 82:2582–2590.

Priestley, C.H.B., and Taylor, R.J. (1972). On the assessment of surface heat flux and evaporation using large-scale parameters. *Mon. Wea. Rev.* 100(2):81–92.

Rasmusson, E.M. (1968). Atmospheric water vapor transport and the water balance of North America. *Mon. Wea. Rev.* 96(10):720–734.

Rodriguez-Iturbe, I. (1986). Scale of fluctuation of rainfall models. *Water Resources Res.* 22(9):15S–37S.

Running, S.W., and Nemani, R.R. (1988). Relating seasonal patterns of the AVHRR vegetation index to simulated photosynthesis and transpiration of forests in different climates. *Remote Sens. Envir.* 24:347–367.

Savijärvi, H.I. (1988). Global energy and moisture budgets from rawinsonde data. *Mon. Wea. Rev.* 116(2):417–430.

Schmugge, T. (1987). Remote sensing applications in hydrology. *Rev. Geophys.* 25(2):148–152.

Schmugge, T.J., Wang, J.R., and Asra, G. (1988). Results from the push broom microwave radiometer flights over the Konza Prairie in 1985. *IEEE Trans. Geosci. Remote Sens.* 26(5):590–596.

Seguin, B., and Itier, B. (1983). Using midday surface temperature to estimate daily evaporation satellite thermal IR data. *Int. J. Remote Sens.* 4(2):371–383.

Sellers, P.J. (1985). Canopy reflectance, photosynthesis and transpiration. *Int. J. Remote Sens.* 6(8):1335–1372.

Sellers, P.J., Mintz, Y., Sud, Y.C., and Dalcher, A. (1986). A simple biosphere model (SiB) for use within general circulation models. *J Atmos. Sci.* 43:505–531.

Sellers, P.J., Rasool, S.I., and Bolle, H.J. (eds.). Satellite data algorithms for studies of the land surface. *Proc. ISLSCP Workshop held at the Jet Propulsion Laboratory, Jan. 1987.* ISLSCP Rep. no 9, Goddard Space Flight Center, Greenbelt, MD.

Serafini, Y.V. (1987). Estimation of the evapotranspiration using surface and satellite data. *Int. J. Remote Sens.* 8(10):1547–1562.

Simpson, J., Adler, R.F., and Negri, A.J. (1988a). On improved validation of rainfall estimates from geosynchronous IR products. pp. D-125–D-140. In *Validation of Satellite Precipitation Measurements for the Global Precipitation Climatology Project.* World Climate Programme Res. Rep. of Int. Wksp., held in Washington, DC, Nov. 17–21, WCRP-1, WMO/TD-No. 203. World Meteorological Organization, Geneva, Switzerland.

Simpson, J.A., Adler, R.F., and North, G.R. (1988b). A proposed tropical rainfall measuring mission (TRMM) satellite. *Bull. Amer. Meteorol. Soc.* 69(3):278–295.

Sinton, W.M., and Strong, J. (1960). Radiometric observations of Mars. *Astrophys. J.* 131:459–469.

Sivapalan, M., Beven, K., and Wood, E.F. (1987). On hydrologic similarity 2. A scaled model of storm runoff production. *Water Resources Res.* 23(12):2266–2278.

Smith, W.L., Leslie, L.M., Diak, G.R., Goodman, B.M., Velden, C.S., Callan, G.M., Raymond, W., and Wade, G.S. (1988). The integration of meteorological satellite imagery and numerical dynamical forecast models. *Phil. Trans. Roy. Soc. London A* 324:317–323.

Soer, G.J.R. (1980). Estimation of regional evapotranspiration and soil moisture conditions using remotely sensed crop surface temperatures. *Remote Sens. Envir.* 9:27–45.

Taconet, O., Bernard, R., and Vidal-Madjar, D. (1986a). Evapotranspiration over

an agricultural region using a surface flux/temperature model based on NOAA-AVHRR data. *JCAM* 25:284–307.
Taconet, O., Carlson, T., Bernard, R., and Vidal-Madjar, D. (1986b). Evaluation of a surface/vegetation parameterization using satellite measurements of surface temperature. *JCAM* 25:1752–1767.
Tarpley, J.D., (1988). Some climatological aspects of satellite-observed surface heating in Kansas. *J. Appl. Meteorol.* 27:20–29.
Thornthwaite, C.W., and Mather, J.R. (1985). The water balance. *Publ. Climatol.* 8:1.
Trenberth, K.E., Christy, J.R., and Olson, J.G. (1987). Global atmospheric mass, surface pressure, and water vapor variations. *J. Geophys. Res.* 92(D12):14,815–14,826.
Tucker, C.J., and Choudhury, B.J., (1987). Satellite remote sensing of drought conditions. *Remote Sens. Envir.* 23:243–251.
Van De Griend, A.A., Camillo, P.J., and Gurney, R.J. (1985). Discrimination of soil physical parameters, thermal inertia, and soil moisture from diurnal surface temperature fluctuations. *Water Resources Res.* 21(7):997–1009.
Walsh, J.E., Jasperson, W.H., and Ross, B. (1985). Influences of snow cover and soil moisture on monthly air temperture. *Mon. Wea. Rev* 113:756–768.
Watson, K., and Hummer-Miller, S. (1981). A simple algorithm to estimate the effective regional atmospheric parameters for the thermal-inertia mapping. *Remote Sens. Envir.* 11:455–462.
Wetzel, P.J., Atlas, D., and Woodward, R.H. (1984). Determining soil moisture from geosynchronous satellite infrared data: A feasibility study. *JCAM* 23:375–391.
Wetzel, P.J., and Chang, J. (1988). Evapotranspiration from nonuniform surfaces: A first approach for short-term numerical weather prediction. *Mon. Wea. Rev.* 116:600–621.
Wetzel, P.J., and Woodward, R.H. (1987).Soil moisture estimation using GOES-VISSR infrared data: A case study with a simple statistical method. *JCAM* 26:107–117.
Willmott, C.J., Rowe, C.M., and Mintz, T. (1985). Climatology of the terrestrial seasonal water cycle. *J Climatol.* 5:589–606.
Wood, E.F., Sivapalan, M., Beven, K., and Band, L. (1988). Effects of spatial variability and scale with implications to hydrologic modeling. *J. Hydrol.* 102:29–47.
Woodward, F.I. (1987). *Climate and Plant Distribution.* Cambridge Univ. Press, Cambridge, England.

7. Evaluation of Canopy Biochemistry

Carol A. Wessman

The mass and energy exchanges within an ecosystem ultimately define the efficiency of photosynthetic processes and system productivity. The manner in which these exchanges occur is largely determined by the plant canopy, the major interface with the atmosphere. The rate at which they occur is modified by the integrated effects of environmental conditions such as climate and moisture and nutrient availability (Mooney and Gulmon, 1982; Van Cleve et al., 1983; Coley et al., 1985). Accordingly, carbon gain, nutrient flux, and heat exchange within a given environment will influence (and be influenced by) adaptations of canopy morphology and foliar biochemistry.

Subtle changes in ecosystem functioning may be expressed in the canopy biochemistry as a result of altered carbon allocation patterns, metabolic processes, and nutrient availability. Relative concentrations of carbohydrates and nitrogenous compounds in plant tissue often reflect the partitioning of carbon resources between roots and shoots (Lainson and Thornley, 1982; Chapin et al., 1987). Foliar nitrogen concentration frequently increases with increased nitrogen availability (Vitousek et al., 1982; Binkley and Reid, 1985; Birk and Vitousek, 1986), whereas reduction in nitrogen supply promotes increased secondary wall thickening and lignification (Gartlan et al., 1980; Waring et al., 1985; Chapin et al., 1986). Storage carbohydrates such as starch, and defensive compounds, such as polyphenols and fibers, also vary predictably with resource availability (Mooney and Gulmon, 1982; Bryant et al., 1983; Coley et al., 1985).

Ecosystem carbon and nitrogen cycles are mutually linked because the quality (organic chemical composition) and quantity of litter supplied by the canopy modulate the processes of decomposition, mineralization, and nitrification (Fogel and Cromack, 1977; Meentemeyer, 1978; Melillo et al., 1982; Pastor and Post, 1986; Aber et al., this volume). These processes, in turn, strongly regulate system productivity through their influence on nitrogen availability (Vitousek, 1982; Nadelhoffer et al., 1983; Pastor et al., 1984). Significant changes in foliar lignin to nitrogen ratios may indicate corresponding changes in decomposition rates affecting nutrient cycling and trace gas fluxes (Delwich et al., 1978; Vitousek, 1983; Goodroad and Keeney, 1984).

The capability to detect changes in canopy biochemistry using remote sensing would provide a means of assessing spatial extent and variation of carbon/nutrient sources and sinks crucial to understanding gas exchange between vegetation and the atmosphere. Present knowledge is limited to small-scale, site-specific studies. Research in analytical chemistry has demonstrated that concentrations of constituents within organic mixtures can be evaluated from near-infrared (IR) reflectance spectra of those mixtures (Wetzel, 1983; Weyer, 1985). These procedures have been successfully extended to laboratory evaluation of foliar material (Marten el al., 1985; Card et al., 1988; Wessman et al., 1988a). With the advent of imaging spectrometry (the acquisition of contiguous, narrow-band spectral measurements in an image format), research has focused on the application of spectroscopy principles to remotely sensed data in the effort to resolve information on subtle spectral features relevant to ecosystem functioning.

The potential to estimate canopy constituents remotely rests on (1) the influence of individual or functional groups of foliar constituents on the overall canopy reflectance curve, and (2) the development of high spectral resolution instruments that will measure the reflectance signal at sufficient detail and quality to document subtle changes in spectral shape and allow reduction of environmental/sensor effects using spectral analysis techniques. This chapter describes the present state of knowledge regarding spectral properties of vegetation, the spectral analysis of organic mixtures, and the role future instruments may play in our assessment of canopy chemistry.

Spectral Properties of Vegetation

The Visible Region (400 to 700 nm)

Absorption by photosynthetic pigments (chlorophyll, xanthophyll, and carotene) dominates the visible wavelengths. Each of the pigments has absorption maxima in the 300 to 500 nm region; however, only chlorophyll absorbs in the red wavelengths (Salisbury and Ross, 1969). Principal

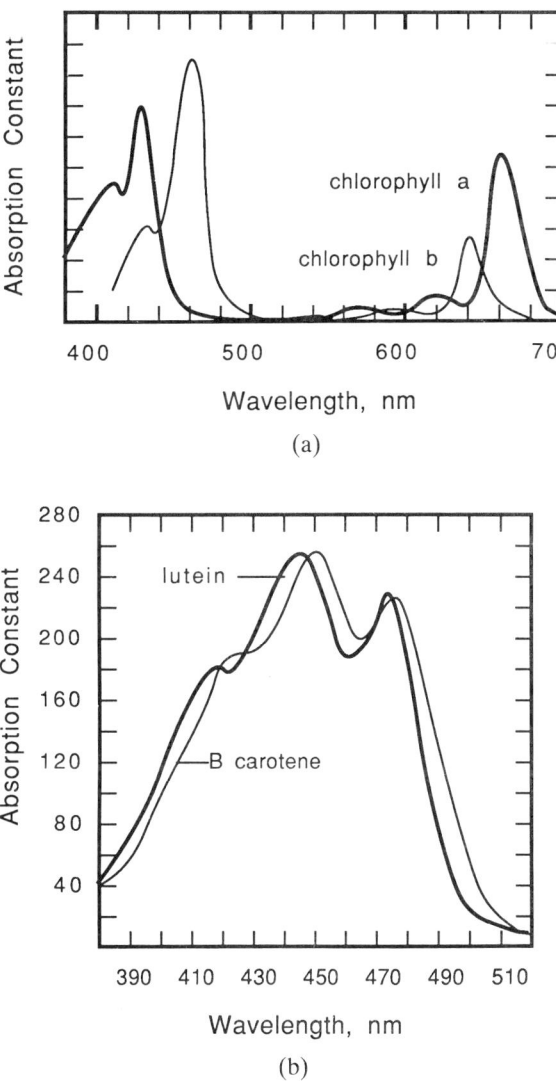

Figure 7.1. (a) Absorption spectra of chlorophylls a and b dissolved in diethyl ether, and (b) absorption spectra of β-carotene in hexane and lutein (a xanthophyll) in ethanol. The absorption constant equals the absorbance (optical density) given by a solution at a concentration of 1 g/l with a thickness (light path length) of 1 cm. [From Salisbury and Ross (1969). Reproduced with permission.]

absorption peaks of extracted chlorophyll a occur at 430 and 660 nm and those of chlorophyll b at 455 and 640 nm (Figure 7.1). When measured in vivo, these peaks shift approximately 20 nm toward the longer wavelengths owing to the difference in refractive indices between the extract solvent and leaf water (Mackinney, 1938; Setak et al., 1971). Changes in chlorophyll concentration with phenological development produce apparent spectral shifts (on the order of 5 to 20 nm) of the absorption edge near 700 nm (Gates et al., 1965; Horler et al., 1983). Environmental stresses that result in chlorophyll loss cause narrowing of the absorption band in the red region and a shift of the red edge to shorter wavelengths. This so-called "blue shift" has been reported for studies of vegetation exposed to heavy metal stress (Collins, 1978; Chang and Collins, 1983), ozone (Ustin and Curtiss, 1988; Westman and Price, 1988), and acidic deposition (Rock et al., 1988).

The relationship between leaf reflectance in the visible region and leaf chlorophyll and nitrogen concentrations has been demonstrated by a number of researchers (Thomas and Oerther, 1972; Thomas and Gausman, 1977; Tsay et al., 1982; Everitt et al., 1985; Nelson et al., 1986). Systematic changes in multispectral crop reflectance are associated with phenological development over the growing season (Collins, 1978; Pinter et al., 1981; Kollenkark et al., 1982; Hinzman et al., 1986). Studies of spectral response to nitrogen fertilization of corn, wheat, and grass pasture have successfully distinguished nitrogen application levels by measuring broad-band canopy reflectance (Walburg et al., 1982; Richardson et al., 1983; Hinzman et al., 1986). The effects of moisture stress and nutrient deficiencies on canopy reflectance are profound, but have not yet been fully quantified. Nitrogen deficiencies will be most evident in the visible region owing to nitrogen's close affiliation with chlorophyll. While moderate- to high-level moisture stress should be readily evident in the IR region, low-level stress may influence chlorophyll state, thereby affecting visible reflectance.

The Short-Wave IR Region, SWIR (700 to 2,500 nm)

Reflectance characteristics of vegetation in the region from 700 to 2,500 nm can be generalized as exhibiting high reflectance in the near IR (700 to 1,300 nm) and high absorption in the middle IR (1,300 to 2,500 nm). The near-IR wavelengths are greatly influenced by cellular structure and refractive index discontinuities within the leaf (Knipling, 1970; Gausman, 1977). Minor water absorption features near 960 and 1,200 nm vary significantly in shape and depth and may be related to both cellular arrangement within the leaf and hydration state (Gates, 1970; Gausman et al., 1978; Goetz et al., 1983). The mid-IR region is dominated by leaf water absorption (Gates et al., 1965, Knipling, 1970) and has been related to plant water status through indices combining these and near-IR bands (Tucker, 1980; Hunt et al., 1987). The region intermediate to the water absorption maxima at

1,450 and 1,940 nm may be strongly influenced by cell structure, morphology, and tissue constituents (Kleman and Fagerlund, 1987; Wessman et al., 1988a).

Near-IR spectroscopy (NIRS) demonstrates that this spectral region (comparable to the SWIR as defined in this chapter) contains quantitative information on constituents within organic mixtures (Marten et al., 1985; Weyer, 1985; McDonald, 1986). NIRS analysis relies on high instrumental signal-to-noise ratios and wavelength reproducibility, one or both of which are difficult conditions to meet in remote sensing instrumentation. Moreover, NIRS techniques follow a highly regimented treatment of foliar samples whereby oven drying and grinding decrease moisture effects and produce uniform particle size. In situ canopy reflectance, on the other hand, has high variablity resulting from environmental and sensor effects, which may introduce enough noise to limit interpretation of the SWIR signal.

Nevertheless, advances in NIRS have simulated work with whole leaf reflectance that promises improved understanding of foliar optical characteristics. The approach suggests that, whereas constituent spectra may not be immediately apparent in the composite spectrum, they are decidedly important factors in its shape. Remotely sensed measurements of vegetation canopy reflectance will certainly not be as sensitive as those of a laboratory spectrophotometer to foliar chemical constituents, but, if sampled at sufficient spectral detail, can indicate those constituents that strongly influence the shape of the spectra.

While there are no major absorption features aside from those of water in the SWIR region, the spectra of organic compounds in this region are a mixture of harmonic overtones and combinations[1] that are mainly caused by stretching and bending vibrations[2] of strong molecular bonds between atoms of low weight (Wetzel, 1983; Weyer, 1985). The molecular functional groups contributing to this region are primarily limited to C–H, O–H, and N–H, whose fundamental absorption bands are no higher than 5,000 to 8,000 nm, depending on their intensity. Overtones are specific to each component of an organic compound and are, with combination bands, more sensitive to changes in the environment of the absorbing molecules

---

[1] When IR radiation is absorbed by a molecule, individual bonds will vibrate in a manner similar to a simple harmonic oscillator. However, unlike ideal oscillators, molecular vibrations can undergo transitions between more than one energy level. These transitions give rise to the overtone absorption bands at approximately 1/2, 1/3, 1/4, ... , and so on, the wavelength of the fundamental absorption, depending on the energy separation between levels. Another type of overtone, called a "combination band", occurs when a single photon has the precise amount of energy to excite two vibrations at once. The energy of the combination band is the sum of the two independent absorptions.
[2] Polyatomic molecules exhibit two distinct types of vibration. Bending vibrations occur when atoms move in or out of plane relative to the molecule. Stretching vibrations occur as lengthening and shortening movement along the bond.

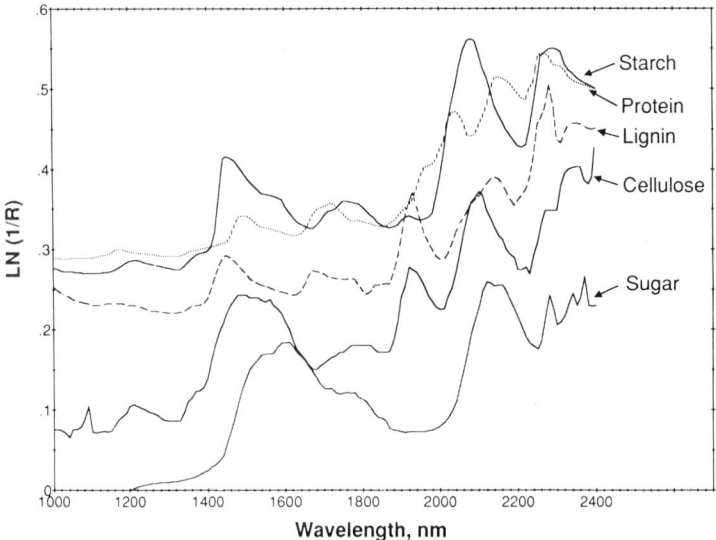

Figure 7.2. Absorption of five biochemical compounds found in leaves plotted as ln(1/R) where R is the reflectance acquired from pure powdered material in a spectrophotometer. (Samples were run at the NASA Ames Research Center; data courtesy of D. Card and D.L. Peterson; milled wood enzyme lignin sample courtesy of John Obst, U.S. Dep. Agric. Forest Products Laboratory, Madison, WI.)

than the fundamental of the same vibration. Slight disturbances in the bonding scheme will produce only small changes in the fundamental band while absorption bands in the SWIR region will experience large shifts in frequency and amplitude.

Extractions of important foliar constituents (primarily in an agricultural context) have been spectrally characterized (Rotolo, 1979; Wetzel, 1983; Weyer, 1985). Carbohydrate absorption spectra are well represented by the starch spectrum, which has a strong O–H combination band at 2,100 nm, an O–H stretch first overtone at 1,460 nm, and a C–H stretch combination band at 2,330 nm (Figure 7.2; Weyer, 1985). Absorption by cellulose closely follows that of starch with an O–H stretch first overtone at 1,490 nm, a C–H stretch first overtone and an O–H stretch combination band at 1,780 nm, an O–H stretch second overtone at 1,820, and an O–H stretch overtone at 2,270 nm. Protein has an N–H stretch first overtone at 1,510 nm; a series of N–H combination bands at 1,980, 2,060 and 2,180 nm; and a C–H bend second overtone at 2,300 nm. The first overtone of the aromatic C–H absorptions (e.g., lignin) is near 1,685 nm and the second is at 1,143 nm. Combination bands occur near 1,420 to 1,450, 2,150, and 2,460 nm.

Water has an important combination band at approximately 1,940 nm,

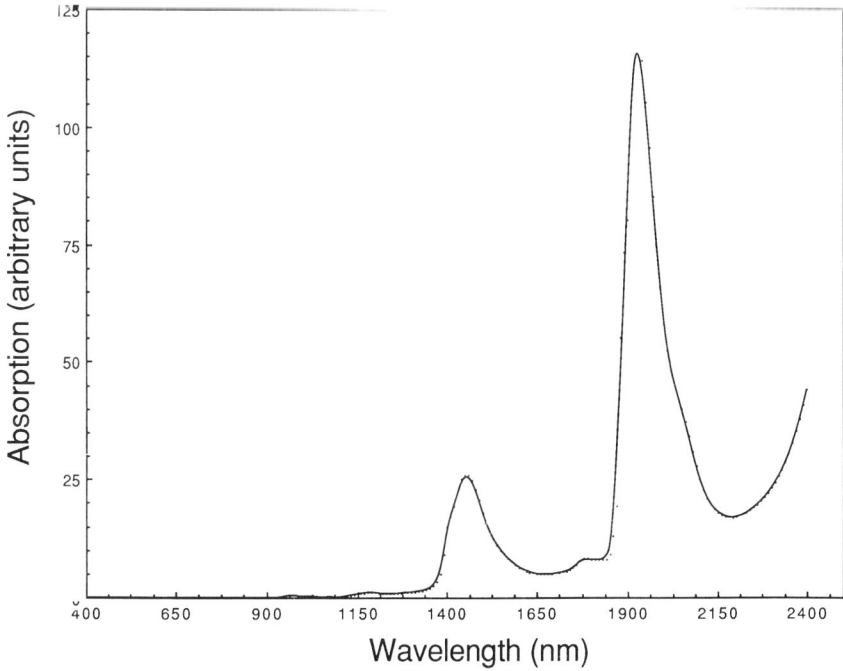

Figure 7.3. Absorption spectrum of liquid water at 20 C. [After Curcio and Petty (1951).]

an O–H stretch first overtone at 1,450 nm, weaker combination bands near 1,200 nm, and an O–H stretch second overtone at 960 nm (Figure 7.3; Curcio and Petty, 1951; Weyer, 1985). The strength of absorption in the 1,450 and 1,940 nm bands dominates broad-band reflectance in the 1,300 to 2,500 nm region (Gates et al., 1965; Knipling, 1970; Tucker, 1980). Changes in leaf moisture content relate strongly to shifts in reflectance amplitude within this region (Olson, 1967; Rohde and Olson, 1971; Gausman et al., 1978) and can complicate analyses for other constituents. Variable moisture content in NIR spectroscopy samples produces less than optimum results in lysine (Rubenthaler and Bruinsma, 1978) and protein (Winch and Major, 1981) predictions using untransformed reflectance data. NIR predictions of acid detergent fiber (ADF) in sorghum overestimate actual ADF concentrations as moisture content increases, but a second derivative transformation of the reflectance data offsets the curve displacement (Fales and Cummins, 1982).

The nature of absorptions in the SWIR are weak and complex since they consist of overlapping overtone and combination bands. Yet, the origins of the observed vibrations are limited. Study of the fine-scale structure of laboratory reflectance spectra using analytical techniques such as derivative spectroscopy, curve fitting, and spectral mixing will provide information on leaf optical properties. Knowledge of absorption characteristics of major leaf constituents (e.g., cellulose, starch and protein) or their effective combinations (e.g., cellulose + starch) may permit remote assessment of canopy level concentrations.

Spectral Analysis of Organic Mixtures

In spectrophotometry, the ratio of incident flux intensity I_o before entering a sample to that of the light flux I after leaving the sample is determined. According to Beer-Lambert's law, which states that concentration is proportional to the logarithm of transmittance [and which for plants is highly correlated with reflectance (c.f. Tucker and Garratt, 1977)], the following relationship holds:

$$I/I_o = e^{-C\ell\alpha} \tag{1}$$

where I/I_o is the measured response, C is the concentration of the sample, ℓ is the path length, and α is the absorption coefficient at wavelength λ. Beer-Lamberts's law describes pure absorption and serves as an adequate model for spectrophotometer data acquired in the laboratory. However, it does not include the appropriate scattering and absorption coefficients necessary when modeling radiation propagating through a canopy (Fukshansky, 1981; Horler et al., 1983). Scattering caused by the architecture of the canopy, as well as the internal structure of each leaf, will lead to increased optical depth and, as a result, increased absorption (Willstatter and Stoll, 1913; Gates et al., 1965). Pure scattering will linearly attenuate light with depth, unlike the exponential effect of absorption (Fukshansky, 1981). Additional complications in the relationship between foliar constituent concentrations and canopy reflectance will result from background spectral contributions, atmospheric effects, and sun-sensor geometry.

The spectral behavior of mixtures, either of single surfaces with inherent complexity (e.g., a leaf) or of a complex of many surfaces (e.g., a landscape), is a function of the type and quantity of reflecting components and their relative influence on the measured response. Absorption bands caused by electronic transitions and bond vibrations may assist in identifying concentrations of foliar constituents using local (derivatives) or global (curve-fitting) analyses of laboratory-acquired spectral data (e.g., Card et al., 1988; Wessman et al., 1988a). Such information, uncomplicated by the atmosphere and illumination geometry, can be used to interpret more complex spectral mixtures acquired with airborne and satellite sensors. Tracking spectral features in reflectance measurements made in the laboratory up to those made at the pixel level should provide some indication of the transfer of spectral information across scales.

Derivative Spectrometry

Derivative spectrometry is commonly employed to resolve or enhance absorption features that are masked by interfering background absorptions and/or by noise (Talsky et al., 1978; Dixit and Ram, 1985). The technique aids in separating overlapping bands and isolating shoulders and weak signals from unwanted background. For a constant intensity I_o over the whole wavelength range (as measured by a spectrophotometer), the first derivative of Equation 1 is obtained as:

$$\frac{(dI/d\lambda)}{I} = -C\ell \left(\frac{d\alpha}{d\lambda}\right) \quad (2)$$

and will be linearly proportional to concentration (Talsky et al., 1978). The sensitivity of the measurements will be high in inflection areas. The second derivative reads as:

$$\frac{(d^2I/d\lambda^2)}{I} = C^2\ell^2 \left(\frac{d\alpha}{d\lambda}\right)^2 - C\ell \left(\frac{d^2\alpha}{d\lambda^2}\right) \quad (3)$$

where direct proportionality to concentration exists only if $d\alpha/d\lambda$ equals zero. If $d^2\alpha/d\lambda^2$ has an extremum value, then sensitivity is very high.

Key properties of derivative spectra are that (1) broad bands are suppressed relative to sharper bands to an extent that increases with derivative order (Figure 7.4), and (2) overlapping bands are resolved even if the shoulders are formed by band maxima separated by less than the largest half-width (Figure 7.5). Higher-order derivatives will also eliminate background functions of higher-order and shape complexity (Talsky et al., 1978). Thus resolution is improved by the sharpening of signals. However, derivative spectra are only useful if features of component spectra are separable from neighboring features by a certain fraction of their average width, γ_{ij}:

$$\gamma_{ij} = 0.5(\gamma_i + \gamma_j) \quad (4)$$

where γ is the full width of the feature at half-height. Higher-order derivatives are required to increase band narrowing and, hence, separability. Noise will increase by a factor of ten with each increase in order unless special measures are taken to improve the signal.

Derivative transformations can be applied to remotely sensed data to reduce baseline shifts (albedo variations) resulting from surface topography, illumination conditions, and/or lack of appropriate calibration information (Dixit and Ram, 1981; Wessman et al., 1987). Derivative spectra from laboratory- and field-acquired measurements are useful for the characterization of shifts in the chlorophyll absorption edge (Horler et al.,

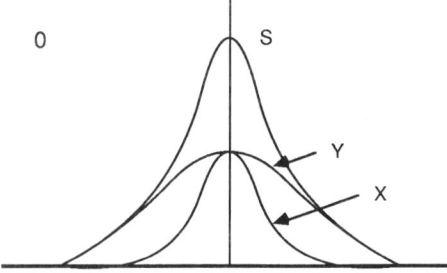

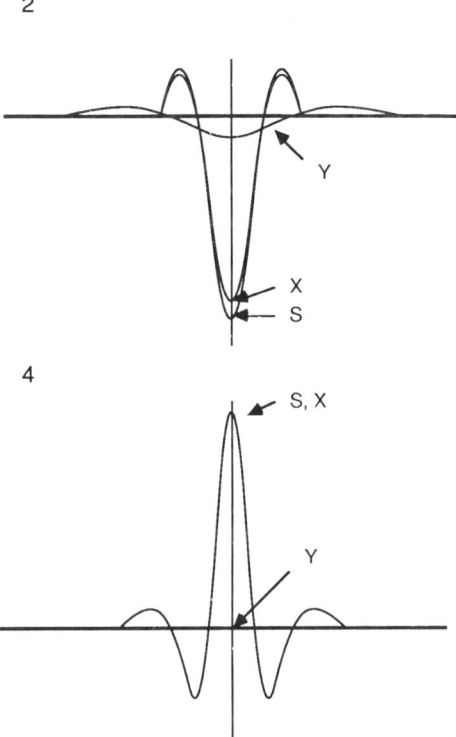

Figure 7.4. Relative amplitudes of two coincident Gaussian bands (bandwidth ratio 1:3) in the zero, second, and fourth derivative orders, illustrating the progressive suppression of the broader band. S is the resultant amplitude. [After Dixit and Ram (1985), p. 319.]

1983; Lichtenthaler et al., 1987; Schmuck et al., 1987; Rock et al., 1988; Ustin and Curtiss, 1988). A derivative transformation of Airborne Imaging Spectrometer (AIS) imagery over temperate deciduous and coniferous forests reduces apparent brightness differences attributable to canopy architecture and shifts in albedo between flight lines (Wessman et al., 1989). Correlative analysis of the transformed spectral data and canopy lignin concentrations suggests that absorption characteristics of lignin or a

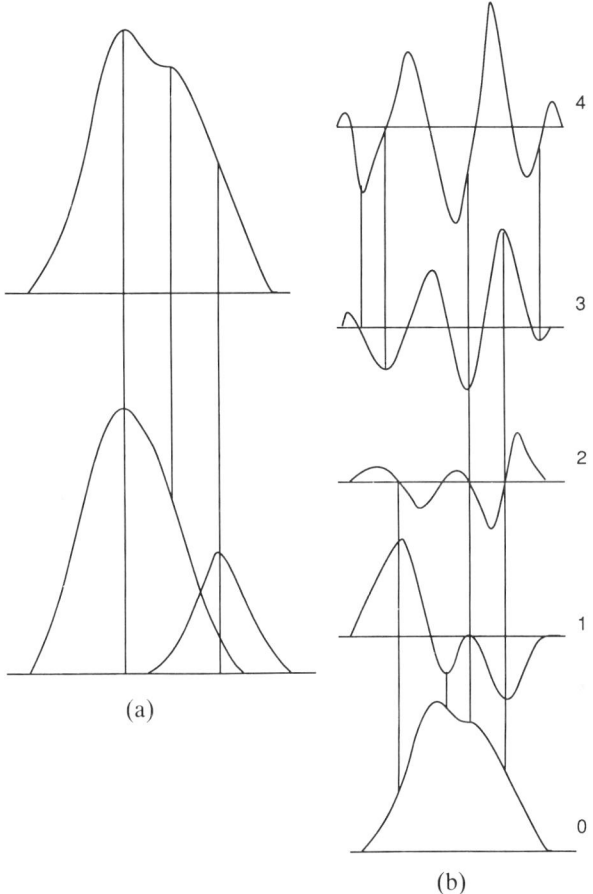

Figure 7.5. (a) Superposition of two unequal Gaussian curves and (b) differentiation of the two curves: fundamental curve and first to fourth derivatives. [After Talsky et al. (1978), p. 787. Reprinted by permission.]

closely associated canopy property influences reflectance in a predictable fashion (Wessman et al., 1988b).

Spectral Mixtures and Curve Fitting

Spectral decomposition techniques have often been applied to mixture spectra in the effort to separate overlapping bands of the component spectra. These techniques have involved fitting Gaussian or Lorentzian curves by a least-squares criterion in order to determine the number and frequency of the bands contributing to the mixture spectrum (Fraser and Suzuki, 1969; Gold et al., 1976). Such standard curve-fitting routines tend to break down because the correct number of constituent bands is required to model accurately the mixture spectra and the absorption bands of many organics are not Gaussian or Lorentzian. Alternative mixture modeling approaches successfully reproduce the mixture spectrum when the spectra of all pure components are known and combine linearly (Blackburn, 1965; Antoon et al., 1977). Essentially, these models are multivariate applications of the Beer-Lambert law and assume the absence of nonlinear effects. A model that linearly combines selected pure foliar constituent spectra (e.g., cellulose, starch, protein, lignin, water) multiplied by their concentration within a given leaf successfully simulates a whole-leaf reflectance spectrum (D. Card, personal communication). The model is easily inverted and demonstrates that interfering absorption features are separable (dependent on noise level) despite significant differences in leaf constituent absorption coefficients.

Unfortunately, pure component spectra frequently are not available, and may not even be representative of constituent spectra under the influence of the mixture's matrix effects. Problems with pure components are incurred when (1) the component spectra are not pure; (2) the components vary between samples, for example, types of protein; (3) the effect of an extraction procedure on the spectral characteristics of a component is unknown; and (4) the component spectrum changes with concentration level (Hruschka and Norris, 1982). In fact, changes in symmetry, chemisorption, physisorption, and hydrogen bonding cause most mixture spectra to be nonlinear combinations of their components (Figure 7.6; Honigs et al., 1984). Consequently, methods that address the nature of matrix interactions may be more representative of component structure and concentration than those utilizing linear combinations of pure component spectra.

Multivariate techniques for mixture analysis have been developed that do not require reference (pure) spectra. These methods rely on the multivariable nature of the data to determine quantitatively one or more variables (i.e., foliar constituent concentrations) from measured values of two or more predicting variables (i.e., spectral reflectance values). Most multivariate methods attempt to model how the chemical constituent influences the spectrum and assume (Naes and Martens, 1987):

$$\mathbf{X} = g(\mathbf{C}) + \mathbf{E} \qquad (5)$$

where g defines the mathematical model, $\mathbf{E}$ is the residual, and $\mathbf{X} = \{x_{ik}\}$: matrix of spectral data for I samples and K wavelengths, and $\mathbf{C} = \{c_{ij}\}$: matrix of chemical data for I samples and J chemical constituents.

Beer's law thus can be expressed as:

$$\mathbf{X} = \mathbf{CK} + \mathbf{E} \qquad (6)$$

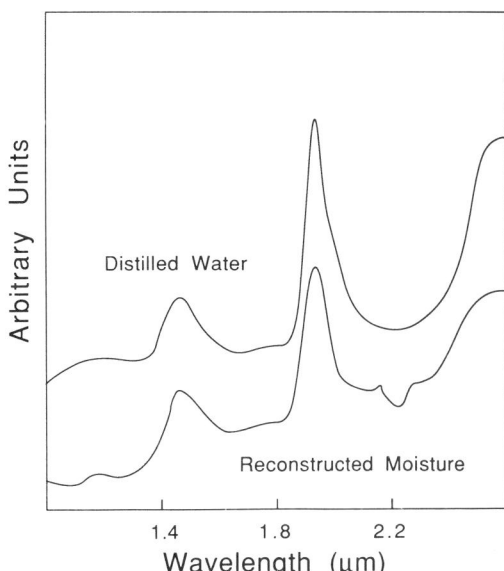

Figure 7.6. An absorbance spectrum of pure H_2O and reconstructed spectrum of moisture in wheat. Broadening of 1.45 and 1.94 μm bands is indicative of variations in hydrogen bonding caused by the protein matrix. [From Honigs et al. (1984), p. 321.]

where **K** is the matrix of unit spectra. From known data for **C** and **X**, the **K**-matrix is estimated from calibration with reference spectra. Concentrations for new samples are then predicted by fitting their data **X** to **K** by some mathematical projection procedure.

Traditionally, the model is inverted to define the chemical constituents as a function of the spectral variables plus noise:

$$\mathbf{C} = h(\mathbf{X}) + \mathbf{F} \qquad (7)$$

where h is the model function and **F** is the residual. Multiple linear and stepwise regression methods give direct estimates of h. Interfering factors, such as other constituents and systematic errors, are compensated for, and not modeled, in this method.

The use of factor analysis combines all the chemical and instrumental variables into a few factor or latent variables that express independent sources of variability within the data (Malinowski and Howery, 1980). In terms of the models described above, a low-dimensional table of factor scores, **T**, models both **X** and **C** as (Naes and Marten, 1987):

$$\begin{aligned} \mathbf{X} &= g(\mathbf{T}) + \mathbf{E} \\ \mathbf{C} &= h(\mathbf{T}) + \mathbf{F} \end{aligned} \qquad (8)$$

where g and h define the relation (commonly defined as linear) between the variables and the scores. The important point is that **C** is function of **T**, which describes some variation in **X**, but **C** is independent of **E**, the error in **X**. Factor analysis has been used to analytically extract and quantify the concentration of component spectra in complex organic mixture spectra measured in the laboratory (Knorr and Futrell, 1979; Windig and Meuzelaar, 1984) and cotton canopy reflectance as measured in the field (Huete, 1986).

Techniques of decomposing spectra have been applied to remotely sensed imagery to determine how many linearly independent sources of variation (called end-members) exist in spectral mixtures, the mixing model required (linear, nonlinear or a combination), and the spectral regions optimal for discriminating end-members (Adams and Adams, 1984; Smith et al., 1985; Adams et al., 1986). End-members are taken from spatially contiguous areas that share the same set of end-members. They can be derived in primarily three ways: (1) as "pure" members from groups of pixels containing only one material; (2) as intersections that result from two units sharing a single end-member; and (3) by determination of pixels at major vertices of the convex hull defined by the image data (Curtiss and Boardman, 1988). Each surface can be considered to contain at least two end-members, representing fully illuminated and shaded conditions, thereby defining the continuum of possible reflectances of a surface on a spectral mixing line. For n spectral end-members, mixing is described within an $(n-1)$ dimensional volume.

First-order mixing models have been found to describe adequately soil-plant spectral interactions within a canopy (Huete, 1986, 1987; Ustin and Curtiss, 1988). The separation of soil influences from the vegetation spectral response using decomposition techniques provides a "cleaner" vegetation signal for characterization of plant canopy spectral properties. Comparisons of decomposed spectra from remote sensing image data with laboratory-acquired end-members representing chemical/biophysiological range extremes may direct analyses of foliar spectral characteristic contributions to canopy reflectance. The end-member concept has been applied in a laboratory setting by Hruschka and Norris (1982), who approximated the near-IR reflectance spectra of ground wheat samples by a linear combination of spectra representing the range of known chemical data. The approximation satisfied a least-squares criterion. Coefficients of these linear combinations were then correlated with the protein and moisture content of the wheat samples. The use of sample spectra as mixture components improved standard errors from correlations of curve-fit coefficients with chemical data over those generated when pure constituent spectra were used as components.

Use of the full spectral range to explore the makeup of mixture spectra will aid in defining the influential constituents. Laboratory spectrophotometer measurements simplify the analysis to those properties primarily related to leaf optical characteristics. However, the fine structure of a leaf spectrum as measured in the laboratory is insufficient to account for the reflectance of plant canopies. Attenuation and multiple scattering of radiation within a plant canopy, as well as the influence of soil, shadow, and other background spectral properties, complicate the vegetation response. These factors must be addressed in order to understand fully the integrated response of whole canopies. Advances in analytical chemistry and their successful application to quantitative foliar analysis indicate that we can, at the very least, improve our understanding of the mechanisms of reflectance by leaves. Should these principles translate in their complete or reduced form to canopy and pixel levels, opportunities will be available for unprecedented ecosystem monitoring.

Current and Potential Use of Imaging Spectrometry

The nature of absorptions within organic mixtures are weak and complex since they consist of overlapping overtone and combination bands. The origins of the observed vibrations are limited and they are all associated with primary constituents of vegetation. Knowledge of absorption characteristics of each of the major leaf constituents (e.g., cellulose, starch, and protein) may permit remote assessment of canopy-level concentrations if high spectral resolution reflectance information is acquired. Direct assessment of low-level constituents is unlikely; chances increase with the predominant materials such as cellulose, lignin, and protein. It may be necessary to create a simpler taxonomy of constituents and their effective spectral and ecological combinations. For example, structural materials such as cellulose and lignin may influence the canopy spectrum in such similar ways as to be indifferentiable. Nonetheless, an estimate of their combined concentration may be very useful for large-scale ecological applications. High spectral resolution measurements will be needed to study characteristics of the canopy reflectance curve and to better separate background factors from the vegetation response.

The potential to estimate canopy chemistry relies to a great extent on the development of imaging spectrometer systems. Broad-band sensors provide adquate contrast between the near-IR plateau and chlorophyll absorption wells, but information is lost in the integration over broad spectral regions. Research has shown, primarily through correlative means and with varying degrees of success, strong relationships between broad-band canopy reflectance and vegetation parameters such as absorbed photosynthetically active radiation (APAR) (Asrar et al., 1984; Hatfield et al., 1984); leaf area index (LAI) (Asrar et al., 1986; Peterson et al., 1988); and water content (Tucker, 1980; Hunt et al., 1987). In each of these studies, the normalized difference (NDVI) and other vegetation indices are affected by illumination conditions, background reflectance, sun-sensor geometry, and so on. The contiguous, narrow spectral measurements made by imaging spectrometers can be mathematically transformed to reduce and/or separate multiple-order effects that are a background to the spectrum of interest. The additional wavebands provided by a spectrometer are requisite for the application of spectral decomposition techniques when more than two to three reflecting components exist in a remotely sensed landscape (Huete and Jackson, 1987).

Recent work by Sellers (1985, 1987) theoretically develops the association between canopy reflectance and biophysical quantities using simple reflectance, photosynthesis, and resistance models. Sellers (1985) demonstrates that plant canopy reflectance data are indicative of instantaneous limiting biophysical rates (photosynthesis, transpiration) rather than any state (biomass, LAI) associated with the canopy. The close connection of photosynthetically active radiation absorbed by the leaf to the chlorophyll density, which can be estimated remotely, leads to near-linear relationships among canopy properties of APAR, photosynthetic capacity (P_c), and minimum canopy resistance (r_c), and the simple ratio vegetation index (SR), and more curvilinear relationships with the normalized difference vegetation index (NDVI). However, large differences in daily net photosynthesis between C_3 and C_4 physiological groups (Brown, 1978; Osmond et al., 1982) may create ambiguities in photosynthetic estimates from vegetation indices (Choudhury, 1987). Studies of landscape variations in nitrogen use efficiency within tall- and shortgrass prairie suggests that natural heterogeneity and management effects can introduce substantial noise into calculations based on NDVI (D. Schimel, personal communication). High correlations between photosynthetic capacity and plant nitrogen couple plant biochemistry to biophysical exchange. A parameter relating NDVI-like indices to flux will be correlated to nitrogen-use efficiency (net carbon assimilated per unit of nitrogen), to the extent that chlorophyll density is correlated to nitrogen content (Schimel et al., 1988); typically 75% of plant nitrogen is contained in the chloroplasts (Chapin et al., 1987). Contiguous, high spectral resolution measurements may provide more accurate estimates of chlorophyll density, canopy nitrogen concentrations, and, as a consequence, the related biophysical rates.

The technology to acquire high spectral resolution images is limited to only a few airborne systems (e.g., the Airborne Visible/Infrared Imaging Spectrometer, AVIRIS). These will provide the prototype data for proposed instruments on the Earth Observing System (Eos), an orbiting platform of sensors designed to provide continuous remotely sensed data at regional to global scales. The High Resolution Imaging Spectrometer (HIRIS) will acquire 192 contiguous bands of spectral information at 10-nm spectral and 30-m spatial resolution. High spectral resolution in the range of 0.4 to 2.4 μm will permit consideration of spectral shape and detail. With the high data rates generated from such resolution, HIRIS imagery will be used to target areas of specific interest and can serve to monitor slowly varying ecosystem parameters (e.g., nitrogen content) over the growing season. HIRIS will act as the intermediate stage in a hierarchical arrangement of imaging spectrometers (e.g., AVIRIS, HIRIS, MODIS).

The MODerate resolution Imaging Spectrometer (MODIS) will measure numerous spectral bands between 0.4 and 12 μm at spectral resolutions of 10 nm and greater and with spatial resolutions of 500 and 1,000 m. High-frequency temporal information can be acquired by MODIS as a consequence of its lower data rates. The acquisition of temporal variability in intercepted PAR, from which maximum canopy photosynthesis and minimum canopy resistance can be estimated, can provide a phenological time series on a regional basis. HIRIS can be used to characterize spatial variability not captured by MODIS and can, in turn, be used for analysis of spectral mixtures at the spatial scales provided by both sensors.

We already recognize the need for temporal and spatial information on biospheric functioning, and the prohibitive logistics of acquiring such information other than through the use of remote sensing technology. Moreover, we recognize that patterns of regional and global interactions will most likely be imperceptible from our present vantage point. Current assessments of global primary productivity using remote sensing (e.g., Justice et al., 1986; Tucker et al., 1986) will increase in usefulness as we become more capable of utilizing such information for modeling and monitoring global-level processes. Estimates of canopy biochemistry through direct assessment of spectral reflectance features or inferred through relationships with other factors contributing to canopy reflectance (e.g., water content) will provide further insights into the nature of biosphere function response to environmental change.

Acknowledgments

I thank J. Aber, D. Peterson, D. Card, and N. Swanberg for past discussions on remote sensing, near and far, of foliar chemistry; P. Matson, S. Running, and P. Vitousek for encouragement in developing these studies; P. Sellers and an anonymous reviewer for helpful comments on the manuscript; and A. Goetz and the Center for the Study of Earth from Space/Cooperative Institute for Research in Environmental Sciences at the University of Colorado for their support.

References

Adams, J.B., and Adams, J.D. (1984). Geologic mapping using Landsat MSS and TM images: removing vegetation by modeling spectral mixtures. *Proc. Int. Symp. Remote Sens. Envir., Third Thematic Conf. Remote Sens. for Exploration Geology*, pp. 615–622.

Adams, J.B., Smith, M.O., and Johnson, P.E. (1986). Spectral mixture modeling: a new analysis of rock and soil types at the Viking Lander I site. *J. Geophys. Res.* 80(B8):8098–8112.

Antoon, M.K., Koenig, J.H., and Koenig, J.L. (1977). Least-squares curve-fitting

of fourier transform infrared spectra with applications to polymer systems. *Appl. Spectr.* 31:518–524.
Asrar, G., Fuchs, M., Kanemasu, E.T., and Hatfield, J.L. (1984). Estimating absorbed photosynthetic radiation and leaf area index from spectral reflectance in wheat. *Agron. J.* 76:300–306.
Asrar, G., Kanemasu, E.T., Miller, G.P., and Weiser, R.L. (1986). Light interception and leaf area estimates from measurements of grass canopy reflectance. *IEEE Trans. Geosci. Remote Sens.* GE-24(1):76–81.
Binkley, D., and Reid, P. (1985). Long-term increase in nitrogen availability from fertilization of Douglas fir. *Canad. J. Forest Res.* 15(4):723–724.
Birk, E.M., and Vitousek, P.M. (1986). Nitrogen availability and nitrogen use efficiency in loblolly pine stands. *Ecology* 67:69–79.
Blackburn, J.A. (1965). Computer program for multicomponent spectrum analysis using least-squares method. *Anal. Chem.* 37:1000–1003.
Brown, R.H. (1978). A difference in N use efficiency in C_3 and C_4 plants and its implications in adaptation and evolution. *Crop Sci.* 18:93–98.
Bryant, J.P., Chapin, F.S. III, and Klein, D.R. (1983). Carbon/nutrient balance of boreal plants in relation to vertebrate herbivory. *Oikos* 40:357–368.
Card, D.H., Peterson, D.L., Matson, P.A., and Aber, J.D. (1988). Prediction of leaf chemistry by the use of visible and near infrared reflectance spectroscopy. *Remote Sens. Envir.* 26:123–147.
Chang, S-H., and Collins, W.E. (1983). Confirmation of the airborne biogeophysical mineral exploration technique using laboratory methods. *Econ. Geol.* 78:723–736.
Chapin, F.S. III, Bloom, A.J., Field, C.B., and Waring, R.H. (1987). Plant responses to multiple environmental factors. *Bioscience* 37:49–57.
Chapin, F.S. III, McKendrick, J.D., and Johnson, D.A. (1986). Seasonal changes in carbon fractions in Alaskan tundra plants of differing growth form: implications for herbivory. *J. Ecol.* 74:707–731.
Choudhury, B.J. (1987). Relationships between vegetation indices, radiation absorption, and net photosynthesis evaluated by a sensitivity analysis. *Remote Sens. Envir.* 22:209–233.
Coley, P.D., Bryant, J.P., and Chapin, F.S. III. (1985). Resource availability and plant antiherbivore defense. *Science* 230:895–899.
Collins, W.E. (1978). Remote sensing of crop type and maturity. *Photogramm. Eng. Remote Sens.* 44(1):43–55.
Curcio, J.A., and Petty, C.C. (1951). The near infrared absorption spectrum of liquid water. *J. Opt. Soc. Amer.* 41(5):302–305.
Curtiss, B., and Boardman, J. (1988). The determination of end-member spectral signatures in imaging spectrometer data. Presented at Computer-Enhanced Analytical Spectroscopy, Second Hidden Peak Symp., Snowbird, UT.
Delwiche, C.C., Bissell, S., and Virginia, R. (1978). Soil and other sources of nitrogen oxide. pp. 459–476. In D.R. Nielsen and J.G. Macdonald (eds.), *Nitrogen in the Environment*. Academic Press, NY.
Dixit, L., and Ram, S. (1985). Quantitative analysis by derivative electronic spectroscopy. *Appl. Spectr. Rev.* 21(4):311–418.
Everitt, J.H., Richardson, A.J., and Gausman, H.W. (1985). Leaf reflectance-nitrogen-chlorophyll relations in buffelgrass. *Photogramm. Eng. Remote Sens.* 51(4):463–466.
Fales, S.L., and Cummins, D.G. (1982). Reducing moisture-induced error associated with measuring forage quality using near infrared reflectance. *Agron. J.* 74:38–41.
Fogel, R., and Cromack, K. (1977). Effects of habitat and substrate quality on

Douglas-fir litter decomposition in western Oregon. *Canad. J. Bot.* 55:1632–1640.
Fraser, R.D.B., and Suzuki, E. (1969). Resolution of overlapping bands: functions for simulating band shapes. *Anal. Chem.* 41:37–39.
Fukshansky, L. (1981). Optical properties of plants. pp. 21–40. In H. Smith (ed.), *Plants and the Daylight Spectrum*. Academic Press, London, England.
Gartlan, J.S., McKey, D.B., Waterman, P.G., Mbi, C.N., and Struhsaker, T.T. (1980). A comparative study of the phytochemistry of two African rain forests. *Biochem. Syst. Ecol.* 8:401–422.
Gates, D.M. (1970). Physical and physiological properties of plants. pp. 224–252. In *Remote Sensing with Special Reference to Agriculture and Forestry*. Natl. Acad. Sci., Washington, DC.
Gates, D.M., Keegan, H.J., Schleter, J.C., and Weidner, V.R. (1965). Spectral properties of plants. *Appl. Opt.* 4(1):11–20.
Gausman, H.W. (1977). Reflectance of leaf components. *Remote Sens. Envir.* 6:1–9.
Gausman, H.W., Escobar, D.E., Everitt, J.H., Richardson, A.J., and Rodriguez, R.R. (1978). Distinguishing succulent plants from crop and woody plants. *Photogramm. Eng. Remote Sens.* 44(4):487–491.
Goetz, A.F.H., Rock, B.N., and Rowan, L.C. (1983). Remote sensing for exploration: an overview. *Econ. Geol.* 78:573–590.
Gold, H.S., Rechsteiner, C.D., and Buck, R.P. (1976). Generalized spectral decomposition method applied to infrared, ultraviolet, and atomic emission spectrometry. *Anal. Chem.* 48:1540–1546.
Goodroad, L.L., and Keeney, D.R. (1984). Nitrous oxide emission from forest, marsh, and prairie ecosystems. *J. Envir. Qual.* 13(3):448–452.
Hatfield, J., Asrar, G., and Kanemasu, E.T. (1984). Intercepted photosynthetically active radiation estimated by spectral reflectance. *Remote Sens. Envir.* 14:65–75.
Hinzman, L.D., Bauer, M.E., and Daughtry, C.S.T. (1986). Effects of nitrogen fertilization on growth and reflectance characteristics of winter wheat. *Remote Sens. Envir.* 19:47–61.
Honigs, D.E., Hieftje, G.M., and Hirschfeld, T. (1984). A new method for obtaining individual component spectra from those of complex mixtures. *Appl. Spectr.* 38:317–322.
Horler, D.N., Dockray, M., and Barber, J. (1983). The red edge of plant leaf reflectance. *Int. J. Remote Sens.* 4(2):273–288.
Hruschka, W.R., and Norris, K.H. (1982). Least-squares curve fitting of near infrared spectra predicts protein and moisture content of ground wheat. *Appl. Spectr.* 36:261–265.
Huete, A.R. (1986). Separation of soil-plant spectral mixtures by factor analysis. *Remote Sens. Envir.* 19:237–251.
Huete, A.R. (1987). Soil-dependent spectral response in a developing plant canopy. *Agron. J.* 79:61–70.
Huete, A.R., and Jackson, R.D. (1987). Suitability of spectral indices for evaluating vegetation characteristics on arid rangelands. *Remote Sens. Envir.* 23:213–232.
Hunt, E.R. Jr., Rock, B.N., and Nobel, P.S. (1987). Measurement of leaf relative water content by infrared reflectance. *Remote Sens. Envir.* 22:429–435.
Justice, C.O., Townshend, J.R.G., Holben, B.N., and Tucker, C.J. (1985). Analysis of the phenology of global vegetation using meteorological satellite data. *Int. J. Remote Sens.* 6(8):1271–1318.
Kleman, J., and Fagerlund, E. (1987). Influence of different nitrogen and irrigation

treatments on the spectral reflectance of barley. *Remote Sens. Envir.* 21:1–14.
Knipling, E.B. (1970). Physical and physiological basis for the reflectance of visible and near-infrared radiation from vegetation. *Remote Sens. Envir.* 1:155–159.
Knorr, F.J., and Futrell, J.H. (1979). Separation of mass spectra of mixtures by factor analysis. *Anal. Chem.* 51:1236–1241.
Kollenkark, J.C., Daughtry, C.S.T., Bauer, M.E., and Housely, T.L. (1982). Effects of cultural practices on agronomic and reflectance characteristics of soybean canopies. *Agron. J.* 74:751–758.
Lainson, R.A., and Thornley, J.H.M. (1982). A model for leaf expansion in cucumber. *Ann. Bot.* 50:407–425.
Lichtenthaler, H.K., and Buschmann, C. (1987). Reflectance and chlorophyll fluorescence signatures of leaves. *Proc. Int. Geosci. Remote Sens. Symp. (IGARRS'87).* IEEE 87CH2434-9. 2:1201–1206.
Mackinney, G. (1938). Applicability of Kundt's rule to chlorophyll. *Plant Physiol.* 13:427–430.
Malinowski, E.R., and Howery, D.G. (1980). *Factor Analysis in Chemistry.* Wiley, NY.
Marten, G.C., Shenk, J.S., and Barton, F.E. II. (eds.) (1985). *Near Infrared Reflectance Spectroscopy (NIRS): Analysis of Forage Quality,* Agric. Hndbk. no. 643, U.S. Dep. Agric., Washington, DC.
McDonald, R.S. (1986). Review: infrared spectrometry. *Anal. Chem.* 58:1906–1925.
Meentemeyer, V. (1978). Macroclimate and lignin control of litter decomposition rates. *Ecology* 59:465–472.
Melillo, J.M., Aber, J.D., and Muratore, J.F. (1982). Nitrogen and lignin control of hardwood leaf litter decomposition dynamics. *Ecology* 63(3):621–626.
Mooney, H.A., and Gulmon, S.L. (1982). Constraints on leaf structure and function in reference to herbivory. *BioScience* 32(3):198–206.
Nadelhoffer, K.J., Aber, J.D., and Melillo, J.M. (1983). Leaf litter production and soil organic matter dynamics along a nitrogen gradient in southern Wisconsin (USA). *Canad. J. Forest Res.* 13:12–21.
Naes, T., and Martens, H. (1987). Multivariate calibration: quantification of harmonies and disharmonies in analytical data. pp. 121–141. In H.L.C. Meuzelaar and T.L. Isenhour (eds.) *Computer-Enhanced Analytical Spectroscopy.* Plenum Press, NY.
Nelson, V.L., Gjerstad, D.H., and Glover, G.R. (1986). Determining nitrogen status of young loblolly pine by leaf reflectance. *Tree Phys.* 1:333–339.
Olson, C.E. Jr. (1967). Optical remote sensing of the moisture content in fine forest fuels. Final Rep. No. 8036-1F, Univ. of Michigan, Ann Arbor.
Osmond, C.B., Winter, K., and Ziegler, H. (1982). Functional significance of different pathways of CO_2 fixation in photosynthesis. In O. Lange, P. Nobel, C. Osmond, and H. Ziegler (eds.), *Encyclopedia of Plant Physiology* (New Series), Vol. 12B. Springer-Verlag, Berlin.
Pastor, J., Aber, J.D., McClaugherty, C.A., and Melillo, J.M. (1984). Aboveground production and N and P cycling along a nitrogen mineralization gradient on Blackhawk Island, Wisconsin. *Ecology* 65(1):256–268.
Pastor, J., and Post, W.M. (1986). Influence of climate, soil moisture, and succession on forest carbon and nitrogen cycles. *Biogeochem.* 2:3–27.
Peterson, D.L., Aber, J.D., Matson, P.A., Card, D.H., Swanberg, N.A., Wessman, C.A., Spanner, M., and Hlavka, C. (1988). Remote sensing of forest canopy and leaf biochemical contents. *Remote Sens. Envir.* 24:85–108.
Pinter, P.J. Jr., Jackson, R.D., Idso, S.B., and Reginato, R.J. (1981). Multidate spectral reflectance as predictors of yield in water stressed wheat and barley. *Int. J. Remote Sens.* 2:43–48.

Richardson, A.J., Everitt, J.H., and Gausman, H.W. (1983). Radiometric estimation of biomass and nitrogen content of Alicia grass. *Remote Sens. Envir.* 13:179–184.

Rock, B.N., Hoshizaki, T., and Miller, J.R. (1988). Comparison of in situ and airborne spectral measurements of the blue shift associated with forest decline. *Remote Sens. Envir.* 24:109–127.

Rohde, W.G., and Olson, C.E. (1971). Estimating foliar moisture content from infrared reflectance data. pp. 144–164. In *Third Biennial Workshop, Color Serial Photography in the Plant Sciences.* Amer. Soc. Photogramm., Falls Church, VA.

Rotolo, P. (1979). Near infrared reflectance instrumentation. *Cereal Foods World* 24(3):94–98.

Rubenthaler, G.L., and Bruinsma, B.L. (1978). Lysine estimation in cereals by near-infrared reflectance. *Crop Sci.* 18:1039–1042.

Salisbury, F.B., and Ross, C. (1969). *Plant Physiology.* Wadsworth, San Francisco, CA, pp. 262, 265.

Schimel, D., Kittel, T., and Parton, W. (1988). Ecosystem controls over atmosphere-biosphere exchange: data from FIFE. *Eos (Proc. Amer. Geophys. Union)* 69:1058.

Schmuck, G., Lichtenthaler, H.K., Kritikos, G., Amann, V., and Rock, B. (1987). Comparison of terrestrial and airborne reflection measurements of forest trees. *Proc. IGARSS'87 Symp.* Ann Arbor, MI, pp. 1207–1212.

Sellers, P.J. (1985). Canopy reflectance, photosynthesis and transpiration. *Int. J. Remote Sens.* 6(8):1335–1372.

Sellers, P.J. (1987). Canopy reflectance, photosynthesis and transpiration. II. The role of biophysics in the linearity of their interdependence. *Remote Sens. Envir.* 21:143–183.

Setak, S., Catsky, J., and Jarvis, P.G. (1971). Determination of chlorophylls a and b. pp. 673–701. In *Plant Photosynthetic Production: Manual of Methods.* Junk, The Hague.

Smith, M.O., Johnson, P.E., and Adams, J.B. (1985). Quantitative determination of mineral types and abundances from reflectance spectra using principal components analysis. *Proc. 15th Lunar and Planetary Sci. Conf. II.*, J Geophy Res., 90 (supple.):C797–C804.

Talsky, G., Mayring, L., and Kreuzer, H. (1978). High-resolution, higher-order UV/VIS derivative spectrophotometry. *Angew. Chem.* 17(11):785–874.

Thomas, J.R., and Gausman, H.W. (1977). Leaf reflectance v. leaf chlorophyll and carotenoid concentrations for eight crops. *Agron. J.* 69:799–807.

Thomas, J., and Oerther, G. (1972). Estimating nitrogen content of sweet pepper leaves by reflectance measurements. *Agron. J.* 64:11–13.

Tsay, M.L., Gjerstad, D.H., and Glover, G.R. (1982). Tree leaf reflectance: a promising technique to rapidly determine nitrogen and chlorophyll content. *Canad. J. Forest Res.* 12:788–792.

Tucker, C.J. (1980). Remote sensing of leaf water content in the near infrared. *Remote Sens. Envir.* 10:23–32.

Tucker, C.J., Fung, I.Y., Keeling, C.D., and Gammon, R.H. (1986). Relationship between atmospheric CO^2 variations and satellite-derived vegetation index. *Nature* 319:195–199.

Tucker, C.J., and Garratt, M.W. (1977). Leaf optical system modeled as a stochastic process. *Appl. Optics* 16:635–642.

Ustin, S.L., and Curtiss, B. (1988). Spectral characteristics of ozone treated conifer species. EPA Final Rep., Contract No. 7B0008NTEX.

Van Cleve, K., Oliver, L., Schlentner, R., Viereck, L.A., Dyrness, C.T. (1983). Productivity and nutrient cycling in taiga forest ecosystems. *Canad. J. Forest*

Res. 13:747–767.

Vitousek, P.M. (1982). Nutrient cycling and nutrient use efficiency. *Amer. Naturalist* 119:553–572.

Vitousek, P.M. (1983). The effects of deforestation on air, soil, and water. pp. 223–245. In B. Bolin and R.B. Cook (eds.)., *The Major Biogeochemical Cycles and Their Interactions*. Wiley, NY.

Vitousek, P.M., Gosz, J.R., Grier, C.C., Melillo, J.M., and Reiners, W.A. (1982). A comparative analysis of potential nitrification and nitrate mobility in forest ecosystems. *Ecol. Mono.* 52(2):155–177.

Walburg, G., Bauer, M.E., Daughtry, C.S.T., and Housley, T.L. (1982). Effects of nitrogen nutrition on the growth, yield, and reflectance characteristics of corn. *Agron. J.* 74:677–683.

Waring, R.H., McDonald, A.J.S., Larsson, S., Ericsson, T., Wiren, A., Arwidsson, E., Ericsson, A., and Lohammar, T. (1985). Differences in chemical composition of plants grown at constant relative growth rates with stable mineral nutrition. *Oecologia* 66:157–160.

Wessman, C.A., Aber, J.D., and Peterson, D.L. (1987) Estimation of forest canopy characteristics and nitrogen cycling using imaging spectometry. pp. 114–118. In G. Vane (ed.), *Proc. SPIE Int. Soc. for Opt. Eng.*, *Imaging Spectrometery II, Vol. 834*. SPIE—Int. Soc. for Opt. Eng., Bellingham, WA.

Wessman, C.A., Aber, J.D., and Peterson, D.L. (1989). An evaluation of imaging spectrometry for estimating forest canopy chemistry. *Int. J. Remote Sens.* 10(8):1293–1316.

Wessman, C.A., Aber, J.D., Peterson, D.L., and Melillo, J.M. (1988a). Foliar analysis using near infrared reflectance spectroscopy. *Canad. J. Forest Res.* 18:6–11.

Wessman, C.A., Aber, J.D., Peterson, D.L., and Melillo, J.M. (1988b). Remote sensing of canopy chemistry and nitrogen cycling in temperate forest ecosystems. *Nature* 335:154–156.

Westman, W.E., and Price, C.V. (1988). Spectral changes in conifers subjected to air pollution and water stress: experimental studies. *IEEE Trans. Geosci Remote Sens.* 26(1):11

Wetzel, D.L. (1983). Near-infrared reflectance analysis: sleeper among spectroscopic techniques. *Anal. Chem.* 55:1165A–1171A.

Weyer, L.G. (1985). Near-infrared spectroscopy of organic substances. *Appl. Spectr. Rev.* 21(1&2):1–43.

Willstatter, R., and Stoll, A. (1913). *Untersuchungen über Chlorophyll*. Springer-Verlag, Berlin.

Winch, J.E., and Major, H. (1981). Predicting nitrogen and digestibility of forages using near infrared reflectance photometry. *Canad. J. Plant Sci.* 61:45–51.

Windig, W., and Meuzelaar, H.L.C. (1984). Nonsupervised numerical component extraction from pyrolysis mass spectra of complex mixtures. *Anal. Chem.* 56:2297–2303.

8. Remote Sensing and Trace Gas Fluxes

Pamela A. Matson and Peter M. Vitousek

Most of the chapters in this volume emphasize the dynamics of energy, water, and carbon at the surface of the earth. These are major processes in the metabolism of the planet, and it is reasonable to expect that thermal and optical remote sensing could detect at least energy transformations rather directly. Although their overall magnitude is smaller, fluxes of trace gases represent another important interaction between the biosphere and atmosphere. At present, these fluxes cannot be directly measured remotely, but we suggest that current remote sensing approaches are nonetheless useful in the analysis of trace gas flux, and that remote sensing techniques that are now under development will make a still greater contribution.

The major trace gases of interest include methane (CH_4), carbon monoxide (CO), nonmethane hydrocarbons (NMHC), nitrous oxide (N_2O), other oxides of nitrogen (NO_x), ammonia (NH_3), and various sulfur-containing trace gases (Mooney et al., 1987). They are important for two major reasons. First, most have significant effects in the atmosphere. The longer lived gases, CH_4 and N_2O, are greenhouse gases that are much more efficient absorbers per molecule than is CO_2. Their concentrations are demonstrably increasing in the troposphere, very rapidly in the case of CH_4. They are now less important overall than is CO_2 (because of their lower concentrations), but their relative importance is increasing. On the other hand, the more reactive gases play major roles in the chemistry of the troposphere; in particular, NMHC and NO_x interact in the formation of tropospheric ozone (O_3), the most serious air pollutant globally.

Second, trace gas fluxes provide information on ecosystem function. While trace gas fluxes are spatially and temporally variable, they arise from relatively well-defined biological processes within terrestrial ecosystems. Most of those processes are not simply correlated with energy capture, but rather reflect interactions of the major biogeochemical cycles. The presence and magnitude of trace gas fluxes therefore can provide insight into the metabolism of terrestrial ecosystems. Indeed, to the extent that trace gas fluxes from biosphere to atmosphere can be measured accurately, we believe that such measurements will drive and constrain our understanding of terrestrial biogeochemistry in the same way that the "watershed ecosystem approach" (Likens et al., 1977; Bormann and Likens, 1979) has guided the current generation of ecosystem studies.

Ultimately, remote measurements are essential to the development of regional estimates of trace gas flux; there will never be enough chamber or cuvette measurements to account for the notorious spatial and temporal variability in flux (Keller et al., 1983, 1986). Our purpose here is to illustrate ways in which remote sensing can be used now to assist in measurements of trace gas emissions from terrestrial ecosystems, and to discuss ways in which techniques that are currently under development may become applicable. We separate the major approaches to utilizing remote sensing into five major categories—remote classification of ecosystems and measurement of flux on the ground in each type; remote estimation of the driving variables for models of trace gas flux; ground-based systems for remote, direct measurements of flux; direct aircraft-based systems; and direct satellite-based systems—and, where possible, give examples, possibilities, and/or limitations of each.

Classification-Based Estimates

The least adventurous way to use remote sensing in developing estimates of trace gas flux is to classify ecosystems into functionally different units that can be distinguished remotely, then to measure fluxes of trace gases on the ground in each of those units, and, finally, to make an areal estimate of flux by multiplying the cover of each type (sensed remotely) by the flux from that type. Where classifications are based on characteristics that are the same factors that control trace gas flux (i.e., hydrology, soil fertility), this approach should yield results that are much better than those obtained by assuming that a particular site is "representative" of a region, or those obtained by sampling a number of available sites and taking an average.

Despite the availability of the methodology, this straightforward use of remote sensing to estimate trace gas flux is more widely discussed than practiced. Matson et al. (submitted) recently used this approach to calcu-

Figure 8.1. Spatial distribution of annual nitrous oxide flux in sagebrush steppe based on vegetation classification and ground-based measurements. Brightest shading has highest fluxes.

late nitrous oxide fluxes from Wyoming sagebrush-steppe ecosystems. Differences in vegetation there are caused by landscape position and winter snow depth; they are highly correlated with rates of production and nitrogen cycling. Reiners et al. (in press) made a supervised classification of the vegetation in a 15 × 15-km area using Landsat Thematic Mapper (TM) data. Matson et al. (submitted) then measured nitrous oxide flux over a two-year period from the major vegetation types within the scene. The product of the classification and the fluxes is an areal estimate of flux from the shrub-steppe landscape (Figure 8.1); it illustrates that one vegetation type (*Artemesia tridentata* var. *vaseyana*), which covers around 15% of the area in this particular image, is responsible for over 35% of nitrous oxide flux. Bartlett et al. (submitted) used a similar approach to estimate methane fluxes from the Everglades. The region was stratified into major wetland types remotely (using TM), and fluxes within representatives of each of the major types were measured through an annual cycle.

This approach can also be applied to evaluating the effects of human disturbance on trace gas fluxes. Matson et al. (in press) made ground-based measurements of nitrous oxide fluxes from an area of tropical forest near Manaus, Brazil, in the intensive study area of the aircraft-based Amazon Boundary Layer Experiment (ABLE). They found wide variation in nitrous oxide fluxes among intact forests that differed in soil fertility (Matson and Vitousek 1987) and, more important, greatly elevated fluxes from pastures that had been cleared from primary forest (Luizao et al., in press).

They estimated that the 11% of the intensively sampled area that was in pasture accounted for more than 40% of nitrous oxide fluxes from the area (Matson et al., in press).

Overall, this approach has a number of strengths. First, it makes use of the variation among ecosystems in the development of areal flux estimates, rather than basing estimates on "representative" sites or attempting to account for real variations by simply averaging many ground-based estimates. Second, it draws attention to the most active sites of flux—sites that are more important than most in atmosphere-biosphere exchange, and that therefore may be more susceptible to disruption. Finally, estimates that are based on accurately characterized differences among ecosystems can be more reasonably extrapolated beyond the immediate sites in which they were collected. For example, areal fluxes of nitrous oxide from regions of Amazonia with greater or lesser deforestation than the ABLE intensive study area can be estimated using classification-based approaches. The approach also has limitations; classifications that do not reflect characteristics that also affect trace gas fluxes may not be useful beyond very narrow limits, and the enormous spatial and temporal variability in trace gas fluxes within a class must still be sampled on the ground, almost invariably inadequately. Most important, the approach cannot account for year-to-year variation in trace gas flux and its controls within a site or type.

Remote Sensing for Driving Variables of Models

A second approach is to use remote sensing to drive biogeochemical models of terrestrial ecosystems, which, in turn, predict fluxes of trace gases. Trace gas production is biologically controlled; it is driven by variations in temperature, moisture, light absorption (for some gases such as nonmethane hydrocarbons), and the chemical characteristics of substrates. To the extent that these can be determined remotely, ecosystem-level models of the processes controlling trace gas flux can be parameterized and run on the basis of information obtained from remote sensing.

In theory, this approach encompasses a wide range of possibilities, from classifications of terrestrial ecosystems that are based on characteristics known to be related to trace gas flux, through successional models that use remote sensing to locate and track disturbances, to process-level models that use remotely sensed information to initialize and update whole-system simulation models. In practice, there are a number of excellent, detailed models of terrestrial biogeochemistry in operation (McGill et al., 1981; Parton et al., 1988b), and substantial effort is going into adapting for remotely sensed driving variables. For example, the Century model of grassland productivity, nutrient cycling, and soil development (Parton et al., 1988b) makes use of information on climate and soil parent material to examine grasslands on a continental scale. It has been modified to predict fluxes of nitrogen-containing trace gases (Parton et al., 1988a,b; Schimel et al., in press); modifications that incorporate some remote sensing inputs have also been carried out (Reiners et al., in press). The Century model's climatic forcing functions are potentially adapted to remote sensing inputs, but the information on soils that it requires still must be obtained by another pathway. Eventually, remotely sensed canopy chemistry may be used to calculate such soil characteristics indirectly (Wessman, this volume), or perhaps an alternative formulation that is driven or modified by canopy rather than soil properties (Running, this volume) will be more useful.

Overall, the strengths of this approach are that it builds upon a long-developing understanding of the biogeochemical regulation of trace gas flux, that it is dynamic in the sense that a change in forcing functions (from day to day or year to year) can yield a reasonable change in trace gas flux, and that it can be applied on spatial scales beyond immediate study areas. A current disadvantage is that the existing models largely look at ecosystems from the bottom up (using soils as a starting point) rather than from the top down, and a large amount of work will be necessary in order to revise existing models or to develop new ones that thoroughly incorporate a top-down perspective (one based on vegetation canopies, which can be seen remotely).

Ground-Based Approaches

A number of ground-based techniques can be used to determine the concentrations or fluxes of trace gases; these include tower-based eddy correlations systems (Baldocci et al., 1988), light detection and ranging (lidar) laser systems (Sachse et al., 1988), the multipass tunable diode laser system, and Fourier transform infrared (FTIR) spectroscopy (Gosz et al., 1988). We will not discuss these approaches in detail, in part because several of the techniques are also utilized in aircraft-based sampling.

The ground-based systems have a number of different advantages. Some can be used to estimate several trace gases simultaneously (FTIR, multiple-pass tunable diode laser), others (such as lidar) can be used to calculate concentration gradients of particular gases remotely, and still others (eddy correlation methods) can be used to calculate whole-system fluxes of particular gases. The last two are particularly useful in that they allow estimation of trace gas fluxes without the often-confounding effects of chambers or other enclosures. All of the ground-based remote techniques are relatively expensive to purchase and maintain, and in operation they are generally confined to a single point. Therefore, they are not as useful for regional or global estimates of flux as are similar aircraft-based sensors.

Aircraft-Based Remote Sensing

We believe that aircraft-based systems currently offer the greatest potential for the application of remote sensing to estimating fluxes of trace gases. Aircraft have the mobility and flexibility to sample important areas at the appropriate time, and a number of new and exciting sensors are now mounted on aircraft.

A number of aircraft-based systems are in operation. One approach is based on treating the planetary boundary layer (PBL), the layer of the troposphere in contact with the surface, as a chamber within which changes in trace gas concentrations can be measured (Harriss et al., 1988; Matson and Harriss, 1988). The National Aeronautics and Space Administration (NASA) Global Troposphere Experiment (GTE), a program concerned with the regulation of tropospheric trace gas concentrations, has carried out field campaigns using this approach over the warm tropical ocean (Atlantic Boundary Layer Experiment or ABLE 1), central Amazonia during the dry (ABLE 2A) and wet (ABLE 2B) season, and arctic Alaska (ABLE 3). The program uses an ultraviolet lidar system mounted on a Lockheed Electra (a four-engine propeller aircraft) to measure the structure of the atmosphere, particularly the diurnal dynamics of the planetary boundary layer. The lidar measures the return of laser light scattered by aerosols (Browell et al., 1988); in this way, aerosol profiles are measured five times a second for each 15-m increment from the aircraft to the surface, with real-time data displayed on the aircraft. The planetary boundary layer is relatively aerosol rich, so its dynamics can be detected readily. Vertical profiles of many trace gas concentrations are then determined with reference to atmospheric structure by on-board analytical instruments (Gregory et al., 1986; Harriss et al., 1988).

The results of the Amazonian dry season field campaign (ABLE 2A) have been published as a collection of papers in the 1988 *Journal of Geophysical Research*. A typical daytime flight during this study period involved a takeoff shortly after sunrise and a climb to near 4,000 m in altitude, followed by vertical profiles, during which on-board investigators could measure vertical distribution of trace gas concentrations within the free troposphere and the planetary boundary layer using in situ instruments and, in the case of ozone, a differential absorption lidar (DIAL) system. As the boundary layer developed through the morning (Figure 8.2), the measured volume of the boundary layer was used to account for changes in the concentrations of any chemical species attributable to mixing with the free troposphere. Any additional change then represented advection, reaction, or flux to or from the surface. In a very large, undisturbed system such as upland Amazonian forests far from rivers or human activity, horizontal advection probably averages out, while chemical reactions can be accounted for by models of tropospheric chemical reactions, at least for the less reactive species. Accordingly, atmosphere/surface exchange (of some

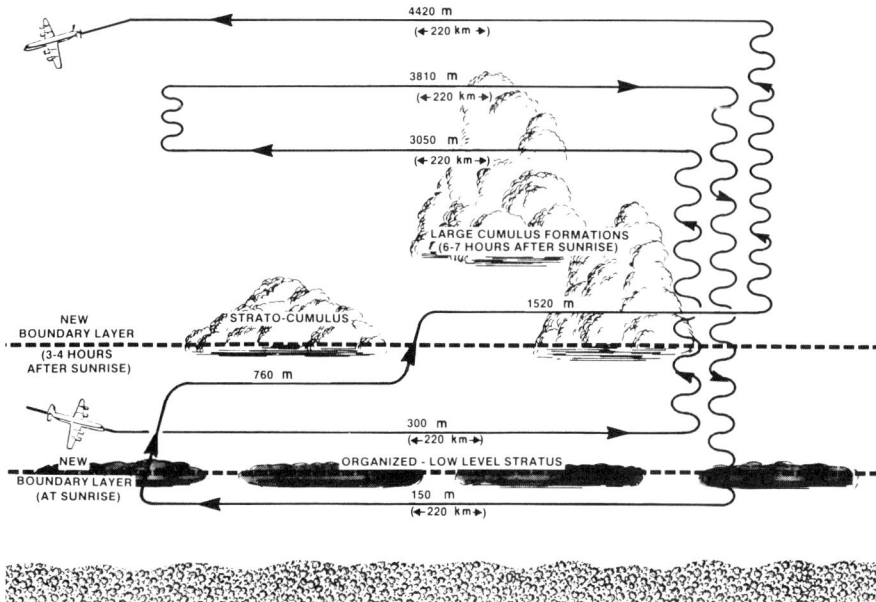

Figure 8.2. Representative flight plan for profile sampling in the planetary boundary layer. [From Matson and Harriss (1988).]

species, under some conditions) can be measured directly on spatial scales from tens to thousands of square kilometers.

In the Amazon, this system was used to measure carbon dioxide exchange between large areas of undisturbed forest and the atmosphere (Wofsy et al., 1988). The spatial and temporal distributions of a number of other trace gases, including carbon monoxide, nitric oxide, and dimethyl sulfide (Sachse et al., 1988; Andreae and Andreae, 1988), were also measured in this way. Additionally, lidar was used to identify haze layers resulting from biomass burning, and the influence of biomass burning on atmospheric chemistry at a distance from source areas was determined (the sources were subsequently traced by satellite-based remote sensing) (Andreae et al., 1988).

The strengths of this overall approach are, first, that it allows direct measurements of trace gas fluxes on spatial scales between those accessible to an investigator on the ground and those involved in Global Circulation Models (GCMs). As such, aircraft-based measurements can provide a check on both local flux measurements and calculations based on global circulation. Moreover, as lidar systems are developed to measure additional gases (CO_2, CO, CH_4, N_2O, NH_4), the approach will become increasingly useful. This approach has the weakness that it is directly applicable only to situations in which the planetary boundary layer is stable, and where horizontal advection within the boundary layer can be accounted for. Even in less stable or less homogeneous circumstances, the approach is useful in that coarse-scale movements of air masses (and their chemistry) can be traced, and areas of convection indentified and sampled. However, it cannot then be used to measure atmosphere-biosphere fluxes directly.

Where the planetary boundary layer cannot be treated as a chamber, aircraft-based eddy correlation systems (Desjardins et al., 1982; Sellers et al., 1988; Sellers, this volume) can be used to measure atmosphere-biosphere exchange. The eddy-correlation approach is discussed in Sellers (this volume), and it will not be described in detail here. For its application to trace gases, the development of rapid-response tunable diode laser systems has allowed measurements of many trace gases to be carried out with sufficient speed and precision for eddy-correlation estimates of flux.

Eddy correlation from aircraft is widely applicable to determining trace gas fluxes, at least for the chemically more stable or better-known species for which conservation of mass (or known rates of reaction) can be assumed. Moreover, the combination of aircraft-based lidar and eddy-correlation systems offers considerably more promise than either technique alone; lidar can be used to determine the coarse-scale structure of the atmosphere (and its chemistry) and to direct sampling, while eddy correlation can be used for finer-scale flux measurements.

Satellite-Based Flux Measurements

While satellite-based remote sensing is now extremely useful for deriving correlates of trace gas flux and drivers for models of trace gas flux, it cannot be used to measure fluxes directly. Perhaps the closest approach to direct measurements of flux has come from the total ozone mapping system (TOMS) measurements of integrated ozone concentrations. Although most of the total signal is contributed by stratospheric ozone, the quantity of stratospheric ozone varies little (spatially) over tropical regions. It may be possible to interpret variation in the TOMS signal as reflecting variation in tropospheric ozone. This ozone is produced chemically within the troposphere, but over the tropics its production is generally limited by NO_x supply (from biomass burning, soil emissions, and lightning) (Jacob and Wofsy, 1988); hence it may be possible to estimate NO_x flux from remote ozone measurements.

Ultimately, satellite-based lidar (or other) systems may allow a broader application of remote sensing to the measurement of trace gas flux—but for the immediate future (including the planned Earth Observing System, or Eos), satellite-based remote sensing will likely contribute indirectly by providing reasonable classifications, driving variables for models, and global circulation patterns rather than by direct measurements of concentrations of flux.

Conclusions

To understand the magnitude and regulation of trace gas fluxes, we need a combination of process-level understanding of the mechanisms involved, models of different spatial and temporal scales that can be driven with remotely sensed data, regional measurements of flux, and the global perspective provided by long-term observations and GCMs. Clearly, the alternative approaches to determining trace gas fluxes outlined in this chapter are complementary, not competitive; they also complement ground-based sampling of the mechanisms regulating flux, long-term observations of atmospheric chemistry, and the further development of GCMs. The research community now has incomplete but useful information about the processes that produce many (not all) of the important trace gases. We know a good deal about their consequences in the atmosphere. We also have good models of nutrient cycling that, in some cases, include predictions of trace gas flux, and we have long-term measurements of the tropospheric concentrations of the long lived gases in a few sites. However, most of the existing biogeochemical models were not developed for use with remote sensing. As such models are developed, the ability to measure the structure of terrestrial ecosystems and canopy chemistry with remote sensing (Sader, 1987; Wessman et al., 1988; Wessman, this volume) will greatly increase their applicability. Additionally, regional-level flux measurements are almost wholly lacking, yet they are essential if we hope to scale up information from flux measurements in chambers or from towers, and apply it on regional and global scales. Both the application of experimental sensors that can drive ecosystem models and the measurement of regional trace gas fluxes will require a much greater utilization of aircraft-based sampling than has hitherto been true of ecological studies.

References

Andreae, M.N., and Andreae, T.W. (1988). The cycle of biogenic sulfur compounds over the Amazon Basin: 1. Dry season. *J. Geophys. Res.* 93(D2):1487–1498.

Andreae, M.O., Browell, E.V., Garstang, M., Gregory, G.L. Harriss, R.C., Hill G.F., Jacob, D.J., Pereira M.C., Sachse, G.W., Setzer, A.W., Silva Dias, P.O., Talbot, R.W., Torres, A.L., and Wofsy, S.C. (1988). Biomass-burning emissions and associated haze layers over Amazonia. *J. Geophys. Res.* 93(D2):1509–1527.

Baldocchi, D.D., Hicks, B.B., and Meyers, T.P. (1988). Measuring biosphere-atmosphere exchanges of biologically related gases with micrometerological methods. *Ecology* 69(5):1331.

Bartlett, D.S., Bartlett, K.B., Hartmen, J.M., Harriss, R.C., Sebacher, D.I., Pelletier-Travis, R., Dow, D.D., and Brannon, D.P. Methane emission from natural wetlands: Associations with surface features and inventory using remote sensing. *Global Biogeochem. Cycles*, submitted.

Bormann, F.H., and Likens, G.E. (1979). *Pattern and Process in a Forested Ecosystem*. Springer-Verlag, NY.

Browell, E.V., Gregory, G.L., Harriss, R.C., and Kirchoff, V.W.J.H. (1988). Tropospheric ozone and aerosol distributions across the Amazon Basin. *J. Geophys. Res.* 93(D2):1431–1451.

Desjardins, R.L., Brach, E.J., Alvo, P., and Schuepp, P.H. (1982). Aircraft monitoring of surface carbon dioxide exchange. *Science* 216:733–735.

Garstang, M., Scala, J., Greco, S., Harriss, R., Beck, S., Browell, E., Sachse, G., Gregory, G., Hill, G., Simpson, J., Tao, W., and Torres, A. (1988). Trace gas exchange and convective transports over the Amazonian rain forest. *J. Geophys. Res.* 93(D2):1528–1550.

Gosz, J.R., Dahm, C.N., and Risser, P.G. (1988). Long-path FTIR measurement of atmospheric trace gas concentrations. *Ecology* 69(5):1326–1330.

Gregory, G.L., Harriss, R.C., Talbot, R.W., Rasmussen, R.A., Garstang, M., Andreae, M.O., Hinton, R.R., Browell, E.V., Beck, S.M., Khalil, M.A.K., Ferek, R.J., and Harriss, S.V. (1986). Air chemistry over the tropical forest of Guyana. *J. Geophys. Res.* 91:8603–8612.

Harriss, R.C., Wofsy, S.C., Garstang. M., Browell, E.V., Molion, L.C.B., McNeal, R.J., Hoell, J.M., Jr., Bendura, R.J., Beck, S.M., Navarro, R.L., Riley, T.J., and Senll, R.L. (1988). The Amazon boundary layer experiment (ABLE 2A): Dry season 1985. *J. Geophys. Res.* 93(D2):1351–1360.

Jacob, D.S., Wofsy, S.C. (1988). Photochemistry of biogenic emissions over the Amazon forest. *J. Geophys. Res.* 93(D2):1477–1486.

Keller, M., Goreau, T.J., Wofsy, S.C., Kaplan, W.A., McElroy, M.B. (1983). Production of nitrous oxide and consumption of methane by forest soils. *Geophys. Res. Lett.* 10:1156–1159.

Keller, M., Kaplan, W.A., and Wofsy, S.C. (1986). Emissions of N_2O, CH_4, and CO_2 from tropical soils. *J. Geophys. Res.* 91:11791–11802.

Likens, G.E., Bormann, F.H., Pierce, R.S., Eaton, J.S., and Johnson, N.M. (1977). *Biogeochemistry of a forested ecosystem*. Springer-Verlag, NY.

Luizao, R., Matson, P.A., Livingston, G., Luizao, R., and Vitousek, P.M. Nitrous oxide flux following tropical land clearing. *Global Biogeochem. Cycles*, in press.

Matson, P.A., and Harriss, R.C. (1988). Prospects for aircraft-based gas exchange measurements in ecosystem studies. *Ecology* 69:1318–1325.

Matson, P.A., and Vitousek, P.M. (1987). Cross-system comparisons of soil nitrogen transformations and nitrous oxide flux in tropical forest ecosystems. *Global Biogeochem. Cycles* 1:163–170.

Matson, P.A., Vitousek, P.M., Livingston, G.P., and Swanberg, N. Sources of variation in nitrous oxide flux from Amazonian forests: effects of soil fertility and disturbance. *J. Geophys. Res.*, in press.

Matson, P.A., Volkmann, C., Coppinger, K. and Reiners, W. Nitrous oxide flux from sagebrush steppe: Variation within and among community types. Submitted.

McGill, W.B., Hunt H.W., Woodmansee, R.G., and Reuss, J.O. (1981). A model of the dynamics of carbon and nitrogen in grassland soils. *Ecol. Bull.* (Stockholm) 33:49–116.

Mooney, H.A., Vitousek, P.M., and Matson, P.A. (1987). Exchange of materials between terrestrial ecosystems and the atmosphere. *Science* 238:926–932.

Parton, W.J., Mosier, A.R., and Schimel, D.S. (1988a). Rates and pathways of nitrous oxide production in a shortgrass steppe. *Biogeochemistry* 6:45–58.

Parton, W.J., Stewart, J.W.B., and Cole, C.V. (1988b). Dynamics of C, N, P, and S in grassland soils: A model. *Biogeochemistry* 5:109–131.

Reiners, W.A., Strong, L., Matson, P.A., Burke, I., and Ojima, D. Estimating biogeochemical fluxes across sagebrush-steppe landscapes with thematic mapper imagery. *Remote Sens. Envir.*, in press.

Sachse, G.W., Harriss, R.C., Fishman, J., Hill, G.F., and Cahoon, D.R. (1988). Carbon monoxide over the Amazon Basin during the 1985 dry season. *J. Geophys. Res.* 93(D2):1422–1430.

Sader, S.A. (1987). Forest biomass, canopy structure, and species composition relationships with multipolarization L-band synthetic aperture radar data. *Photogramm. Engi. Remote Sens.* 53(2):193–202.

Schimel, D.S., Parton, W.J., and Woodmansee, R. Biogeochemistry of great plains grasslands. *Climatic change*, in press.

Sellers, P.J., Hall, F.B., Asrar G., Strebel, D.E., and Murphy, R.E. (1988). The first ISLSCP field experiment (FIFE). *Bull. Amer. Meteorol. Soc.* 69(1):22–27.

Wessman, C.A., Aber, J.D., Peterson D.L., and Melillo, J.M. (1988). Remote sensing of canopy chemistry and nitrogen cycling in temperate forest ecosystems. *Nature* 335:154–156.

Wofsy, S.C., Harriss, R.C., and Kaplan, W.A. (1988). Carbon dioxide in the atmosphere over the Amazon Basin, *J. Geophys. Res.* 93(D2):1377–1388.

9. Satellite Remote Sensing and Field Experiments

Piers J. Sellers, F.G. Hall, D.E. Strebel, G. Asrar, and R.E. Murphy

The past few years have seen an increasing pressure brought to bear on biologists to apply their research efforts to larger spatial scales. The motivation for this pressure has scientific as well as sociopolitical roots. First, from the science viewpoint, it has become apparent that satellite-based and atmospheric observing techniques can allow researchers a different view of biospheric functioning by spatially integrating the effects of biospheric states and fluxes, respectively, at moderate cost; see, for example, the studies of Tucker et al. (1986), Goward et al. (1985), Justice et al. (1985), Fung et al. (1987), Goward and Dye (1987), and Houghton (1987). The new opportunities opened up by the use of these techniques compel the whole biological community to use a new means of studying life on the earth. Second, from the sociopolitical viewpoint, there is a growing perception in public and government circles that the global environment is changing, partly as a result of man's industrial and agricultural activities. The mechanisms of global change are only just beginning to be investigated and understood, see Rotty (1983), Trabalka (1985), and the review of Schlesinger and Mitchell (1987), but biospheric processes may play a strong role in

---

This chapter is based in part on Sellers, P.J., Hall, F.G., Asrar, G., Strebel, D.E., and Murphy, R.E. (1988). The first ISLSCP field experiment (FIFE). *Bull. Amer. Meteorol. Soc.* 69(1):22–27. Adapted with permission of the American Meteorological Society.

any developing scenario. Of particular relevance is the so-called "greenhouse effect" (Schlesinger and Mitchell, 1987; Hansen et al., 1981), which hypothesizes that the build up of anthropogenically generated CO_2 in the atmosphere may bring about a warming in the troposphere and a cooling of the stratosphere. (To date, however, none of the greenhouse effect simulations performed with numerical models of the earth's atmosphere have addressed the direct effects of increased CO_2 or temperature on the physiology of the terrestial biota). In any case, more and better information is needed about global change so that its effects can be anticipated or avoided. Together, these two pressures have motivated a large number of biologists to realign their research efforts toward the study of large-scale processes and atmosphere-biosphere interactions.

Modeling Biosphere-Atmosphere Interactions

The study of land surface biota on global and regional scales automatically implies a study of the interactions between the biosphere and the atmospheric environment. Figure 9.1 shows the temporal and spatial domains of various biological and meteorological phenomena, from which it can be seen that the preferred areas of research for biologists and meteorologists do not match in terms of time and space scales. Biologists have had their greatest successes in the study of small-space-scale, short-time-scale phenomena—for example, photosynthesis and physiology—with proportionately fewer solid gains at larger scales. Meteorologists, on the other hand, have accomplished most in the regional to global space domain over time periods ranging from a few hours to days; for example, the use of geophysical fluid dynamical theory to describe the planetary atmospheric circulation has resulted in the routine prediction of global weather to provide forecasts that are usable out to several days. In the description of processes acting on smaller space scales, such as turbulence, or larger time scales, such as atmospheric chemistry, there have been successes but comprehensive theories have yet to be widely accepted.

The mismatch of scales between the biological and meteorolgical communities had brought about a clean division of biosphere-atmosphere modeling studies into two distinct categories. The first is the modeling of predominantly biophysical interactions between the terrestrial surface and the atmosphere (e.g., Dickinson, 1984; Sellers et al., 1986), where the principal objective is to simulate realistically the biological, physical, and dynamical processes that govern the motion of the atmosphere and thereby calculate the time evolution of the three-dimensional fields of temperature, humidity, wind speed, radiative flux divergence, and precipitation; in short, to simulate the global weather or climate starting from some specified initial condition. The primary tool for this kind of study is the atmospheric General Circulation Model (GCM), which incorporates

9. Satellite Remote Sensing and Field Experiments 171

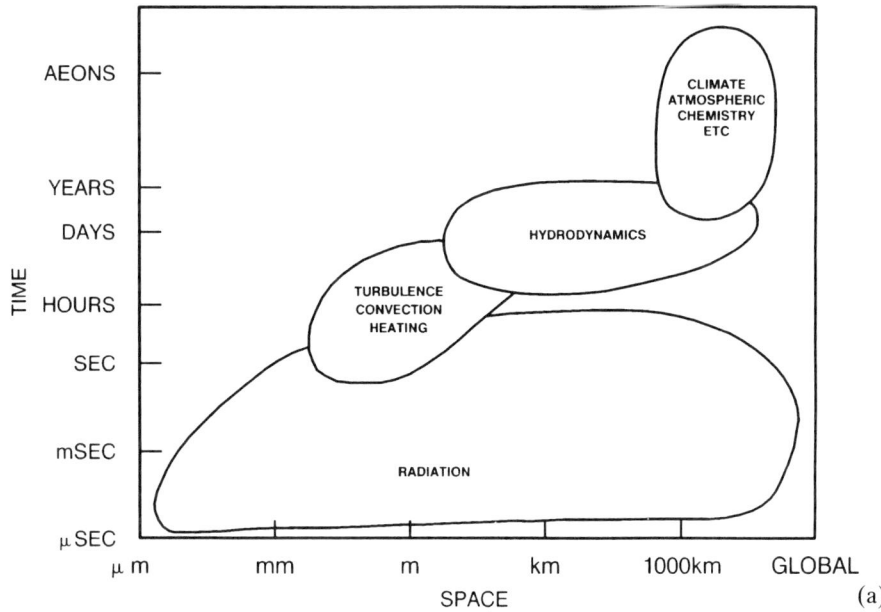

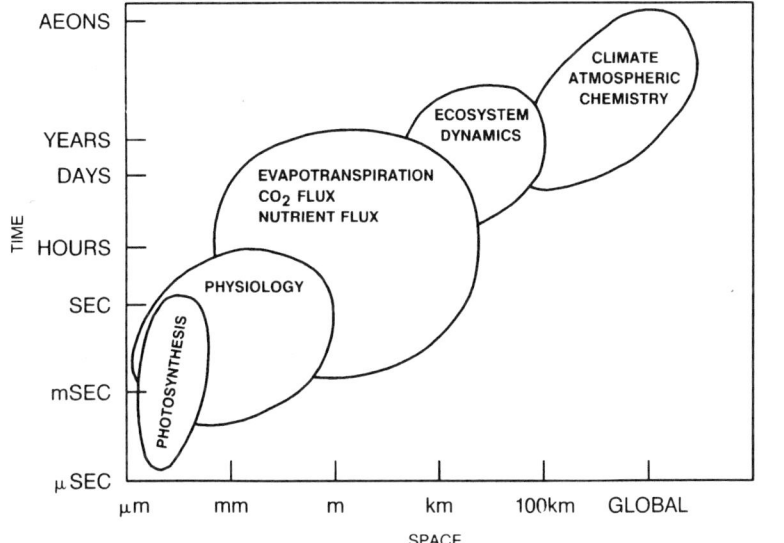

Figure 9.1. Time and space scales commonly considered by (a) meteorologists and (b) biologists. Usually, each area of interest—for example, hydrodynamics in (a) or evapotranspiration in (b)—has its own set of consistent models with less rigorous links to processes up- and down-scale.

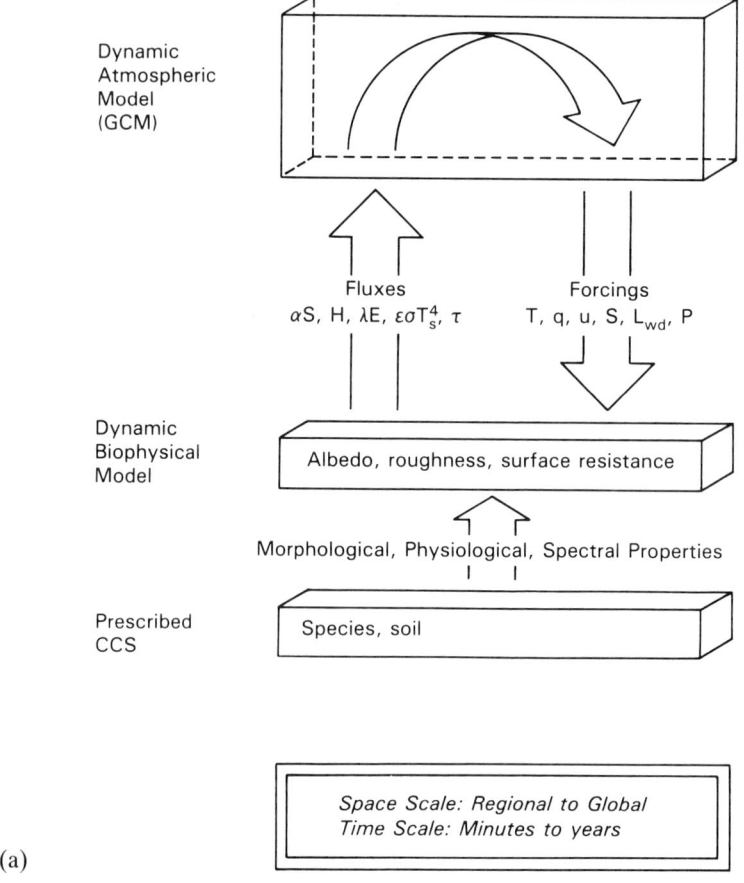

(a)

Figure 9.2. Atmosphere-biosphere models. (a) *Biophysical models*. The distribution of species, or community composition and structure (CCS), is prescribed and supplies the biophysical parameter set that controls exchanges of heat, mass, radiation, and momentum between the surface and atmosphere. In operation, the dynamic atmospheric model applies forcings of temperature T, humidity q, wind speed u, insolation S, downward long-wave radiation L_{wd}, and precipitation P. The biophysical surface model uses these forcings to calculate the returning fluxes of reflected short-wave radiation αS, sensible heat flux H, latent heat flux λE, upward long-wave flux $\epsilon \sigma T_s^4$, and shear stress or aerodynamic drag τ. These fluxes affect the subsequent dynamic development of the atmosphere and thus the future forcings. (b) *Ecosystem dynamics models*. The atmospheric condition is usually prescribed from climate records or GCM output. The meteorological forcings of T, q, u, S, L_{wd}, and P are then used to determine survival and growth rates of a species mixture via a series of physiological models. The results of these calculations affect subsequent successional changes and alterations in the CCS.

9. Satellite Remote Sensing and Field Experiments 173

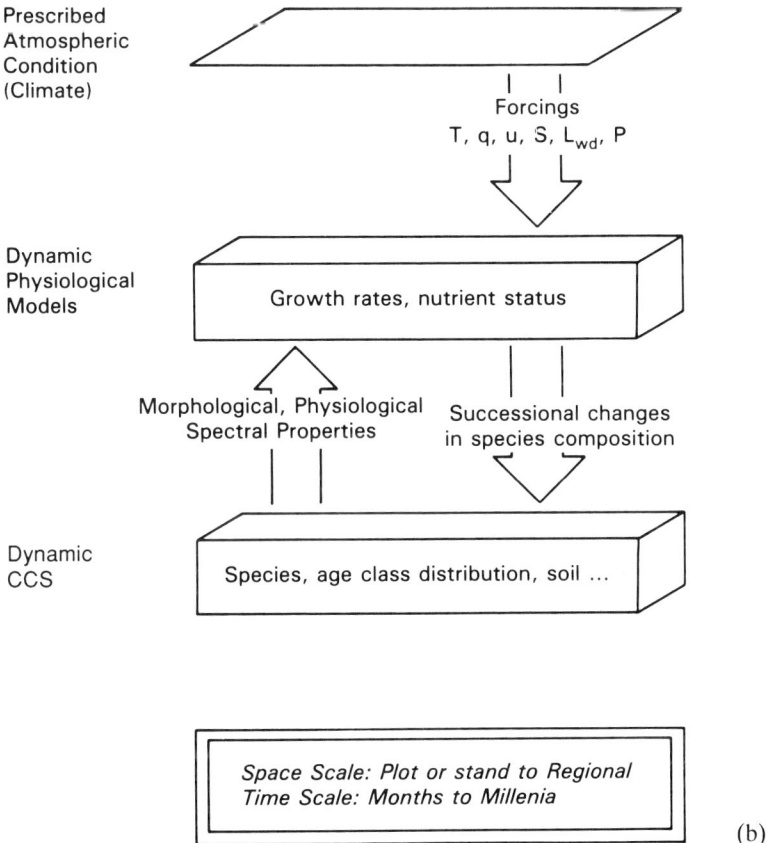

Figure 9.2. *Continued.*

the primitive equations describing the motion of the atmosphere in terms of fluid dynamics, and mathematical descriptions of the important one-dimensional physical processes—radiative heating, convection, turbulent transport, and latent and sensible heating—which transfer heat, mass, and momentum throughout the atmospheric column. An essential component of such a model is a correct description of the radiation, momentum, and sensible and latent heat fluxes at the surface-atmosphere interface as these may significantly contribute to the internal heating and the drag force exerted on the lower atmosphere.

Over the oceans, the modeling of these fluxes is fairly straightforward, provided the sea-surface temperature (SST) field is known from observation or calculated from an ocean circulation model. For the land, however, the problem is more complex; recent work has shown that the partitioning of absorbed energy at the terrestrial surface may be a strong function of the type, density, and health of the vegetation there; see Sato et al. (1989). In particular, the vegetation's stomatal resistance seems to play a large role in limiting land-surface evapotranspiration rates and hence has a direct feedback effect on the calculated precipitation rates over continental interiors during the summer, as a significant fraction of evaporated moisture may be recirculated back to the land surface as rainfall. More details on the results and potential of such modeling studies are given in a later chapter but it is worth pointing out at this stage that all such modeling approaches assume quasi-stationarity in the vegetation's community composition and structure (CCS); see Figure 9.2a. Thus, for any given simulation run, the spatial distribution of species, or some crude analog thereof, is assumed to be invariant with time. This is hardly a constraint on the modeling studies which can be addressed with such models, as computer power currently limits simulation runs to periods of a few years or decades at most, too short a time for significant changes in CCS to develop or have any effect in any case.

The second category of biosphere-atmosphere model addresses the issue of successive changes in CCS over periods of decades to millennia, illustrated in Figure 9.2b. These models, referred to here as ecosystem dynamics models, simulate the time evolution of CCS within a region starting from some initial conditions and forced by a time series of atmospheric conditions; temperature, humidity, precipitation, and so on (Shugart et al., 1973). Generally speaking, these models regard the atmosphere purely as a source of applied forcing and do not consider any feedback effects from the surface back to the atomsphere. As can be seen from Figure 9.2, therefore, biosphere-atmosphere modeling efforts are divided roughly into those that deal with a dynamic atmosphere and a prescribed field of CCS (Figure 9.2a); and those that deal with a prescribed atmospheric condition, or climate, and a dynamic CCS (Figure 9.2b). The distinction is made all the sharper by considering the simulation time step typically employed in each type of model; the GCMs usually operate on a time step of the order of several minutes whereas ecosystem succession models employ time steps of a few weeks to months. As yet, few direct or indirect connections have been made between the two classes of models, although some efforts have been made to use the simulated climate generated by GCMs initialized with higher, usually doubled, atmospheric CO_2 concentrations to force biogeochemical-cycle models, (Parton et al., 1987), and to explore the possible changes in biome distribution due to atmospheric warming effects, (Solomon, 1986).

The Role of Remote Sensing

Satellite remote sensing can play a significant role in the investigation of biosphere-atmosphere interactions, for both biophysical models and ecosystem succession models. Both categories of model require observations for initialization and for validation, and in the case of biophysical models, the requirements demanded of an observational system are fairly extreme. To be truly useful, the observational system must satisfy the following criteria:

1. It should provide information on the thermodynamic, biological, and hydrological state of the land surface.
2. It should be capable of providing global coverage at high temporal and spatial resolutions; that is, several times a day and around 1 km, respectively.
3. It should utilize a consistent observational/interpretational technique everywhere.
4. It should be relatively economic.

Only satellite remote sensing coupled with interpretational algorithms can satisfy all of the above criteria. As yet, however, the processing of raw sensor counts, the initial output from the satellite instruments, into useful parameters for land surface studies is in a fairly primitive stage (Sellers et al., 1988b). This is so because each of the processing steps—calibration, atmospheric-geometric correction, radiance-to-parameter calculation, and parameter-to-biophysical-quantity calculation—has its own problems and uncertainties, all of which add up to degrade the value of the final product. These problems are briefly discussed as follows:

1. *Calibration*: To convert the digital counts to radiance values, the calibration of the satellite sensor must be known to a fairly high precision. In fact, many sensors are launched without calibration and all sensors are subject to some drift in calibration once in orbit.
2. *Atmospheric-geometric correction*: The effects of atmospheric attenuation, reflection, and emission must be accounted for; this is often done with numerical multilayer atmospheric radiative transfer models (e.g., Wiscombe et al., 1984). Additionally, the radiation field at the surface and within the atmosphere is often highly anisotropic, thus necessitating a full treatment of the effects of sun-target-sensor geometry. Detection and elimination of cloud-contaminated fields is often extremely difficult.
3. *Radiance-to-parameter calculation*: Once these first two problems have been dealt with, the scientist is left with an estimate of a spectral directional radiance, emittance or reflectance from the surface. This must then be interpreted into a physical parameter, such as surface temperature, green leaf area index (LAI), or near-surface soil moisture.
4. *Parameter-to-biophysical-quantity calculation*: The derivation of the physical parameter itself may be the final objective. Often, however, the scientist desires to use a combination or a time series of these parameters to derive another biophysical quantity. For example, a time series of surface temperature measurements may be used to calculate the diurnal surface energy balance and thence the corresponding time

series of sensible and latent heat fluxes from the surface; see Becker et al. (1988) in the review of Sellers et al. (1988b). As another example, near-infrared surface reflectances, which are closely correlated with green vegetation density, are time integrated to provide an estimate of total net primary productivity over a specified period (Goward et al., 1985).

A schematic of this data-processing chain is shown in Figure 9.3, together with the problems and uncertainties associated with each step. As yet, we have few cases where the raw data have been processed all the way through to a final product and this compared with a direct observation at the appropriate scale.

The Need for Experiments

There have been many theoretical studies that demonstrate the utility of satellite data for the study of land-surface processes. As discussed above, however, the actual utility or potential of these data is largely unknown as the products of a full and rigorous processing of the satellite data have seldom been compared with simultaneous and appropriate validation measurements. The key word here is "appropriate," as this implies a spatial scale consistent with the satellite sensor resolution, which is normally on the order of tens of meters to a few kilometers, and plainly much larger than the sites or plots associated with most in situ biological research. In conducting experiments involving remotely sensed data, scientists therefore have to address the following issues:

1. *Scale.* The in-situ validation measurements have to be carried out at a scale comparable to that of the products derived from the remotely sensed data; that is, they have to be conducted at roughly the same scale as the satellite sensor resolution. This alone has serious implications as the means must be found to transfer biological understanding and expertise from the customary meter scale to much larger scales. This process of upscale integration requires innovations in both modeling and measurement techniques.
2. *Simultaneity.* Many of the biophysical parameters of interest exhibit substantial rates of change over a diurnal cycle; for example, surface temperature, or evapotranspiration rate. Clearly, validation measurements must be precisely timed to coincide with the time of acquisition of the satellite data.
3. *Ancillary measurements.* Figure 9.3 and the accompanying text list four steps in processing the satellite data to a biophysical quantity: calibration, atmospheric-geometric correction, radiance-to-parameter calculation, and parameter-to-biophysical-quantity calculation. Ancillary measurements must be carried out to address the first two steps if the

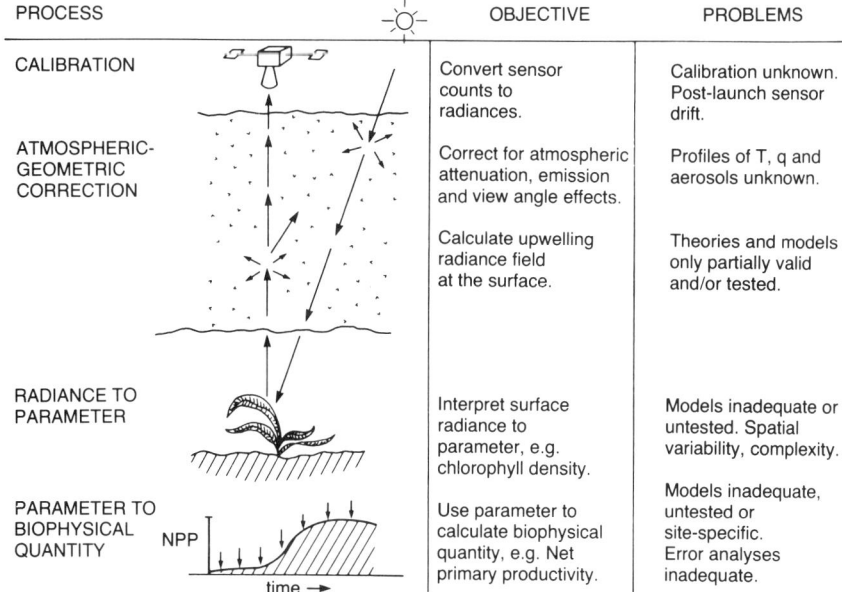

Figure 9.3. Basic processing steps required to convert satellite data to biophysical quantities.

potential of the satellite data is to be fully realized. For calibration purposes, observations should be taken over a test site of known reflectance properties (Slater et al., 1986), and for the atmospheric correction step, measurements of the temperature, humidity, and aerosol profiles should be made over the test site at the time of satellite data acquisition. These measurements are ancillary to the main validation measurements made at or near the surface but are nonetheless vital for the interpretation of the satellite data.

These requirements for large-scale, simultaneous surface validation measurements combined within an intensive ancillary measurement effort represent substantial scientific and financial investments that act as powerful disincentives to actually carrying out such field experiments. Nevertheless, they must be done if we are ever to make use of the satellite data in a quantitative and scientifically defensible way. It can also be argued that when the cost of executing such experiments is scaled against the cost of the satellite observation systems, the return on the investment can be seen to represent good value.

Up to now, we have discussed large-scale experiments in the abstract: what they are, why they are needed, and roughly what is involved in constraining their design. In the next section, we review one such field experiment, the First ISLSCP[1] Field Experiment (FIFE). FIFE was primarily concerned with the use of satellite data to calculate land-surface energy and mass balances and therefore falls within the first category of biophysical experiments we have discussed; see Sellers et al. (1988a). For an example of a long-time scale ecosystem dynamics study, the reader is referred to the forest dynamics modeling effort of Hall et al. (1987), which investigated change within a forested landscape over a period of several years.

The First ISLSCP Field Experiment (FIFE)

Theoretical Background

The experiment was designed to determine the extent to which satellite data and modeling could yield information on the energy and mass balance of a vegetated land surface. An explicit recognition of the role of vegetation in the land-surface energy balance was central to the design and execution of the experiment.

The simplest realistic treatments of land-surface–atmosphere interactions, as used in GCMs, deal with the radiation balance first and the partitioning of the absorbed energy second.

The surface radiation balance may be written as:

$$R_n = S(1 - \alpha) + L_{wd} - \epsilon \sigma T_s^4 \qquad (1)$$

where R_n = net radiation, Wm^{-2}; S = total surface insolation, Wm^{-2}; α = broad-band surface albedo; L_{Wd} = downwelling (from sky to surface) long-wave radiation, Wm^{-2}; T_s = surface temperature, K; ϵ = surface emissivity; and σ = Stefan-Boltzman constant, $Wm^{-2} K^{-4}$.

The absorbed radiation, R_n, is then partitioned into the following flux terms:

$$R_n = G + P_s + H + \lambda E \qquad (2)$$

where G = ground heat flux, Wm^{-2}; P_s = energy for photosynthesis, Wm^{-2}; H = sensible heat flux, Wm^{-2}; λE = latent heat flux, Wm^{-2}; E = evaporation rate, $kg\ m^{-2}\ s^{-1}$; and λ = latent heat of vaporization, $J\ kg^{-1}$.

Generally speaking, the energy absorbed for photosynthesis, P_s, is of the order of 1% or less of R_n and the ground heat flux term, G, is usually less than 10% of R_n when averaged over a diurnal cycle. This means that the bulk of the absorbed energy is partitioned into the sensible, H, and latent, λE, heat fluxes, which are returned to the atmosphere by turbulent

---

[1] ISLSCP (International Satellite Land Surface Climatology Project) is sponsored by the World Meteorological Organization (WMO), the Internation Commission of Scientific Unions (ICSU), and the United Nations' Environmental Program.

diffusion. The ratio of H to λE is of particular importance to atmospheric processes as it has substantial effects on the timing and location of the internal heating of the atmosphere.

The partitioning of absorbed energy into H, λE, and G may be described by three equations:

$$H = \frac{[T_s - T_a]}{r_a} \rho c_p \tag{3}$$

where T_s = surface temperature, K; T_a = air temperature at reference height z_r, K; ρ, c_p = density, specific heat of air; kg m^{-3}, J Kg^{-1} K^{-1}; and r_a = aerodynamic resistance between the surface and reference height z_r, s m^{-1};

$$\lambda E = \frac{[e^*[T_s] - e_a] \rho c_p}{r_a + r_{surf}} \frac{}{\gamma} \tag{4}$$

where $e^*(T_s)$ = saturated vapor pressure at surface temperature, T_s, kP_a; γ = psychrometric constant, kP_a K^{-1}; e_a = air vapor pressure at z_r, mb; and r_{surf} = surface resistance to vapor transfer, s m^{-1}; and

$$G = f\left[\frac{dT_s}{dt}, W\right] \tag{5}$$

where W = soil wetness.

A schematic of this transfer scheme is shown in Figure 9.4. Equations (3) and (4) are comparable to the Ohm's law description of electrical current flow where

$$I = \frac{V}{R} \tag{6}$$

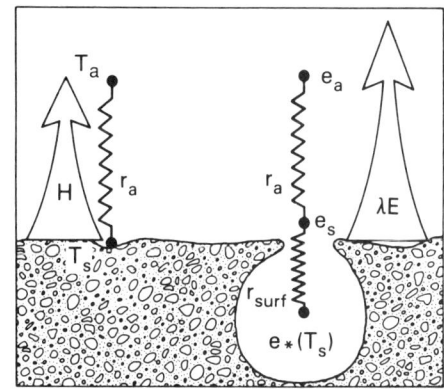

Figure 9.4. Schematic of simple biophysical heat flux resistance model. Note that the source for sensible heat flux, H, is the surface skin at temperature, T_s. For the latent heat flux, λE, the source for vapor is conceptualized as the saturated plant interior with a vapor pressure of $e^*(T_s)$.

(Where I = current, amperes; V = potential difference, volts; and R = resistance, ohms) in that the fluxes H and λE are equivalent to the current I; and the top and bottom lines on the right-hand sides of Equations (3) and (4) are equivalent to the voltage drop V and the resistance R respectively. The relative sizes of the resistance terms r_a and r_{surf} clearly control the ratio H to λE. r_a, the aerodynamic resistance, is usually described using an eddy diffusion model, whereby:

$$r_a = \int_{surface}^{Z_r} \frac{1}{K_{h,v}} dz \qquad (7)$$

where $K_{h,v}$ = eddy diffusion coefficient for heat and water vapor transfer, $m^2\ S^{-1}$.

Monteith (1973) explains in detail how Equation (7) may be solved with some simple assumptions about how turbulent transport operates close to the surface. When the sensible heat flux is relatively small (i.e., near-neutral conditions), this procedure yields

$$r_a = \frac{1}{u_r}\left[\frac{1}{k}\log\left(\frac{Z_r}{Z_o}\right)\right]^2 \qquad (8)$$

where u_r = wind speed at Z_r, $m\ s^{-1}$; Z_o = roughness length of the surface, meters; and k = von Karman's constant, $\simeq 0.41$.

From Equation (8), we can see that transfer from the surface to the atmosphere will be most efficient when the surface is rough, that is, when Z_o is large, and when the wind speed is high. The surface resistance term is a biophysical concept (Monteith, 1973); it can be thought of as the integrated impedance to the release of moisture from within the soil or plant tissues to the adjacent external air layer. In vegetated areas, the major component of r_{surf} is the stomatal resistance of plant leaves to the diffusion of saturated water vapor from the mesophyll to the leaf boundary layer. In nature, many plants seem to control their stomatal functioning so as to maximize their assimilation rates for a minimal loss of water vapor, and to achieve this, the stomatal functioning is sensitive to variations in the photosynthetically active radiation (PAR) flux, 0.4 to 0.7 μm (micrometers), the leaf temperature, the humidity of the surrounding air, and the moisture content of the leaves, itself a function of evaporation rate and soil moisture content (Farquhar and Sharkey, 1982; Jarvis, 1976).

The set of Equations (1) through (8) forms the basis of many biophysical models of the surface energy balance, including the descriptions of Dickinson (1984) and Sellers et al. (1986). Table 9.1 outlines procedures for estimating terms in the radiation and energy balances from remotely sensed data. These are discussed in more detail below.

First, satellite data may be used to estimate different components of the surface radiation balance, see Equation (1). In principle, the insolation S, the surface albedo α, the downward long-wave flux L_{wd}, and the surface temperature T_s, and hence long-wave loss, may be calculated from existing

Table 9.1. Procedure for Calculating Surface Energy Balance from Satellite Data. Derivations of Numbered Equations Are Described in the Text

Derived Quantity	Satellite Observation	Satellite
Insolation S	Radiation Balance TOA reflectances*	GOES, NOAA, Meteosat
Albedo, α	TOA reflectances	GOES, NOAA, Meteosat
Upward long-wave flux, $\epsilon \sigma T_s^4$ surface temperature, T_s	TOA emission, sounder profiling	GOES, NOAA, Meteosat
Downward long-wave flux, L_{wd}	Sounder profiling	NOAA

$$[\text{Net radiation: } R_n = S(1-\alpha) + L_{wd} - \epsilon \sigma T_s^4] \quad (1)$$

Surface State

Vegetation state, SVI	Vegetation indexes from TOA, I_N, I_V radiances	NOAA, SPOT, Landsat
Surface temperature, T_s	TOA emission sounder profiling	GOES, NOAA, Meteosat
Soil moisture, W	Microwave brightness temperature, T_B	Nimbus-7

Calculating Surface Heat Fluxes

$$[\text{Energy balance: } R_n = H + \lambda E + G + P_s] \quad (2)$$

I. $\lambda E = R_n - H - G - P_s$
    Equations 1, 2, 3, and 5
II. $\lambda E = f(T_s, SVI, W)$
    Equations 4, 9, and 10

*TOA = top of the atmosphere.

satellite radiometers and sounders; see Diak and Gautier (1983), Dedieu et al. (1987), Morcrette and Deschamps (1986), and Price (1986), respectively, or a summary of all methods in Sellers et al. (1988b). By inserting these estimates in Equation (1), we obtain an estimate of the net radiation R_n by summation.

Once R_n has been estimated it may be partitioned into its component heat fluxes using indirect modeling methods that utilize remotely sensed data; see the review of Becker et al. (1988). Two methodologies are commonly proposed. In the first, a satellite-based measurement of T_s is combined with estimates of T_a and r_a obtained from meteorological data or models to solve Equation (3) to yield H. This and an estimate of G based on a time series of T_s observations and some simple thermal diffusion modeling [see Equation (5)], are subtracted from R_n in Equation (2) to yield an estimate of λE. However, this method has been found to be fairly sensitive to uncertainties in the terms on the right-hand side of Equation 3 (Abdellaoui et al., 1986). In the second method, Equation (4) is solved using estimates of e_a and r_a obtained from meteorology, T_s from thermal IR satellite data, and an estimate of r_{surf} from either vegetation index observations (Tucker et al., 1981) or soil moisture measurements from microwave observations (Njoku and Patel, 1986).

The vegetation index observations offer particular promise. Theoretical (Sellers, 1985, 1987; Hope, 1987) and empirical (Asrar et al., 1984; Tucker et al., 1981) investigations have shown that the ratio of the above-canopy near-IR and visible reflectances is almost linearly related to the fraction of photosynthetically active radiation absorbed by the plant canopy (APAR). Theoretical work indicates that APAR is in turn related to the area-averaged canopy photosynthetic capacity, P_c^*, which is correlated with the minimum canopy resistance, r_c^*; see Figure 9.5.

Thus,

$$r_{surf} \simeq \frac{r_c^*}{[f(T)f(\delta e)f(\psi_l)]} \tag{9a}$$

and

$$r_c^* \, \alpha \, \frac{I_N}{I_V} \tag{9b}$$

where r_c^* = minimum canopy resistance as determined by PAR flux, s m^{-1}; $f(T), f(\delta e), f(\psi_l)$ = stress terms that act to increase canopy resistance due to the effects of temperature, vapor pressure deficit, or leaf water potential, respectively—$0 \leq f(x) \leq 1$; and I_N, I_V = upwelling radiances (or sensor counts) observed in near-IR (0.7 to 1.1 μm) and visible (0.4 to 0.7 μm) wave bands, respectively; W m^{-2} sr^{-1}.

In sparsely vegetated regions, an estimate of the near-surface soil moisture content might also be interpreted into a value of r_{surf}. Thus,

$$r_{surf} = f(W), \quad W = f(T_B) \tag{10}$$

where W = soil wetness and T_B = brightness temperature as observed in the microwave region, (0.1 to 50 cm), K.

The theoretical bases for most of the above methods have been widely known for many years, but very few quantitative tests have been carried out under field conditions. The aim of the FIFE project was to collect the satellite observations, the ancillary data, and estimates of the surface fluxes

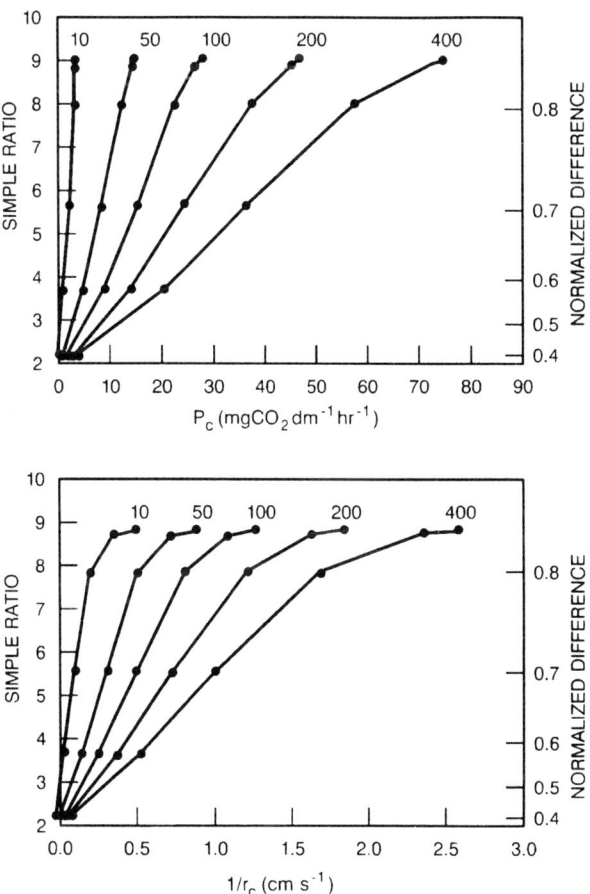

Figure 9.5. Simulated relationship between spectral vegetation indices, SVI, stress-free canopy photosynthetic rate P_c, and canopy resistance r_c. Simulation was conducted for a maize canopy. The SVIs are defined as follows:

$$\text{Simple ratio} = \frac{I_N}{I_N}$$

$$\text{Nomalized difference} = \frac{I_N - I_V}{I_N + I_V}$$

See Equation (9) in text.

and surface biophysical states to allow a thorough evaluation of all of these methods with the eventual objective of calculating the surface energy balance from satellite and meteorological data.

Experiment Design

Three important issues framed the design of FIFE:

- The size of the site
- The location of the site
- The duration of the observational effort

The Size of the Site

Two conflicting criteria constrained the size of the site. It was desirable to have as small a site as possible to allow a concentration of the surface measurement network and thereby accurate validation of the models. On the other hand, the site had to be large enough to be observable from orbit so that a number of pixels could be placed within its boundaries with confidence. Additionally, it was decided at an early stage to make use of airborne eddy-correlation equipment, which implies a minimum length scale of 15 km; this is roughly ten times the size of the length scale of the larger planetary boundary eddies that transport the fluxes away from the surface and thus represents a minimal representative sample of these turbulent structures. The site was thus designed around an intensive surface observational network located within a square 15 km on a side.

The Location of the Site

As FIFE represented the first effort of its kind, it was decided that the experiment location should favor a conservative experimental approach, particularly with respect to sampling density. The following criteria were put forward to define the site selection:

- *Surface homogeneity*. As far as possible, the site was to have uniform vegetation and moderate terrain throughout its area.
- *Grassland*. Measurement of fluxes and radiances over forests is technically difficult to say the least: towers and special platforms have to be erected on site and surface anisotropic effects can be fairly extreme. For purely practical purposes, grassland areas are easier to work on.
- *Strong seasonal cycle*. A mid-continent site with a strong seasonal cycle was preferable to allow a wide range of physiological and climatic conditions to be observed.
- *Logistics*. The site had to be in the United States and had to be close to an academic institution capable of supplying scientific and technical site

support. Also, it was highly desirable to have airfields close by for basing the research aircraft.
- *Research archive.* It was desirable that the site had been the subject of previous research efforts. The resulting archive would be valuable for the detailed experiment design.

A grassland area in Nebraska and an area including the Konza Prairie reserve and adjacent pastures in Kansas were listed for the study. The Konza Prairie site was eventually chosen as the logistic support available in the area was optimum.

Color Plate 2 shows how close the site and airfields are to the town of Manhattan, Kansas, where accommodations and technical/administrative support provided by Kansas State University were available.

The Duration of the Observational Effort

Clearly, an annual cycle is the minimum duration required for some aspects of the observational effort, to allow an understanding of changes induced by different climatic and physiological conditions. However, it was impractical to monitor the swiftly changing biophysical variables, such as surface temperature, surface heat fluxes, and upwelling radiances, over such a long period. The site observations were therefore split into two categories: a semicontinuous *monitoring* effort and a series of *intensive field campaigns* (IFCs). These are discussed in the following.

The monitoring effort was charged with the collection of the satellite and meteorological data necessary to drive the biophysical models described in the previous section, with only a few validation measurements included where appropriate. These measurements were also to provide an outline of the phenological and climatic conditions at the site throughout the year. The slowly changing variables of soil moisture and some biophysical properties, such as LAI, and biomass, were monitored throughout the complete year of 1987 and continue at the time of writing. Additionally, some 16 Portable Automatic Mesonet (PAM) stations or the equivalent were distributed about the site to measure meteorological variables, including incoming and reflected radiation, air temperature, vapor pressure, wind speed, soil temperature, and precipitation every 15 minutes. Throughout the year, satellite data were collected over the site whenever viewing conditions permitted. At the time of the satellite overpasses, sun photometer observations were taken to permit atmospheric correction of the data.

The IFCs were directed at taking the detailed, labor-intensive, and difficult measurements necessary for validating the satellite data products over a range of spatial scales. Specifically, time series of flux measurements and biophysical state variables were to be obtained, with particular emphasis on the times of the satellite overpasses. The research aircraft were committed to the site during the IFCs to carry out two essential tasks: to make radiometric measurements, some of which would be compared with satellite data, and to make estimates of heat, mass and momentum fluxes over the site. The measurements included:

- *Radiance observations.* Airborne and surface platforms were used to measure the upwelling radiation reflected and emitted by the surface, both to compare with contemporaneous satellite observations and to explore the relationship between radiances and surface conditions.
- *Flux observations.* Three aircraft and 16 surface platforms were used to measure the diurnal variation of the sensible and latent fluxes at the surface and within the boundary layer. Carbon dioxide flux was also measured using a variety of techniques.
- *Biophysical observations.* A whole range of detailed biophysical observations were made to validate parameter-to-radiance models and to understand the biophysical controls on the latent heat and CO_2 fluxes. Measurements of stomatal and canopy resistance, leaf chlorophyll content, leaf water potential, trace gas flux, and many replicates of biometric measurements were made.
- *Ancillary measurements.* Laser equipment, radiosondes, sun photometers, and spectrometers were all used to provide the data necessary for processing and interpreting the satellite and aircraft-based radiance data.

Obviously, all of this involves a considerable amount of resources and effort that could not be held in place for extended periods. Four IFCs were therefore proposed, each of 12 to 16 days' duration, during which these intensive observations were to be made. The IFCs were timed to coincide with the cardinal phases of the vegetation development, see Figure 9.6:

IFC-1; green-up   May 26 to June 6
IFC-2; peak greenness   June 25 to July 11
IFC-3; dry-down   August 6 to August 21
IFC-4; senescence   October 5 to October 16

During each IFC, four or five aircraft and some 150 people were working at or near the site.

The fine detail of the experimental design followed from the need to cover a wide range of time and space scales. The site was stratified according to treatment and slope; the stratification was moderately complex as the grassland is maintained in a shrub-free state by managed burning of the vegetation in April–May at intervals of one to a few years. The final stratification plan allowed the science team to distribute the 16 automatic

9. Satellite Remote Sensing and Field Experiments 187

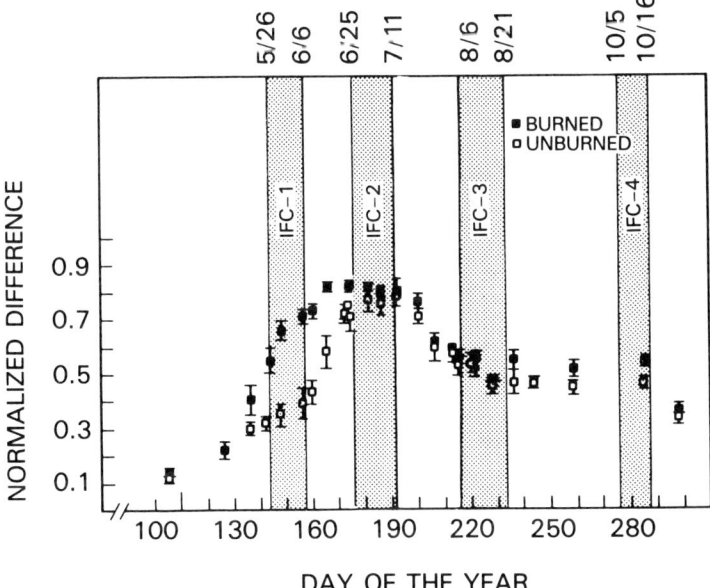

Figure 9.6. Annual cycle of the normalized-difference vegetation index (NDVI) as observed at Konza Prairie using a hand-held radiometer. The NDVI is an indicator of the area-averaged photosynthetic capacity of the vegetated surface. Timings of the intensive field campaigns (IFCs) are shown.

weather stations and the 20-odd flux measurement rigs around the site in such a way as to obtain reasonable estimates of the mean and within-site variability of conditions over the entire (15 km^2) area. This stratification also provided the basis for the soil moisture and biometric observations.

Coordinating the Measurement Program

Having decided on the size and location of the test site and the timing of the observing periods, the operations planning of FIFE was directed at achieving the following goals.

A. The simultaneous acquisition of satellite, atmospheric, and surface data

FIFE aimed to provide a data set that would allow direct comparison between the satellite observations and (near-)surface parameters and processes. The following data were acquired to achieve this objective:

1. *Satellite data.* National Oceanic and Atmospheric Administration (NOAA)–9, NOAA–10, Systeme Probatoire d'Observation de la Terre

(SPOT), Landsat, Geostationary Orbiting Environmental Satellite (GOES).
2. *Airborne radiometric data.* Data from a range of airborne remote-sensing instruments flown over the site during the satellite overpasses to ensure that cloud-free spectral data were acquired and to study the effect of the atmosphere on the radiometric signal.
3. *Surface–near-surface fluxes.* Measurements of latent (evapotranspiration) and sensible heat, CO_2, and momentum fluxes at and above the surface.
4. *Surface–near-surface states.* Measurements of meteorological and atmospheric optical, thermal, and biophysical properties of the surface, and the physical and chemical properties and water content of the soil.

B. *Multiscale observations of biophysical parameters and processes controlling energy and mass exchange at the surface and how these are manifested in "satellite-resolution" radiometric data*

To achieve this second objective, an active effort was made to acquire data over a range of spatial scales. These data are currently being used to test various methods of interpreting small-scale processes (e.g., photosynthesis, transpiration, scattering of light by leaves) up to the scale of satellite pixels of various resolutions. Two issues must be tackled as a direct consequence of this objective. The first is that data must be acquired in order to validate integration procedures; the second is that the effects of coarsening resolution on radiometric data must be studied explicitly. In this sense, the focus of FIFE is directly bound to the problem of studying processes and states over a range of scales, from individual plant leaves up to the entire site. The following data-acquisition strategy was proposed to achieve this objective:

1. *Radiometric data.* These data were collected at the leaf (1 cm^2), canopy (1 m^2), flux-station (10^2 to 10^4 m^2), and satellite-resolution (10^2 m^2, 10^3 m^2, 10^4 m^2, 64 km^2) scales using a range of ground-based, airborne, and satellite instruments.
2. *Flux data.* These data were collected at leaf, canopy, and flux-station scales as defined above and also at scales comparable to the whole site (i.e., [10 to 20 km]^2). Porometers, closed flux meters, and Bowen ratio, surface and airborne eddy-correlation, sound direction and ranging (SODAR), and light direction and ranging (LIDAR) equipment were utilized to perform these tasks. Fluxes of heat, water vapor, and CO_2 were studied at all scales while momentum flux was observed on the flux-station scale and at larger scales.
3. *Biophysical data.* Biophysical data (vegetation physiological, physical, and optical properties; soil physical and chemical properties; soil moisture, etc.) were acquired. These data will be used to check the prognostic variables in various simulation models of surface processes. It is ex-

pected that they will play a vital role in validating the various spatial-integration techniques used to estimate area-averaged quantities.

C. *The provision of integrated analysis through a highly responsive central data system*

A key element of the overall strategy was to design a data system that would provide centralized storage of and cooperative access to all of the data collected during the experiment. The goals of the FIFE Information System (FIS) are, first, to capture and preserve the data and, second, to provide convenient access to the data as rapidly as possible. The database is located at NASA/Goddard Space Flight Center, with electronic communication links to investigators' "home" locations and the experimental site. A database-management system provides on-line access to a complete data inventory and all single-point (nonimage) data. In addition, user support staff is available to preprocess and distribute copies of images and other large data sets. From the outset, it was considered imperative to have the information-system effort directed on a day-to-day basis by a scientist involved in the experiment, supported by scientists drawn from the different disciplines contributing to the data collection and processing effort. The FIS design has been, and continues to be, flexible and evolutionary. The FIFE scientists continue to work with the FIS staff to specify data system needs and processing priorities and to review the system design. As investigators use the system and work with its products, new requirements emerge and the system is modified.

Experiment Execution

The monitoring effort started on the site in May 1987 and continues today. During the IFCs, the surface flux rigs and detailed radiometric measurement and atmospheric sounding equipment (including balloons, SODAR and LIDAR) were moved into the site. Seven aircraft participated in FIFE (see Figure 9.7), but five had major roles in the day-to-day execution of the experiment. These were the following:

1. *The NASA C-130.* This aircraft is equipped with two scanners that collect visible, near-IR, and thermal-IR data over a total of 14 wavelength bands. From its operating height of 16,000 feet above ground level (agl) (4.84 km), the aircraft could "cover" the entire site twice with about 30% overlap using six flight lines. This series of observations routinely provides multiple-view-angle measurements of surface radiance. The aircraft is also equipped with a sun photometer (for atmospheric corrections); a pointable, linear-array, high-spectral-resolution radiometer; and a microwave radiometer for soil-moisture mapping.
2. *Three flux (measurement) aircraft.* The National Center for Atmospheric Research (NCAR) King Air and the University of Wyoming King Air

Figure 9.7. Aircraft and airborne instruments involved in FIFE. *ER-2*: Advanced Very High Resolution Imaging Spectrometer (AVIRIS) and Ocean Color Scanner (OCS). The instruments were used to obtain short-wave reflectances (0.3 to 2.6 μm) over the site on two occasions. *C-130*: Thematic Mapper Simulator (TMS), Thermal Imaging Multichannel Spectrometer (TIMS), and Advanced Solid-State Array (ASAS) Sensor were used to obtain high-resolution, visible, near-IR, and thermal data over the site with spatial resolutions of around 10 to 20 m. The push-broom microwave radiometer (PBMR) was used in soil-moisture surveys. *NCAR King Air*: Flux aircraft equipped for measuring sensible and latent heat flux and momentum flux (gust probe). *NAE Twin Otter*: Equipped like NCAR King Air with additional instrumentation to measure CO_2 flux. *U-Wy King Air*: Equipped like NCAR King Air with x-band Sideways-Looking Airborne Radar (SLAR) for soil-moisture surveys. *NASA Helicopter*: Multichannel Radiometer (MMR) for visible, near-IR, and thermal observations. C-band scatterometer for soil moisture studies. *NOAA Aerocommander*: Gamma-ray equipment for soil-moisture surveys.

Color Plates

Color Plates   III

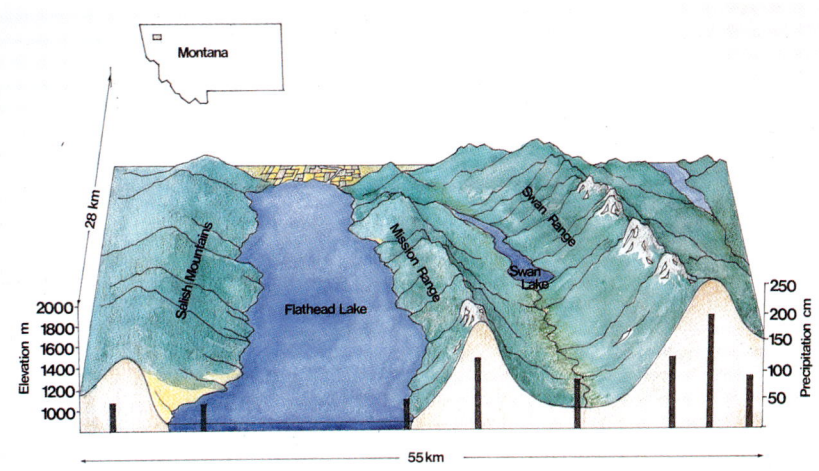

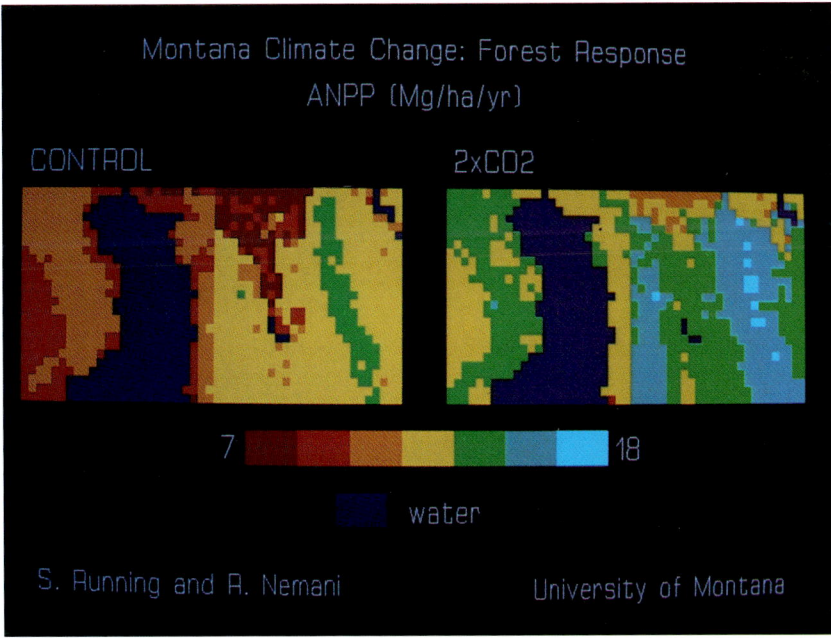

Color Plate 1. A map of simulated forest NPP for a 1,200-km² area of Montana under; first, current conditions, and, second, the response projected given 2× atmospheric CO_2 with a +4-degree air temperature and +10% precipitation and incorporating physiological responses changing water-use efficiency and ecosystem LAI.

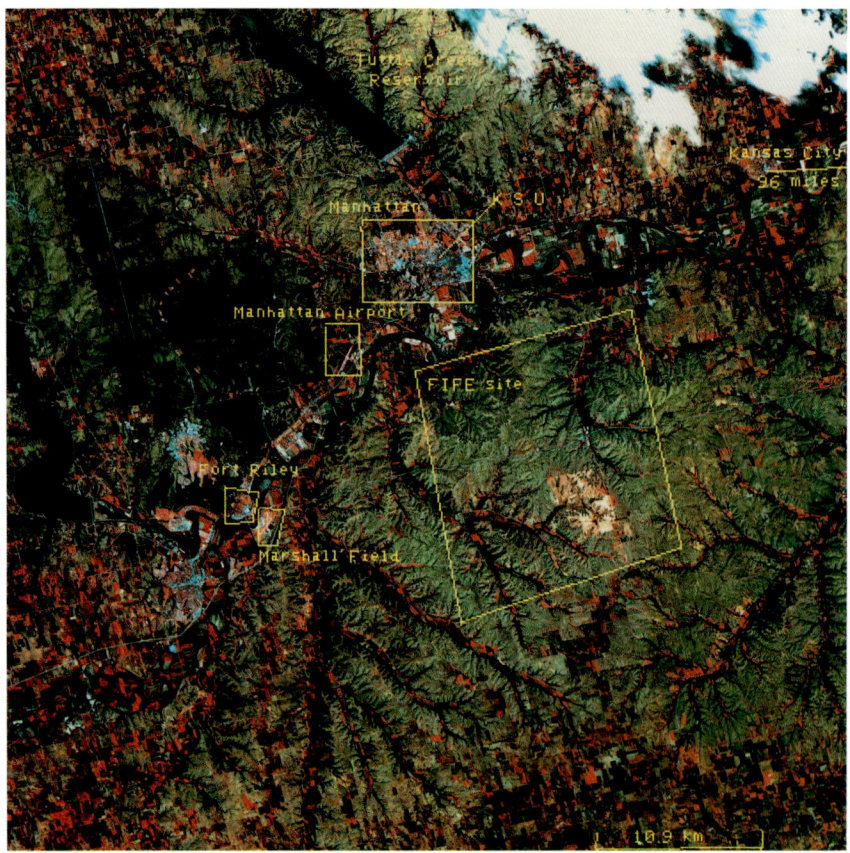

Color Plate 2. SPOT image of the FIFE site near Manhattan, Kansas, taken at 1200 GMT on March 20, 1987. The corner coordinates of the site are as follows:

Corner	Northing	Easting	North latitude*	West longitude*
Northwest	4,333,000	706,000	39 07 24	96 37 02
Northeast	4,333,000	722,000	39 07 10	96 25 56
Southeast	4,317,000	722,000	38 58 32	96 26 15
Southwest	4,317,000	706,000	38 58 46	96 37 19

*Based on Clarke (1866) ellipsoid.

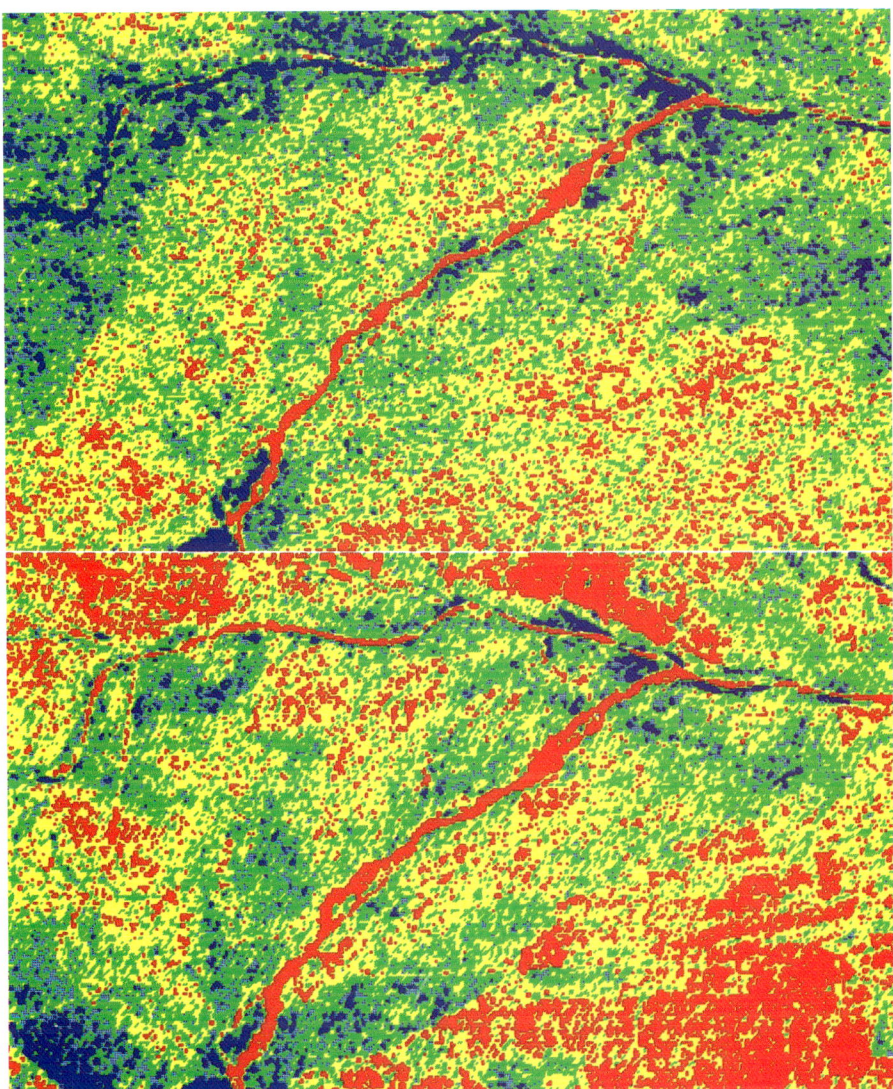

Color Plate 3. Changes in vegetation cover in the Entire Creek area of central Australia over the period 1980 to 1983 determined from Landsat using the index of Pickup and Nelson (1984). The upper image shows conditions in 1980, and the lower image is for 1983. The colors purple, blue, green, yellow, and red represent progressively smaller percentages of vegetation cover.

VI   Color Plates

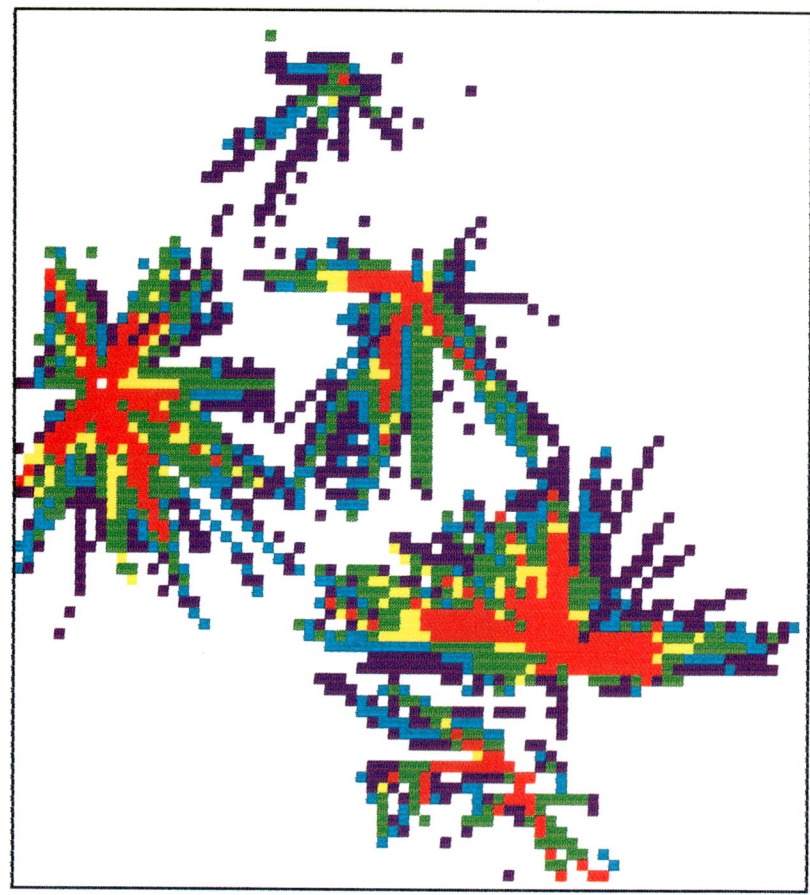

Color Plate 4. Patterns of trampling by cattle in a large arid-zone paddock estimated from animal behavior models calibrated using vegetation cover changes derived from Landsat band 5. The colors purple, blue, green, yellow, and red represent an increasing number of cattle passing through each grid cell.

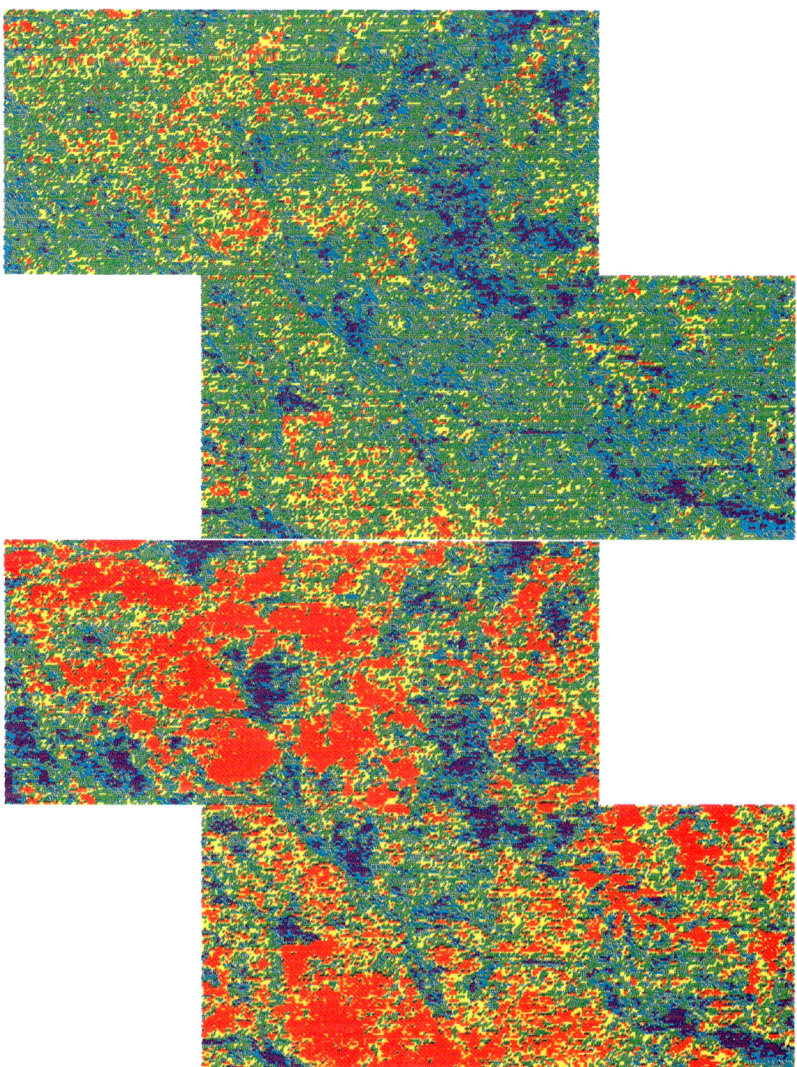

Color Plate 5. Effects of an increase in erosion intensity on spatial patterns of erosion and deposition in a 230-km^2 area of arid central Australia. Red represents erosion, the yellow and green areas are relatively stable, and blue and purple indicate deposition of increasing intensity. The upper image indicates the state of the area in 1972. The lower image is an erosion forecast using a prototype derived from an area of extreme erosion.

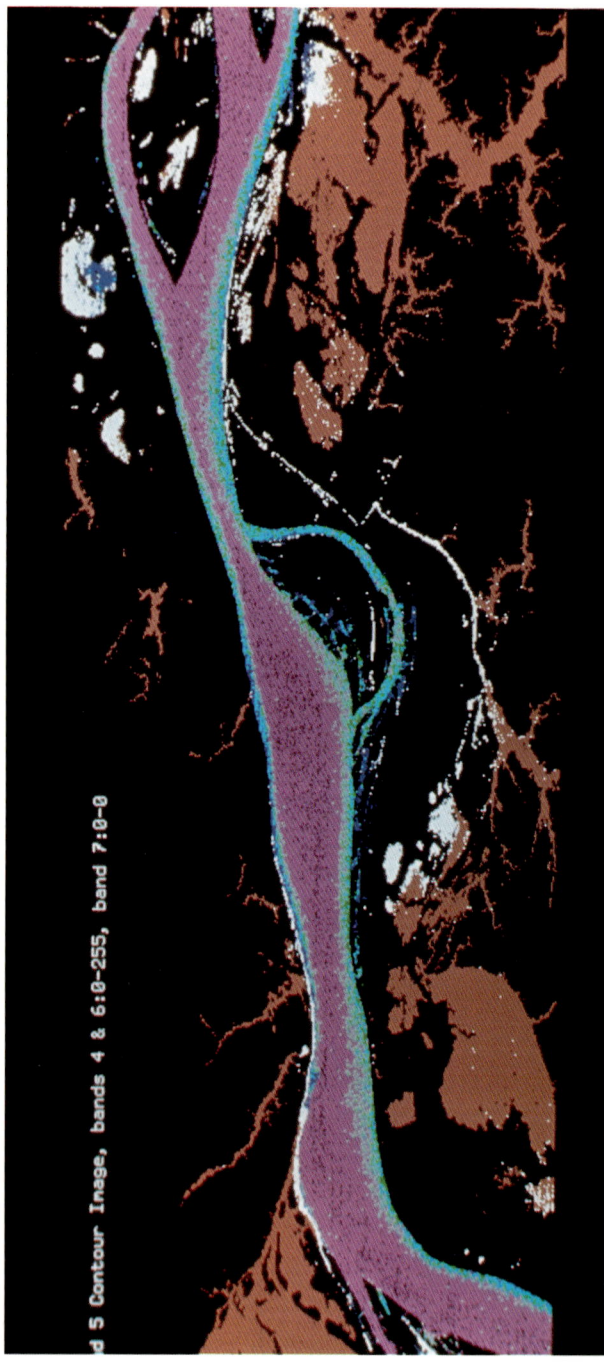

Color Plate 6. MSS image, taken near Manaus, July 31, 1977. The color-enhancement scheme was developed to emphasize the gradient in water types across the floodplain and main channel. Each color band represents a narrow range of brightness values from band 5 (600 to 700 nm). Although the image has not been calibrated, the change from peach to white to blue to green to pink to maroon indicates a gradient from essentially humic-rich, clear water on the floodplain to highly turbid water in the center of the main channel. Essentially, these color bands are a discrete representation of the mixing of these two water types.

are equipped to measure turbulent fluxes of momentum, sensible heat, and latent heat using eddy-correlation equipment. The National Research Council of Canada (NRC) contributed its similarly equipped Twin Otter to FIFE for three IFCs. The Twin Otter also has the unique capability of measuring CO_2 fluxes (area-averaged photosynthesis minus respiration). Usually, two flux aircraft were available during an IFC to allow full coverage of the diurnally changing conditions and also to permit intercomparison among instruments on the different aircraft.
3. *The NASA H-1 (Huey) helicopter.* The H-1 is equipped with a pointable radiometer that records visible, near-IR, and thermal IR radiation reflected or emitted from the surface. A C-band scatterometer is used for soil-moisture surveys. The helicopter's hovering capability provides precisely located nadir and multiangle radiometric data on a nearly simultaneous basis over large areas.

The surface meteorological and flux stations operated more or less continuously during IFCs, as did the atmospheric-sounding programs, soil-moisture and biophysical surveys, and surface radiometry efforts.

The deployment of aircraft during the IFCs sometimes followed a range of idiosyncratic schemes, but three broad plans involving coordinated activities by aircraft and ground teams were followed repetitively. These coordinated mission plans (CMPs) were as follows:

CMP-1: Diurnal Cycle Observations

This involved several subsatellite flights by the C-130 and helicopter and two or three flights by flux measurement aircraft; see Figure 9.8. The aim of CMP-1 was to collect data to allow the calculation of the surface radiation and energy balances. Typically, a CMP-1 involved eight to ten aircraft missions and resulted in continuous 24-hour (or 48-hour) operations by scientists and aircrew. The aircraft were deployed as shown in Figure 9.8, with the following aims: (1) Acquisition of radiometric data to compare with the simultaneously acquired satellite and surface radiometric data to compare with the simultaneously acquired satellite and surface radiometric data. In the course of this effort, multiangle data were acquired from which the surface bidirectional-reflectance-distribution function (BRDF) can be reconstructed. (2) Acquisition of airborne (area-averaged) estimates of the fluxes to compare with the measurements produced by the combined surface networks and estimates derived from the satellite and airborne radiometric data. The experiment objectives required that, in order to describe adequately the biological and physical dynamics of the diurnal cycle, three or four C-130 flights, two or three "flux-aircraft" flights, and two helicopter flights had to be coordinated with each other and with satellite overpasses within a given 24-hour period.

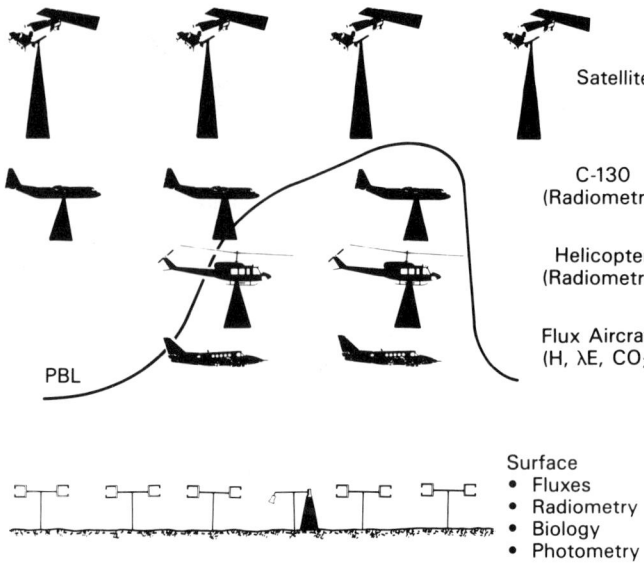

Figure 9.8. Flights and surface activities conducted during intensive field campaigns, diurnal cycle (CMP-1).

CMP-2: Combined Radiometric Mission

Essentially, this involved subsatellite flights with the C-130 and the helicopter, and ground-based instrumentation to allow comparison between satellite observations and the instantaneous surface condition; see Figure 9.9.

CMP-3: Soil-Moisture Survey

The C-130, the helicopter, the radar-equipped University of Wyoming King Air, and the radar-equipped NOAA aircraft were used to conduct soil moisture surveys over the entire site in conjunction with ground-based validation efforts; Figure 9.10.

Table 9.2 shows the number of CMPs conducted during each IFC (note that a CMP-1 can consist of several CMP-2s). Clearly, IFC-1 and IFC-3 were very successful in having a large number of CMP-1s and other missions completed. In IFC-2 and IFC-4, bad weather conditions and instrument failures prevented many flights from being carried out.

Figure 9.11 shows the situation at the time of a satellite overpass (NOAA-9) on June 4, 1987, near the end of IFC-1. The local time is

9. Satellite Remote Sensing and Field Experiments   193

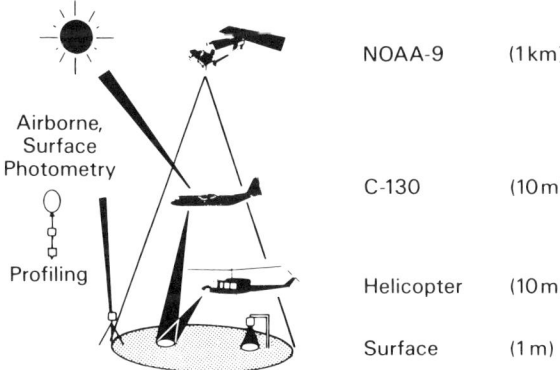

Figure 9.9. Flights and surface activities conducted during intensive field campaigns, multiscale simultaneous radiometric observations (CMP-2).

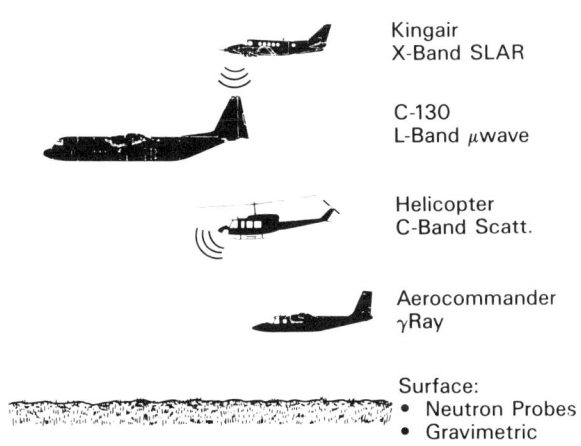

Figure 9.10. Flights and surface activities conducted during intensive field campaigns, soil-moisture survey (CMP-3).

Table 9.2. Individual Aircraft Mission and Combined Mission Plan (CMP) Summaries for Each IFC. For the Aircraft Missions, the First Figure Refers to the Number of Data Flights, the Second to the Number of Hours Flown

	IFC-1	IFC-2	IFC-3	IFC-4	Total
C-130	14/32	16/44	21/56	10/26	61/158
Flux aircraft	8/28	11/43	22/48	16/50	57/169
Helicopter	12/36	12/20	17/23	10/30	51/109
NOAA AeroCommander	3/96	6/16	—	—	9/25
CMP-1	3	1	3	1	10
CMP-2	10	12	14	8	44
CMP-3	4	4	3	2	13

15:17 and the NOAA-9 Advanced Very High Resolution Radiometer (AVHRR) instrument is collecting data with a surface resolution of 1 km over the site as surface photometry, balloons, and SODAR probe the atmospheric optical, thermal, and physical properties. The C-130 is completing a flight line at 16,000 feet (agl) while acquiring reflectance and emittance data with its scanners at a surface resolution of about 10 m by 10 m. A sun-tracking photometer mounted on top of the C-130 monitors the optical thickness of the atmosphere from 16,000 feet on up. The helicopter is acquiring detailed radiometric data from 1,000 feet (agl) over a preselected site while scientists on the surface take radiometric (BRDF, emittance, etc.) and biophysical (chlorophyll density, photosynthesis, leaf transpiration, etc.) observations. The NCAR King Air is in the process of flying an "L-shaped" flight pattern at about 500 feet (agl) along the two downwind sides of the site while taking eddy-correlation data to estimate area-averaged surface fluxes. At the time, over 100 people (scientists, research assistants, air-crew technicians) were working on, near, or above the FIFE site.

In total, nearly 180 data missions were flown by FIFE aircraft over the site, amounting to over 450 hours of flying time. The satellite data acquired over the site during 1987 are estimated to be as follows: 5,000 GOES images, 1,000 NOAA (AVHRR) images, 30 SPOT images, and five Landsat scenes. Individual investigator data are anticipated to be both diverse and voluminous.

The FIFE Information System began processing satellite and meteorological data well before IFC-1. An inventory of these data and some historical point data sets were on-line and accessible from the experiment site during IFC-1. Automatic Meteorological Station data and the various

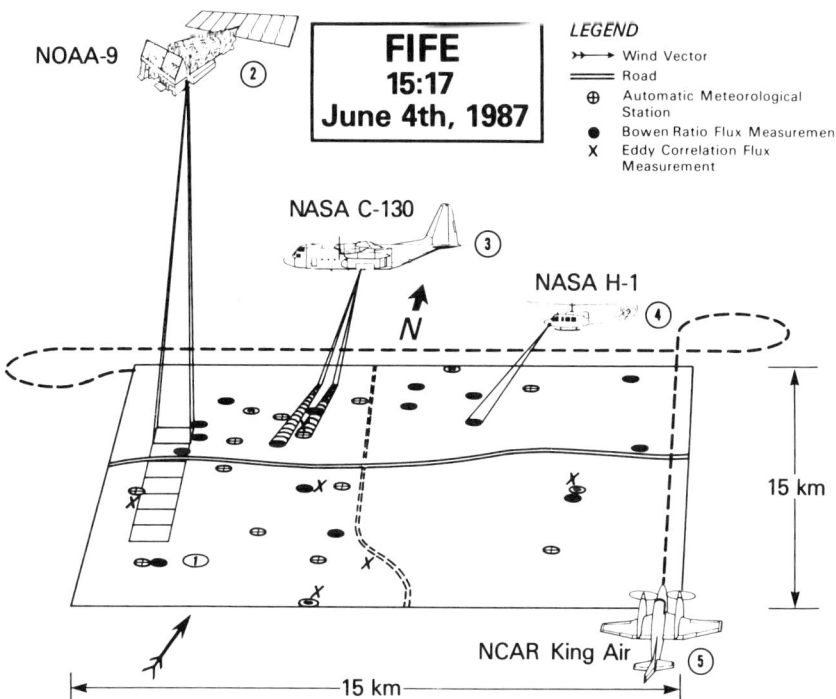

Figure 9.11. Situation at the FIFE site at 1517, June 4, 1987 (NOAA-9 overpass). (1) Surface-flux stations and automatic meteorological stations monitor surface and near-surface fluxes and near-surface meteorological conditions. (2) NOAA-9 satellite scans the site at 1-km resolution. (3) NASA C-130 traverses the site at 16,000 feet agl taking scanner and sun photometer data. (4) The helicopter hovers above preselected site at 1,000 feet agl and acquires radiometric data. (5) NCAR King Air collects eddy correlation data at 500 feet agl. The data-acquisition activities are discussed in the text.

types of support data from IFC-1 were available by the start of IFC-2. By the end of IFC-3, a full user interface was in place and standard data sets from IFC-1 were available in addition to the on-line data. As the data collection was completed, investigators began to send their data to the FIS to be assimilated and available for the analysis phase of the experiment.

Analysis of the Data

The FIFE data set is currently being analyzed by about 50 scientists and research staff and a review of preliminary results should be available before the end of 1989.

In one respect, the project is entering its most critical phase. The data have been collected and the observational effort has been completed without serious mishap. At this point, most of the participating scientists are

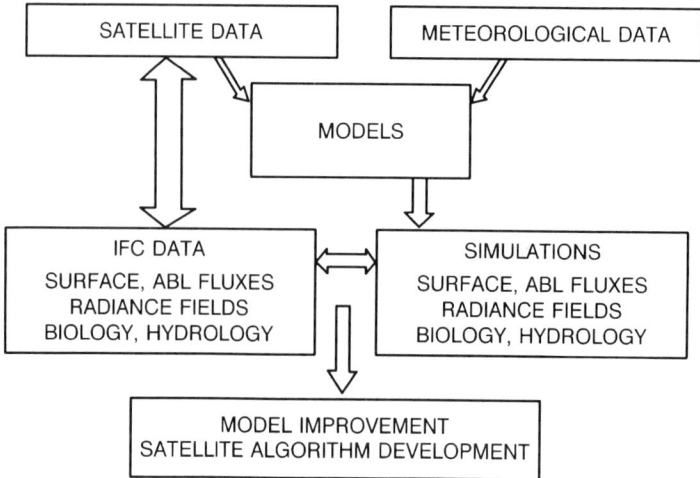

Figure 9.12. Flowchart showing development of FIFE project in analysis phase. (1) Satellite data, meteorological data, and IFC data (observations taken during IFCs) are loaded into the FIFE information system following data-quality checks. (2) Biophysical models and remote sensing algorithms are combined to calculate surface fields of heat and mass fluxes and radiances in simulations. The satellite data and meteorological data are used to drive the models. (3) The simulations are compared with the directly equivalent IFC data. Discrepancies lead to the improvement of models and algorithms.

working at their home institutions on the particular aspects of the data-acquisition effort for which they were primarily responsible; in other words, they are completing the boxes marked "satellite data," "meteorological data," and "IFC data" in Figure 9.12. However, the success of the project can only really be judged when the first two elements are combined with models to produce calculations to match the IFC data, that is, when the satellite data are fully processed, combined with other data, and compared with equivalent field observations. This stage will require the active cooperation of many scientists from different disciplines and the application of existing and novel models and concepts to sometimes unfamiliar scales and applications. The ultimate success of the project can only be assessed after this stage of the research effort has been completed.

Implications for the Future

FIFE is a relatively complex, interdisciplinary, and resource-intensive experiment aimed directly at studying biological controls on the climate system and how we can observe them from orbit. It is complex because it requires studies at all scales (from the leaf scale to the "satellite-pixel" scale); it is interdisciplinary because we need simultaneous measurements of heat, radiation, and mass fluxes and the associated biophysical states from satellite, airborne, and surface platforms; and it is necessarily resource intensive because of the large number of surface stations and aircraft required to cover a site big enough to be observable from orbit. However, the resulting data set is now available from a single combined repository and it is unique for all the reasons previously stated. For the first time, the data will exist that will allow scientists to test models and algorithms on scales consistent with satellite observations but with enough supporting data on finer scales to test the validity of various spatial aggregation techniques. These techniques, which are intimately bound up with the "scale problem," have to be mastered if we are to use satellites for land-surface-climatological studies in a biophysically defensible way. In this respect, FIFE could serve as a useful model for future ecological and climatological remote sensing experiments.

The results of FIFE will also have some immediate and practical consequences. The specification of the planned Earth Observing System (EOS) and its associated data system, which is a satellite-based effort targeted squarely at understanding earth systems science problems related to atmospheric-oceanic-biospheric interactions, is not yet complete. The experience of FIFE should be highly relevant to EOS in terms of specifying scientific objectives, instrument design, orbit configurations, data-handling procedures, analysis techniques, and validation procedures. For example, the planning and results of FIFE to date have emphasized the need for the following attributes of a satellite-based earth science effort: high-temporal resolution, a range of spatial resolutions, and pointable sensors.

In FIFE, the NOAA series of satellites and GOES were used to provide a moderate-temporal-resolution, coarse-spatial-resolution data set with SPOT and aircraft data providing the high-spatial-resolution, pointable-instrument capability. The 18-day overpass frequency of Landsat (which combines fine spatial and spectral-resolution data with poor temporal resolution) has proved to be marginally useful for bioclimatological applications; out of five Landsat overpasses during FIFE IFCs, three occurred when the site was covered in a continuous cloud layer and a fourth when conditions were partially cloudy and the C-130 scanner was inoperable. By contrast, NOAA-9, NOAA-10, and SPOT have been underflown on numerous occasions and their data make up a semicontinuous record of surface conditions. From this, it can be concluded that the ideal remote-sensing system should not compromise the requirement of high time resolution by placing undue emphasis on spatial resolution or instruments of marginal utility.

The interdisciplinary nature of FIFE has compelled the participating scientists to contribute the knowledge of their own discipline toward sometimes unfamiliar scientific areas and a hugely increased range of scales. An integrated data system is a key factor in making this interdisciplinary, multiscale research feasible. As a result, FIFE has created the environment for discussion of all aspects of the land-surface component of EOS and will provide a data set to test hypotheses.

Concluding Remarks

The need for experiments to validate remote sensing hypotheses is more likely to increase rather than decrease in the next few years. There is already a wide gap between the potential of currently available data and its utilization by the scientific community and operational organizations that require such data for routine applications. This gap exists in spite of the apparent benefits to be gained from remote sensing, mainly because of the uncertainties and errors involved in interpreting the raw sensor counts all the way through to surface or atmospheric parameters. Some of these uncertainties are due to technical problems, such as calibration; some are the consequence of our incomplete understanding, such as the atmospheric-geometric correction problem; and some are just there because their solution requires the intense cooperation of scientists from a range of disciplines working within unfamiliar scale domains. This last problem, which is arguably the most difficult to overcome, requires that many different individuals be organized and motivated toward a common goal. In this respect, field experiments offer not only a technical solution to the problem of interpreting satellite data, but also a forum for scientists to bring their individual skills to a difficult problem that is, by definition, interdisciplinary.

The advent of the EOS project, which has the goal of investigating the entire earth system using a large arsenal of novel sensors, presents the biological community with some daunting challenges and some hard choices. Perhaps the greatest challenge is to determine which quantities should be measured and with what precision. Once this question has been settled, the need will be to determine a combination of sensors, operational modes, and algorithms to obtain these quantities, at which point the need for field experiments will become more pressing. Over the next few years, therefore, we can expect to see a series of field validation efforts using airborne prototype instruments combined with surface equipment to determine the final inventory of research instruments for EOS. It is hoped that these efforts will carry biologists over the gap that lies between their current status as hopeful theoreticians and that of confident, practicing scientists working in the global domain.

Acknowledgments

Many of the FIFE team science investigators contributed to the Experiment Plan through their participation in the December, 1986 and March, 1987 FIFE workshops. This chapter is partly based on the FIFE Experiment Plan and the published paper of Sellers et al. (1988a). The SSG members R. Grossman, S. Verma, C. Bruegge, B. Blad, and J. Wang, supported by J. Dozier, F. Davis, J. Norman, D. Schimel, F. Becker, and S. Goward, provided much of the material in these documents. FIFE staff members and others maintained a constant level of support in spite of many other duties, notably, B. Markham, J. Newcomer, G. Beckman, S. Goetz, D. Fadler, R. Kennard, I. McPherson, R. Kelly, G. Alger, J. Killeen, J. Ormsby, F. Wood, E. Peck, and C. Walthall.

Marlene Schlichtig of COLA typed and edited the paper and Carol Kulwich of NASA supervised the artwork.

References

Abdellaoui, A., Becker, F., and Olory-Hechinger, E. (1986). Use of METEOSAT for mapping thermal inertia and evapotranspiration over a limited region of Mali. *J. Clim. Appl. Meteorol.* 25:1489–1506.

Becker, F.B., Camilo, P.J., and Choudhury, B.J., (1988). Surface heat fluxes. In: P.J. Sellers, S.I. Rasool, and H-J. Bolle (eds.), *Satellite Data Algorithms for Land Surface Studies* ISLSCP Rep. no. 9, JPL, Pasadena, CA (available from authors).

Dedieu, G., Deschamps, P.Y., and Kerr, Y.H. (1987). Satellite estimation of solar irradiance at the surface of the earth and of surface albedo using a physical model applied to Meteosat data. *J. Clim. Appl. Meteorol.* 26:79–87.

Diak, G., and Gautier, C. (1983). Improvements to a simple physical model for estimating insolation from GOES data. *J. Clim. Appl. Meteorol.* 22:505–508.

Dickinson, R.E. (1984). Modeling evapotranspiration for three-dimensional global climate models. pp. 58–72. In J.E. Hanson and T. Takahashi, (eds.), *Climate Processes and Climate Sensitivity*. Geophysical Monograph. 29. Amer. Geophys. Union, Washington, DC.

Farquhar, G.D., and Sharkey, T.D. (1982). Stomatal conductance and photosynthesis. *Ann. Rev. Pl. Physiol.* 33:317–345.

Fung, I.Y., Tucker, C.J., and Prentice, K.C. (1987). Application of advanced very high resolution radiometer vegetation index to study atmosphere-biosphere exchange of CO_2. *J. Geophys. Res.* 93 (D3):2999–3015.

Goward, S.N., and Dye, D.G. (1987). Evaluating North American net primary productivity with satellite observations. *Adv. Space Res.* 7 (11):165.

Goward, S.N., Tucker, C.J., and Dye, D.G. (1985). North American vegetation patterns observed with the Nimbus-7 Advanced Very High Resolution Radiometer. *Vegetatio* 64:3–14.

Hall, F.G., Strebel, D.E., Goetz, S.J., Woods, K.D., and Botkin, D.B. (1987). Landscape pattern and successional dynamics in the boreal forest. *Proc. 1987 IGARRS Symp.* IEEE87CH2434-9, pp. 473–483.

Hansen, J., Lacis, A., Lebedeff, S., Lee, P., Rind, D., and Russell, G. (1981). Climate impact of increasing carbon dioxide. *Science*, 213:957–966.

Hope, A.S. (1987). Parameterization of surface moisture availability for evapotranspiration using combined remotely sensed spectral reflectance and thermal observations. PhD thesis, Univ. Maryland, College Park.

Houghton, R.A. (1987). Biotic changes consistent with the increased seasonal amplitude of atmospheric CO_2 concentrations. *J Geophys. Res.*

Jarvis, P.G. (1976). The interpretation of the variations in leaf water potential and stomatal conductance found in canopies in the field. *Phil. Trans. Roy. Soc.* (London), Ser. B., 273:593–610.

Justice, C.O., Townshend, R.G., Holben, B.N., and Tucker, C.J. (1985). Analysis of the phenology of global vegetation using meteorological satellite data. *Int. J. Remote Sens.*, 6:1271–1318.

Monteith, J.L. (1973). *Principles of Environmental Physics.* Edward Arnold, London.

Morcrette, J., and Deschamps, P. (1986). Downward longwave radiation at the surface in clear-sky atmospheres: Comparisons of measured, satellite derived, and calculated fluxes. *ISLSCP. Proc. Int. Conf. held in Rome, Italy, 1985*; ESA-SP 248, ESA, Paris, France, pp. 257–262.

Njoku, E.G., and Patel, I.R. (1986). Observations of the seasonal variability of soil moisture and vegetation cover over Africa using satellite microwave radiometry. *ISLSCP. Proc. Int. Conf. held in Rome, Italy, 1985*; ESA-SP 248, ESA, Paris, France, pp. 349–356.

Parton, W.J., Schimmel, D.S., Cole, C.V., and Ojima, D.S. (1987). Analysis of factors controlling soil organic matter levels in great plains grasslands. *Soil Sci. Soc. Amer. J.* 51:1173–1179.

Price, J.C. (ed.) (1986). Interpretation of thermal infrared data. The heat capacity mapping mission. *Remote Sen. Rev.* 1:N2 (Harwood Academic Publishers).

Rotty, R.M. (1983). Distribution of and changes in industrial carbon dioxide production. *J. Geophys. Res.*, 88:1301–1308.

Sato, N., Sellers, P.J., Randall, D.A., Schneider, E.K., Shukla, J., Kinter III, J.L., Hou, Y-T., and Albertazzi, E. (1989). Effects of implementing the simple biosphere model (SiB) in a general circulation model. *J. Atmos. Sci.* 46(18):2757–2782.

Schlesinger, M.E., and Mitchell, J.F.B. (1987). Climate model circulations of the equilibrium climatic response to increased carbon dioxide. *Rev. Geophys.* 25(4):760–798.

Sellers, P.J. (1985). Canopy reflectance, photosynthesis and transpiration. *Int. J. Remote Sens.* 6(8):1335–1372.

Sellers, P.J. (1987). Canopy reflectance, photosynthesis and transpiration II: The role of biophysics in the linearity of their interdependence. *Remote Sens. Envir.* 21:143–183.

Sellers, P.J., and Hall, F.G. (1987). *The FIFE Experiment Plan*, NASA Internal Document, 623, NASA/GSFC, Greenbelt, MD.

Sellers, P.J., Hall, F.G., Asrar, G., Strebel D.E., and Murphy, R.E. (1988a). The first ISLSCP field experiment (FIFE). *Bull. Amer. Met. Soc.* 69(1):22–27.

Sellers, P.J., Mintz, Y., Sud, Y.C. and Dalcher, A. (1986). A simple biosphere model (SiB) for use within general circulation models. *J. Atmos. Sci.* 43(6):505–531.

Sellers, P.J., Rasool S.I., and Bolle, H-J. (1988b). Satellite data algorithms for land surface studies. ISLSCP Rep. No. 9, JPL, Pasadena, CA.

Shugart, H.H., Crow, T.R., and Hett, J.M. (1973). Forest succession models: A rationale and methodology for modeling forest succession over large regions. *Forest Sci.* 19:203–212.

Slater, P.N. (1986). Variations in in-flight radiometric calibration. *ISLSCP. Proc. Int. Conf. Held in Rome, Italy, 1985*; ESA-SP 248, ESA, Paris, France, pp. 349–356.

Solomon, A.M. (1986). Linking GCM climate data with data from static and dynamic vegetation models. pp. 95–98. In: C. Rosenzweig and R.E. Dickinson (eds.), *Climate-Vegetation Interactions*. UCAR, Boulder, CO.

Trabalka, J.R. (ed.), (1985). *Atmospheric Carbon Dioxide and the Global Carbon Cycle*. U.S. Dep. Energy, Washington, DC. (available as NTIS DOE/ER-0239 from Nat. Tech Inf. Serv., Springfield, VA).

Tucker, C.J., Fung, I.Y., Keeling, C.D., and Gammon, R.H. (1986). Relationship between atmospheric CO_2 variations and a satellite-derived vegetation index. *Nature*, 319:195–199.

Tucker, C.J., Holben, B.H., McMurtrey, J.H.E. (1981). Remote sensing of total dry matter accumulation in winter wheat. *Remote Sens. Environ.* 11:171–190.

Wiscombe, W.J., Welch, R.M., and Hall, W.D. (1984). The effects of very large drops on cloud absorption. Part I: Parcel models. *J. Atmos. Sci.* 41:1336–1355.

10. Remote Sensing of Spatial and Temporal Dynamics of Vegetation

Richard J. Hobbs

Most of the world's vegetation is in a state of flux at a variety of spatial and temporal scales. Plant growth and reproductive patterns respond to seasonal fluctuations in climate. Yearly climatic variations are also responsible for differences in species growth and establishment patterns, leading to changes in species composition and distributions. Over long periods of time, directional vegetational changes may occur through succession. Vegetation changes may take place at extremely small scales, for instance, in canopy gaps created by the death of individual trees (Shugart and West, 1981; Runkle, 1985), or over larger scales where vegetation responds to such disturbances as fires or floods. Species distributions may change rapidly in response to episodic events (e.g., Hobbs and Mooney, 1989), or over longer periods in response to climatic shifts (e.g., Davis, 1986; Delcourt and Delcourt, 1987). Evidence of past vegetational changes resulting from changes in climate during glaciation cycles reinforce the view that major vegetational shifts are possible.

If man-induced global climatic changes occur as predicted (e.g., Broeker, 1987: Bolin et al., 1987), large changes in vegetation are likely, and it is important that we have the ability to measure such changes and to develop predictive models of future change. The Earth System Sciences Committee (1988) recently stated: "Of particular long term concern are changes in vegetation cover, soil moisture, and biome extent, productivity and nutrient cycling. None of these is satisfactorily measurable on a global scale at this time, and a concerted effort is needed to remedy this situation." This chapter examines the role of remote sensing in the study of vegetation dynamics. As techniques of studying vegetation using remote sensing have been covered elsewhere in this volume, problems associated with detecting change are the focus here. First, types of vegetation change are looked at, and then how remote sensing can be used to delineate different vegetation types or communities is considered. Finally, ways of detecting vegetation changes using remotely-sensed data and the related problems are examined.

Patterns of Change

The fact that vegetation is changing at a variety of spatial and temporal scales makes it essential that we take into account variability at one scale when trying to interpret changes at another.

The types of variability that have to be considered are:

1. Seasonal response
2. Interannual variability
3. Directional vegetation change, which may be caused by;
   a. Intrinsic vegetation processes (i.e., succession)
   b. Land-use and other human-induced changes
   c. Changes in global climatic patterns

Seasonal Variations

While the radiation climate varies diurnally and seasonally (owing to differences in angle of illumination and atmospheric variations), so too does the vegetation. Of particular interest here are phenological patterns over the course of a year. Many plant communities have distinct seasonal peaks of growth and flowering activity (e.g., Bell and Stephens, 1984; Mooney et al. 1986) that can markedly affect spectral reflectance (e.g., Warren and Hutchinson, 1984). Different components of the vegetation often grow at different times of year and this also affects the overall reflectance of the community. Examination of year-to-year variability therefore must take seasonal changes into account. An extreme example of this is an annual grassland where live vegetation is present for only part of the year. The change in spectral reflectance between summer and winter may be much greater than that caused by any directional change in vegetation. On the other hand, seasonal variations can be used to differentiate between herbaceous and woody vegetation or among different woody vegetation types with different phenological patterns.

Interannual Variability

Interannual variability in vegetation takes place as a result of climatic variability and its effects on germination and growth. Thus large variations in vegetation composition and growth are seen in arid and semiarid areas where rainfall is sporadic and the response of vegetation to such rainfall is rapid (e.g., Walker, 1979; Ayyad 1981; Griffin and Friedel, 1985). National Oceanic and Atmospheric Administration (NOAA) Advanded Very High Resolution Radiometer (AVHRR) data have been used to document interannual variation in vegetation in sub-Saharan Africa (Tucker, 1986, Tucker et al., 1986). Such data allow entire ecological zones to be monitored, but can also be used for more localized interannual comparisons. Yearly variations in vegetation can take the form of changes in the spatial distribution of plant growth, as in Tucker et al. (1986), or may involve differences in species dominance from year to year, such as in California grasslands (Pitt and Heady, 1978). Again, such interannual variations may result in large changes in spectral reflectance but not indicate any directional change in the vegetation.

Directional Vegetation Changes

Directional change indicates a progressive or irreversible change in the vegetation that results from something other than annual climatic variability. This can be viewed as a change in vegetation at one particular spot or as a change in the vegetation pattern owing to boundary shifts. The definition of what constitutes directional change depends on the scales of observation; for instance, the vegetation in a particular forest canopy gap may be undergoing directional change over a period of a few years, whereas the forest as a whole remains fairly static in terms of overall composition. Similarly, although the forest may appear to be fairly static when viewed over years or decades, it can in fact be seen to have changed markedly when looked at over hundreds or thousands of years. In fact, vegetation change often takes place over relatively long periods, which makes it difficult to obtain good data that will document such change. One must then resort to indirect methods of data acquisition, making use of historical and stratigraphic records (e.g., Davis, 1986; Hamburg and Cogbill, 1988). Only in certain cases can change be observed directly, because of the availability of permanent recording sites, or because of the rapidity with which change occurs. Examples of long-term study plots include those of Watt (1981) and the Rothampstead grassland plots (Silvertown, 1980, Tilman, 1982). Rapid vegetation change is seen where there is a rapid response to changes in climate, level of herbivory, or management regime, or where there is a response to episodic climatic extremes. Hobbs and Mooney (1989) have reviewed the effects of such episodic events on

Mediterranean-type systems, and Williams et al. (1987) give an example of rapid community change as a result of extreme rainfall events.

Directional change may occur because of intrinsic vegetation processes or it may be forced by external events. Change may be due to variations in life history characteristics, physiological status, and competitive abilities of the plants that make up a given community, or it may be initiated due to disturbances such as a fire, a storm, or a flood. Considerable debate has centred on the mechanisms of vegetation change or succession (Connell and Slatyer, 1977; West et al., 1981; Gray et al., 1986; Huston and Smith, 1987; Pickett et al., 1987; Tilman, 1988), and these will not be explored in detail here. It is sufficient to say that our understanding of the processes of vegetation changes is incomplete for many vegetation types and that a variety of mechanisms of change are involved.

Superimposed on natural patterns of vegetation change are changes brought about by man. These include changes in land use and in vegetation cover through deforestation or reforestation, the introduction or removal of herbivores, or the introduction of nonnative species, and the changes attributable to the effects of pollution. In many areas of the world, the natural vegetation is being significantly fragmented by clearance for agriculture and is being markedly changed by current management practices. In particular, stock grazing and altered fire regimes affect many vegetation types (e.g., Adamson and Fox, 1982; Saunders et al., 1987; Hobbs and Hopkins, 1989), while pollution is being implicated in the large-scale decline of forests in the northern hemisphere (e.g., Postel, 1984; Johnson and Siccama, 1984; Schutt and Cowling 1985).

The effects of man on vegetation are the major type of change in evidence on a global scale at the present time. However, superimposed on these effects are the potential effects of global-scale climatic changes that may significantly alter the vegetation of many parts of the world (Bolin et al., 1987; Broeker, 1987). These potential changes are as yet speculative and we lack models and predictive capability to assess their extent or likelihood. We can hypothesize, for instance, that changes in rainfall regime will cause both changes within individual communities and changes in community distributions and boundaries (Figure 10.1), but the changes are likely to be complex, involving many lag and secondary effects. International programs are now getting under way that will target these problems (National Academy Press, 1986; Earth System Sciences Committee, 1988; Mooney, 1988).

Can We Detect Differences in Vegetation?

The ability to use remote sensing to detect vegetation change depends on the initial ability to deal adequately with the static situation. Thus we must be able to detect differences in vegetation composition, structure, produc-

Figure 10.1. Hypothetical response of two plant communities to large-scale climate change, involving movement through the PAM-AN plane (plant available moisture–available nutrients). [Modified from Graetz et al. (1988).]

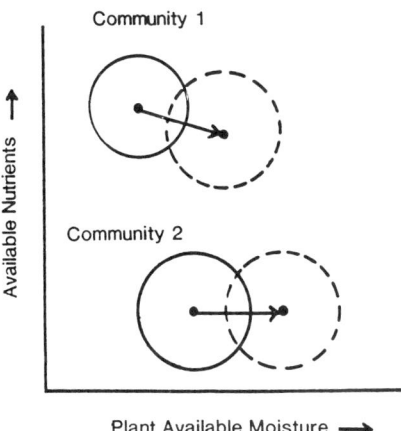

tivity, and "health." There has been much work that shows that remote sensing can quite accurately determine some or all of these characteristics for many vegetation types (Botkin et al., 1984; Committee on Planetary Biology, 1986). This has been done on a variety of scales, ranging from global or continental, using NOAA AVHRR data (Townshend and Tucker, 1984; Goward et al., 1985; Tucker et al., 1985a; Clark et al., 1986; Yates et al., 1986; Dregne and Tucker 1988) down to regional and local, using Landsat Multi-Spectral Scanner (MSS) and Thematic Mapper (TM) data, and SPOT and airborne scanners (e.g., Saxon and Dudzinski, 1984; Morton, 1986; Ustin et al., 1986). Further developments include the use of radar (Synthetic Aperture Radar, Shuttle Imaging Radar, e.g., Green, 1986), which has the advantage of being able to penetrate cloud cover.

The resolution required depends on the type of problem being tackled. For large-scale (continental and global) investigations the AVHRR provides the only viable source of data in terms of data-handling problems. Finer-resolution data furnish more detailed information for smaller-scale studies but the area that can be handled is limited by data storage and processing and cost. Different types of vegetation and different situations will require a variety of approaches. For instance, a single AVHRR data cell in the Western Australian wheat belt is likely to contain 20 to 30 different vegetation types in dozens of native vegetation remnants surrounded by agricultural land (e.g., Hobbs et al., 1989). The AVHRR pixel would not register any of this detail and Landsat MSS or TM data would be required to resolve the native vegetation mosaics. From a global perspective, the AVHRR data provide an accurate-enough assessment of the character-

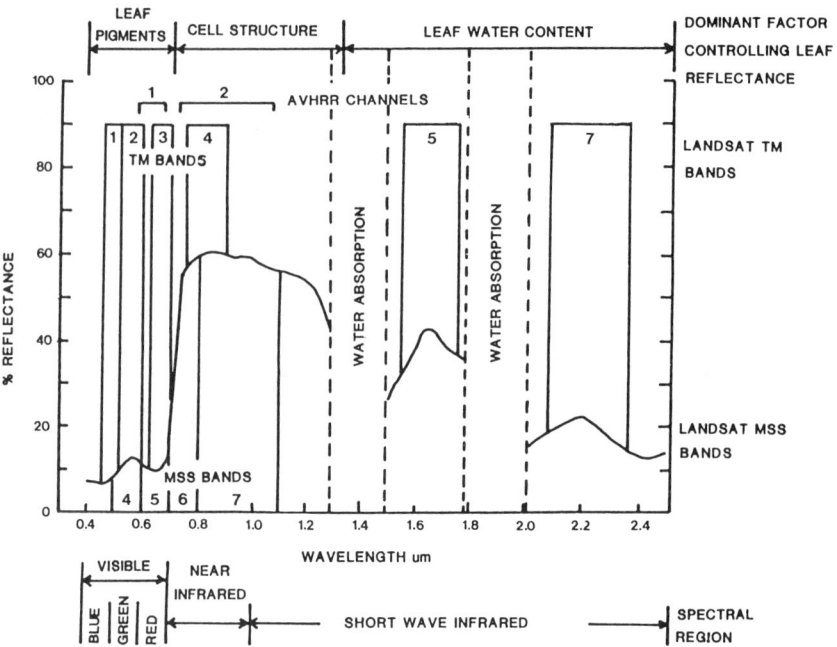

Figure 10.2. Typical vegetation reflectance curve, with AVHRR and Landsat MSS and TM bands superimposed. [Modified by permission of the publisher from Rock et al. (1986), *BioScience* 36, 439–45. Copyright 1986 by American Institute of Biological Sciences.]

istics of the area, but for local land management purposes, more detail is required.

Figure 10.2 illustrates a typical vegetation reflectance curve (Goetz et al., 1983; Rock et al., 1986). Strong absorption by the photosynthetic pigments occurs at around 0.48 and 0.68 μm, while reflectance of green light is evident at 0.52 to 0.6 μm. Strong reflectance in the region of 0.75 to 1.3 μm, or the near-infrared (IR), is characteristic of healthy leaf tissue, and the slope and position of the sharp rise in reflectance between the visible and near-IR have been directly correlated with leaf chlorophyl concentrations (Horler et al., 1980, 1983; Rock et al., 1986). Reflectance values in the 1.65 μm and 2.2 μm regions can provide an accurate indication of leaf water content (Rohde and Olson, 1971; Tucker, 1980). The techniques and problems involved with measuring and interpreting remotely sensed data in these various spectral regions have been discussed by other contributers to this volume and will not be explored further here. Rather this chapter will concentrate on particular problems associated with detecting vegetation change.

Of particular importance in considering vegetation mapping has been the development of vegetation or greenness indices such as the normalized difference vegetation index or NDVI (see Perry and Lanternschlager, 1984). This is a normalized ratio between the visible and near-IR spectral bands that enhances vegetation and reduces variations caused by changes in irradiance, which varies as a function of solar elevation. Photosynthetically active vegetation reflects less radiation in the visible range than in the near-IR range, and thus higher index values indicate where more green vegetation is present. Comparisons of NDVIs from different times can yield information on variations in vegetation productivity and condition (e.g., Goward et al. 1985; Tucker et al., 1985b; Roller and Colwell, 1986; Towhshend and Justice, 1986). However, the index represents only a relatively crude measure of "color" of the land surface, and the relationship between the index and actual changes in vegetation is still being assessed. Particularly important in this respect are large-scale experiments such as those set up under the International Satellite Land Surface Climatology Project (ISLSCP).

More complex techniques of plant community recognition and mapping are available that use all of the spectral data derived from satellites rather than an index derived from a subset of the data. Such techniques involve the determination of a 'spectral signature' for each community and the statistical analysis of multispectral data (e.g., Walker et al., 1986; Yool et al., 1986). Our work in western Australia has shown that Landsat MSS data can be used effectively to classify and map relatively complex vegetation patterns, at least to a broad degree of community separation (Hobbs et al., 1989). The analysis first determines spectral classes (Richards and Kelly, 1984) by the selection of relatively homogeneous training areas and their subsequent ordination. Spectral classes are then derived from the ordination and all pixels allocated to a spectral class based on the probability of membership of each class. However, classification requires a skilled interpreter/classifier and still produces subjective results.

The technique used by Hobbs et al. (1989) was able to separate the broad structural vegetation types (i.e., woodland, mallee, dense shrubland, and heath), and also was able to distinguish heath regenerating after disturbance from the equivalent undisturbed community. It could not, however, distinguish different types of woodland or different dominant species. From the point of view of recognizing vegetation change, therefore, only relatively large structural changes (e.g., change from savanna to woodland) would be detectable. Jupp et al., (1986) and Walker et al., (1986), on the other hand, have been able to detect a gradient of woodland structure from recently cleared areas through various stages of regeneration to intact woodland. This indicates good potential for remote sensing of woodland dynamics. Similarly, Weaver (1987) has shown that airborne Thematic Mapper data can be used to distinguish different stages of heathland canopy development. There are a number of complicating factors that have to be recognized, however. In addition to the problems of variations

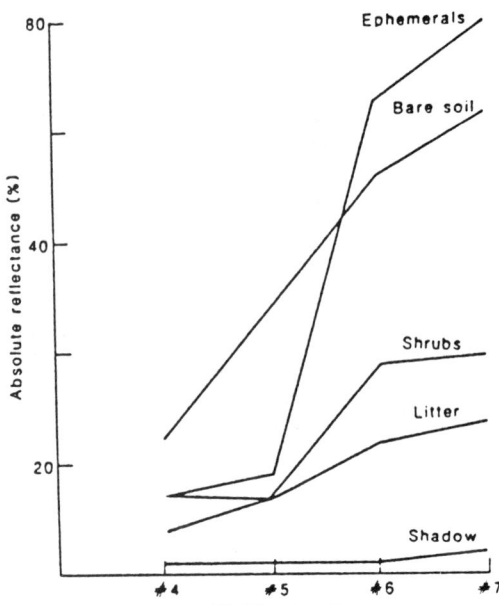

Figure 10.3. Absolute reflectance values for arid vegetation components in the four Landsat MSS wavebands. [Reprinted by permission of the publisher from Graetz (1987), *Remote Sens. Environ.* 23: 313–331. Copyright 1987 by Elsevier Science Publishing Co., Inc.]

in solar angles, shadows, and plant phenology raised earlier (e.g., Wardley et al., 1987), terrain is also important. For instance Walsh (1987) has shown that the same forest type may have quite a different spectral response on a shallow than on a steep slope. Sing (1987) has also found that, while some tropical vegetation types are separable using Landsat MSS data, others are not, owing to the similarity of cover types. Types that were not separable included shifting cultivation versus grassland and scrub versus forest. Similarly, Adomeit et al. (1981) could not separate different vegetation types in eastern Australia using Landsat MSS data. Clearly, the detection of vegetation change in these situations would be difficult without more satisfactory separation of vegetation types.

Utilizing all the spectral information available from Landsat data may not always be advantageous. For instance, under rangeland conditions where vegetation is very sparse (usually less than 25% cover) and there is a constant soil color, there is a high spectral redundancy and four-band Landsat MSS data effectively become two band, that is, red and near-IR (Graetz, 1987). Pixels are then spatially integrated averages of varying proportions of separate landscape components, such as bare soil, litter, and shrubs. While these components have distinct spectral signatures when analyzed separately (Figure 10.3), classification of mixtures based on spectral classes may not be possible (e.g., Graetz et al., 1983). In such cases, simple models to relate satellite data to measured proportions on the ground may be more appropriate (Pech et al., 1986a, 1986b; Pech and

Davis, 1987; Graetz, 1987). Detection of differences is thus possible by estimating the proportion (or percent cover) of each component in each pixel. More subtle differences (e.g., in shrub cover) may therefore be detectible using these methods than would be possible using classification techniques.

Can We Detect Vegetation Change?

Remote sensing has been used successfully to detect large-scale vegetation changes brought about through deforestation (Malingreau and Tucker, 1988) or forest fires (Malingreau et al., 1985). Both studies used AVHRR data, the first utilizing only the 3.5- to 3.9-μm thermal channel, and the second utilizing the NDVI to recognize the incidence of fires (Figure 10.4). Such techniques provide a series of images or index values that can be compared visually or used to derive statistics on extent of change. Further examples include the studies of Tucker et al. (1985b), who investigated the changes in vegetation in sub-Saharan Africa over a number of years. In many cases, a straightforward estimate of changes in leaf area index (LAI) may provide the information required to document vegetation change and concomitant changes in ecosystem processes. Further information on changes in vegetation characteristics are also possible, however.

Detailed investigations of particular components of reflectance curves

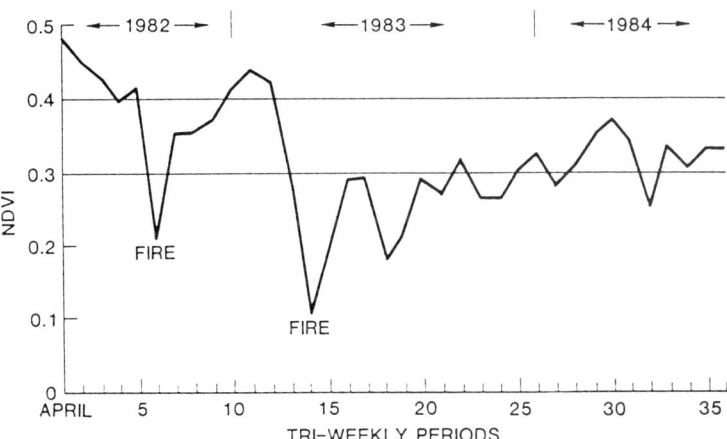

Figure 10.4. Changes in the normalized difference vegetation index (NDVI) of tropical forest in East Kalimantan, as derived from AVHRR data, showing the incidence of forest fires. [Reprinted by permission of the publisher from Malingreau et al. (1985) *Ambio* 14:314–321. Copyright 1985 by Swedish Academy of Sciences.]

(as in Figure 10.2) are now being used to detect forest decline suspected to be the result of air pollution. Airborne spectral measurements have indicated that forest damage can be characterized by changes in the visible, near-, and short-wave IR, and by a shift of the "red edge" toward shorter wavelengths (Rock et al., 1988; Herrmann et al., 1988). Short-wave IR data can also be used to detect various levels of water stress in vegetation (Rock et al., 1986). There is thus potential for assessing the health of vegetation remotely and establishing areas where existing vegetation may be under stress and where future change is likely.

Methods of statistical analysis have been developed to investigate localized vegetation changes. Jensen and Toll (1982) and Sing (1983) have presented a method of change detection by image differencing in which spatially registered images are subtracted to produce a further image that represents the change between the two images. This method involves determining a threshold boundary between change and no-change pixels based on comparison of the subtracted pixel value with a statistical distribution. Pickup (this volume) has discussed problems associated with differencing techniques. Wickware and Howarth (1981) and Christensen et al. (1988) derived independent classifications of data from two recording dates and then evaluated these for change using a postclassification crosstabulation technique. Other methods have been explored by Weismiller et al. (1977), although they found the postclassification technique the most successful. There are a number of problems associated with the use of these techniques. These relate, first, to the separation of true vegetation change from phenological change or change in atmospheric conditions between recording dates, and, second, to problems of misclassification and registration errors (Wickware and Howarth, 1981). Registration errors can be minimized (Christensen et al., 1988), but classification problems remain.

Principal components analysis and transformation have been used to enhance regions of change in multitemporal Landsat data and to remove some of the problems of separating real change from background radiation and atmospheric differences (Byrne et al., 1980; Richards, 1984, 1986; Fung and LeDrew, 1987). Principal components analysis of data from two dates results in the gross differences between dates associated with overall radiation and atmospheric changes being displayed along the first principal component and changes in land cover being displayed along the second and higher components (Figure 10.5). Subsequent pixel classification on the basis of higher-order components can then be used to provide maps of vegetation change between the two recording dates (e.g., Richards, 1984). In this way, confusion of spectral signatures between dynamic and static cover types can be avoided. The technique is valuable for dealing with small areas, but is scene dependent (Fung and LeDrew, 1987).

Image differencing, principal components analysis, and other spectral techniques look at pixels without reference to their neighbors. Vegetation

10. Remote Sensing of Spatial and Temporal Dynamics of Vegetation   213

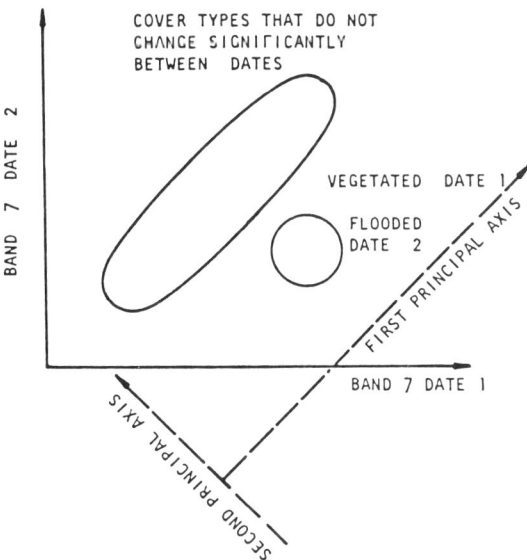

Figure 10.5. Use of principal components analysis to detect change between Landsat scenes from different times. Cover types that do not change significantly are spread along axis 1 of the analysis, while change pixels are separated on the second- and higher-order axes. [Reprinted by permission of the publisher from Richards (1984) *Remote Sens. Environ.* 16:35–46. Copyright 1984 by Elsevier Science Publishing Co., Inc.]

change can often involve large changes in spatial pattern but small changes in the spectral response of individual pixels. Little use has so far been made of spatial variability or texture in assessing vegetation change. Recently, however, Foran (1988) has found it useful in arid landscapes, especially where spectral change is small. Pickup and Chewings (1986, 1988) have examined the spatial changes in arid vegetation associated with erosion. Frank (1984) has also used local variance analysis as a classification aid.

Change detection requires data collected over a certain time period. Often vegetation change is relatively gradual, and the remote sensing record is relatively short. Ecology has suffered from a paucity of long-term studies in the past (e.g., Taylor, 1934; Hobbs and Mooney, 1989), and it is essential that we now make sure that adequate long-term data are collected both on the ground and remotely. In this respect, it is important that data continue to be collected in a standard manner and in a format such that they remain compatible with earlier data. This is not necessarily as easy as it sounds and, for instance, although the Landsat series have been in operation since 1972, the satellites have not provided exactly compatible measurements (e.g., Chaudhury, 1985). The usefulness of current technology for monitoring purposes has also been questioned recently because of the lack of reliable imagery, especially in tropical areas where cloud cover is frequent (Currey et al., 1987). Similarly, the Earth System Sciences Committee (1988) has stated: "The database for documenting even present patterns (of land use) on a global basis, let alone future changes, is totally inadequate, both because of inadequate or distorted coverage and because of a lack of uniform standards for reporting." They suggest that global coverage is probably not possible because of data managment and processing problems. They suggest, instead the establishment of a sample of individual scenes that can be revisited at intervals and subjected to a consistent classification scheme. Thus although we must seek advances in instrumentation, we must also seek long-term consistency in data collection. The choice of scenes to be monitored should be based on an initial screening to detect areas where rapid change is likely, either through land-use changes or through changes in global climate.

Conclusions

The major problems involved with the detection of vegetation change using remote sensing can be summarized as follows:

1. In what aspects of vegetation change are we interested? Short-term phenological and interannual changes have to be taken into account before longer-term changes can be detected. Similarly, large-scale changes require different techniques of detection than changes at smaller spatial scales. The requirement for finer-resolution data has to be traded off against the increased data storage and handling required.
2. How can we best distinguish different types of vegetation at any given time? The existing technology provides a basis for assessing different aspects of the vegetation, ranging from overall productivity to levels of water or pollution stress. We also have methods available that can classify different types of vegetation present at a broad structural level. We need to be able to do this on a global scale in order to obtain reliable estimates of biome extent that will act as a baseline for change detection. At a finer scale, sensor resolution must be matched against the spatial scale of the vegetation mosiac under examination. Mixed pixels can become a problem, especially in sparse vegetation types.
3. How can we detect change from one time to the next? Change detection involves comparisons of data from different times. This requires, first, the availability of compatible data sets collected over a long enough time period, and, second, the availability of methods of comparison. Our success in detecting future vegetation change therefore depends on our ability to maintain data continuity and to develop methods of processing and interpreting the data once we have them. The development of adequate statistical techniques for detecting change is an important priority. A further priority is the establishment of selected areas for long-term monitoring; these should include areas where change is most likely to occur.

References

Adamson, D.A., and Fox, M.D. (1982). Change in Australasian vegetation since European settlement. pp. 109–146. In J.M.B. Smith (ed.), *A History of Australasian Vegetation*. McGraw-Hill, Sydney, Australia.

Adomeit, E.M., Jupp, D.L.B., Margules, C., and Mayo, K.K. (1981). The separation of traditionally mapped land cover classes by LANDSAT data. pp. 150–165. In A.N. Gillison and D.J. Anderson (eds.), *Vegetation Classification in Australia*. Australian Nat. Univ. Press, Canberra.

Ayyad, M.A., (1981). Soil-vegetation-atmosphere interactions. pp. 9–31. In D.W. Goodall, R.A. Perry, and K.M.W. Hones (ed.), *Arid-land Ecosystems: Structure, Function and Management*, Vol. 2. Cambridge Univ. Press, Cambridge, England.

Bell, D.T., and Stephens, L.J. (1984). Seasonality and phenology of kwongan species. pp. 205–226. In J.S. Pate, and J.S. Beard (eds.), *Kwongan Plant Life of the Sandplain*. Univ. of Western Australia Press, Nedlands.

Bolin, B., Doos, B.R., Jager, J., and Warrik, R.A. (eds.) (1987). *The Greenhouse Effect, Climatic Change, and Ecosystems. SCOPE 29*. Wiley, NY.

Botkin, D.B., Estes, J.E., MacDonald, R.M., and Wilson, M.V. (1984). Studying the earth's vegetation from space. *BioScience* 34:508–514.

Broeker, W.S. (1987). Unpleasant surprises in the greenhouse? *Nature* 328:123–126.

Byrne, G.R., Crapper, P.F., and Mayo, K.K. (1980). Monitoring land cover changes by principal components analysis of multitemporal Landsat data. *Remote Sens. Envir.* 10:175–184.

Chaudhury, M.U. (1985). Landsat series: technical properties and application to vegetation studies. pp. 23–29. In *Remote Sensing in Vegetation Studies*. ESCAP-BIOTROP, Bogor, Indonesia.

Christensen, E.J., Jensen, J.R., Ramsey, E.W., and Mackey, H.E. (1988). Aircraft MSS data registration and vegetation classification for wetland change detection. *Int. J. Remote Sens.* 9:23–38.

Clark, C.A., Cate, R.B., Trenchard, M.H., Boatright, J.A., and Bizzell, R.M. (1986). Mapping and classifying large ecological units. *BioScience* 36:476–478.

Committee on Planetary Biology. (1986). Remote Sensing of the Biosphere. Nat. Academy Press, Washington, DC.

Connell, J.H., and Slatyer, R.O. (1977). Mechanisms of succession in natural communities and their role in community stability and organisation. *Amer. Nat.* 111:1119–1144.

Currey, B., Fraser, A.S., and Bardsley, K.L. (1987). How useful is Landsat monitoring? *Nature* 328:587–589.

Davis, M.B. (1986). Climatic instability, time lags, and community disequilibrium. pp. 268–284. In J. Diamond and T.J. Case (eds.), *Community Ecology*. Harper and Row, NY.

Delcourt, P.A., and Delcourt, H.R. (1987). *Long-Term Forest Dynamics of the Temperate Zone. A Case Study of Late-Quaternary Forests in Eastern North America*. Springer-Verlag, NY.

Dregne, H.E., and Tucker, C.J. (1988). Green biomass and rainfall in semi-arid sub-Saharan Africa. *J. Arid Envir.* 15:245–252.

Earth System Sciences Committee (1988). *Earth System Science, a Closer View*. NASA, Washington, DC.

Foran, B.D. (1988). Detection of yearly cover change with Landsat MSS on pastoral landscapes in Central Australia. *Remote Sens. Envir.* 23:333–350.

Frank, T.D. (1984). The effect of change in vegetation cover and erosion patterns on albedo and texture of Landsat images in a semi-arid environment. *Ann. Assoc. Amer. Geogr.* 74:393–407.

Fung, T., and LeDrew, E. (1987). Application of principal components analysis to change detection. *Photogramm. Eng. Remote Sens.* 12:1649–1658.

Goetz, A.F.H., Rock, B.N., and Rowan, L.C. (1983). Remote sensing for exploration: an overview. *Econ. Geol.* 78:573–590.

Goward, S.N., Tucker, C.J., and Dye, D.G. (1985). North American vegetation patterns observed with the NOAA-7 advanced very high resolution radiometer. *Vegetatio* 64:3–14.

Graetz, R.D. (1987). Satellite remote sensing of Australian rangelands. *Remote Sens. Environ.* 23:313–331.

Graetz, R.D., Gentle, M.R., Pech, R.P., O'Callaghan, J.R., and Drewien, G. (1983). The application of Landsat image data to rangeland assessment and monitoring: an example from South Australia. *Aust. Rangel. J.* 5:63–73.

Graetz, R.D., Walker, B.H., and Walker, P.A. (1988). The consequences of climatic change for seventy percent of Australia. pp. 399–420. In G.I. Pearman (ed.), *Greenhouse. Planning for Climatic Change*. CSIRO, Melbourne, Australia.

Gray, A.J., Crawley, M.J. and Edwards, P.J. (1986). *Colonisation, Succession and Stability*. Blackwell, Oxford, England.

Green, G.M. (1986). Use of SIR-A and Landsat MSS data in mapping shrub and intershrub vegetation at Koonamore, South Australia. *Photogramm. Eng. Remote Sens.* 52:659–670.

Griffin, G.F., and Friedel, M.H. (1985). Discontinuous change in central Australia: some major implications of ecological events for land management. *J. Arid Environ.* 9:63–82.

Hamburg, S.P., and Cogbill, C.V. (1988). Historical decline of red spruce populations and climatic warming. *Nature* 331:428–430.

Herrmann, K., Rock, B.N., Ammer, U., and Paley, H.N. (1988). Preliminary assessment of airborne imaging spectrometer and airborne thematic mapper data acquired for forest decline areas in the Federal Republic of Germany. *Remote Sens. Environ.* 24:129–149.

Hobbs, R.J., and Hopkins, A.J.M. (1990). From frontier to fragments: European impact on Australia's vegetation. *Proc. Ecol. Soc. Aust.* 16 (in press).

Hobbs, R.J., and Mooney, H.A. (1989). Effects of episodic events on Mediterranean-climate ecosystems. In F. di Castri, C. Floret, S. Rambal, and J. Roy (eds.), *Timescales of Water Stress Response of Mediterranean Biota* (in press).

Hobbs, R.J., Wallace, J.F., and Campbell, N.A. (1989). Classification of vegetation in the Western Australian wheatbelt using Landsat MSS data. *Vegetatio* 80:91–105.

Horler, D.N.H., Barber, J., and Barringer, A.R. (1980). Effects of heavy metals on the absorbance and reflectance spectra of plants. *Int. J. Remote Sens.* 1:121–136.

Horler, D.N.H., Dockray, M., Barber, J., and Barringer, A.R. (1983). Red edge measurements for remotely sensing plant chlorophyll content. *Adv. Space Res.* 3:273–277.

Huston, M., and Smith, T. (1987). Plant succession: life history and competition. *Amer. Nat.* 130:168–198.

Jenson, J.R., and Toll, D.L. (1982). Detecting residential land use development at the urban fringe. *Photogramm. Eng. Remote Sens.* 48:629–643.

Johnson, A.H., and Siccama, T.C. (1984). Decline of red spruce in the northern Appalachians: Assessing the possible role of acid deposition. *Tappi J.* 67:68–72.

Jupp, D.L.B., Walker, J., and Penridge, L.K. (1986). Interpretation of vegetation structure in Landsat MSS imagery: A case study in disturbed semi-arid eucalypt woodlands. Part 2. Model-based analysis. *J. Envir. Manag.* 23:35–37.

Malingreau, J.P., Stephens, G., and Fellows, L. (1985). Remote sensing of forest fires: Kalimantan and North Borneo in 1982–83. *Ambio* 14:314–321.

Malingreau, J.P., and Tucker, C.J. (1988). Large-scale deforestation in the southeastern Amazon Basin of Brazil. *Ambio* 17:49–55.

Mooney, H.A. (1988). Ecologists and the global change program. *Trends Ecol. Evol.* 3:4–5.

Mooney, H.A., Hobbs, R.J., Gorham, J., and Williams, K. (1986). Biomass accumulation and resource utilisation in co-occurring grassland annuals. *Oecologia* (Berlin) 70:555–558.

Morton, A.J. (1986). Moorland plant community recognition using Landsat MSS data. Remote Sens. Environ. 20:291–298.

National Academy Press (1986). *Global Change in the Geosphere-Biosphere: Initial Priorities for and IGBP.* Nat. Acad. Press, Washington, DC.

Pech, R.P., and Davis, A.W. (1987). Reflectance modeling of semiarid woodlands. *Remote Sens. Environ.* 23:365–377.

Pech, R.P., Davis, A.W., and Graetz, R.D. (1986a). Reflectance modeling and the derivation of vegetation indices for an Australian semi-arid shrubland. *Int. J. Remote Sens.* 7:389–403.

Pech, R.P., Davis, A.W., Lamcraft, R.R., and Graetz, R.D. (1986b). Calibration of Landsat data for sparsely vegetated arid rangelands. *Int. J. Remote Sens.* 8:1829–1850.

Perry, C.R., Jr. and Lanternschlager, L.F. (1984). Functional equivalence of spectral vegetation indices. *Remote Sens. Envir.* 14:169–182.

Pickett, S.T.A., Collins, S.L., and Armesto, J.J. (1987). Models, mechanisms and pathways of succession. *Bot. Rev.* 53:335–371.

Pickup, G., and Chewings, V.H. (1986). Random field modelling of spatial variations in erosion and deposition in flat alluvial landscapes in arid central Australia. *Ecol Model.* 33:269–296.

Pickup, G., and Chewings, V.H. (1988). Forecasting patterns of soil erosion in arid lands from Landsat MSS data. *Int. J. Remote Sens.* 9:69–84.

Pitt, M.D., and Heady, H.F. (1978). Responses of annual vegetation to temperature and rainfall patterns in northern California. *Ecology* 59:336–350.

Postel, S. (1984). Acid pollution, acid rain and the future of forests. *Worldwatch Paper* 58:1–22.

Richards, J.A. (1984). Thematic mapping from multitemporal image data using the principal components transformation. *Remote Sens. Environ.* 16:35–46.

Richards, J.A. (1986). *Remote Sensing Digital Image Analysis.* Springer-Verlag, Berlin.

Richards, J.A., and Kelly, D.J. (1984). On the concept of spectral class. *Int. J. Remote Sens.* 5:987–991.

Rock, B.N., Hohsizaki, T., and Miller, J.R. (1988). Comparison of in situ and airborne spectral measurements of the blue shift associated with forest decline. *Remote Sens. Environ.* 24:109–127.

Rock, B.N., Vogelmann, J.E., Williams, D.L., Vogelmann, A.F., and Hoshizaki, T. (1986). Remote detection of forest damage. *BioScience* 36:439–445.

Rohde, W.G., and Olson, C.E. Jr. (1971). Estimating foliar moisture content from infrared reflectance data. pp. 144–164. In *Third Biennial Workshop: Color Aerial Photography in the Plant Sciences and Related Fields.* Amer. Soc. Photogramm., Falls Church, VA.

Roller, N.E.G., and Colwell J.E. (1986). Course-resolution satellite data for ecological surveys. *BioScience* 36:468–475.

Runkle, J.R. (1985). Disturbance regimes in temperate forests. pp. 17–33. In S.T.A. Pickett and P.S. White (eds), *The Ecology of Natural Disturbance and Patch Dynamics.* Academic Press, NY.

Saunders, D.A., Arnold, G.W., Burbidge, A.A., and Hopkins, A.J.M. (Eds.) (1987). *Nature Conservation: The Role of Remnants of Native Vegetation.* Surrey Beatty, Sydney, Australia.

Saxon, E.C., and Dudzinski, M.L. (1984). Biological survey and reserve design by Landsat mapped ecolines—A catastrophe theory approach. *Aust. J. Ecol.* 9:117–123.

Schutt, P., and Cowling, E.B. (1985). Waldsterben, a general decline: symptoms, development. *Plant. Dis.* 69:548–558.

Shugart, H.H., and West, D.C. (1981). Long-term dynamics of forest ecosystems. *Amer. Sci.* 69:647–652.

Silvertown, J. (1980). The dynamics of a grassland ecosystem: botanical equilibrium in the park grass experiment. *J. Appl. Ecol.* 17:491–504.

Sing, A. (1983). Univariate image-differencing for forest change detection with Landsat. pp. 154–160. In *Remote Sensing for Rangeland Monitoring and Management.* Remote Sensing Society, Reading, England.

Sing, A. (1987). Spectral separability of tropical forest classes. *Int. J. Remote Sens.* 8:971–979.

Taylor, W.P. (1934). Significance of extreme or intermittent conditions in distribution of species and management of natural resources, with a restatement of Leibig's law of minimum. *Ecology* 15:374–379.

Tilman, D. (1982). *Resource Competition and Community Structure.* Princeton Univ. Press, Princeton, NJ.

Tilman, D. (1988). *Plant Strategies and the Dynamics and Structure of Plant Communities.* Princeton Univ. Press, Princeton, NJ.

Townshend, J.R.G., and Justice, C.O. (1986). Analysis of the dynamics of African vegetation using the normalized difference vegetation index. *Int. J. Remote Sens.* 7:1435–1445.

Townshend, J.R.G., and Tucker, C.J. (1984). Objective assessment of advanced very high resolution radiometer data for land cover mapping. *Int. J. Remote Sens.* 5:497–504.

Tucker, C.J. (1980). Remote sensing of leaf water content in the near infrared. *Remote Sens. Environ.* 10:23–32.

Tucker, C.J. (1986). Maximum normalized difference vegetation index images for sub-Saharan Africa for 1983–1985. *Int. J. Remote Sens.* 7:1383–1384.

Tucker, C.J., Justice, C.O., and Prince, S.D. (1986). Monitoring the grasslands of the Sahel 1984–1985. *Int. J. Remote Sens.* 7:1715–1731.

Tucker, C.J., Townshend, J.R.G., and Goff, T.E. (1985a). African land-cover classification using satellite data. *Science* 227:369–375.

Tucker, C.J., Vanpraet, C.L., Sharman, M.J., and Van Ittersum, G. (1985b) Satellite remote sensing of total herbaceous biomass production in the Senegalese Sahel: 1980–1984. *Remote Sens. Envir.* 17:233–249.

Ustin, S.L., Adams, J.B., Elvidge, C.D., Rejmanek, M., Rock, B.N., Smith, M.O., Thomas, R.W., and Woodward, R.A. (1986). Thematic mapper studies of semiarid shrub communities. *BioScience* 36:446–456.

Walker, B.H. (1979). Management principles for semi-arid ecosystems. pp. 379–388. In B.H. Walker (ed.), *Management of Semi-arid Ecosystems.* Elsevier, Amsterdam, Netherlands.

Walker, J., Jupp, D.L.B., Penridge, L.K., and Tian, G. (1986). Interpretation of vegetation structure in Landsat MSS imagery: A case study in disturbed semi-arid eucalypt woodlands. Part 1. Field data analysis. *J. Envir. Manag.* 23:19–33.

Walsh, S.J. (1987). Variability of Landsat MSS spectral responses of forests in relation to stand and site characteristics. *Int. J. Remote Sens.* 8:1289–1299.

Wardley, N.W., Milton, E.J., and Hill, C.T. (1987). Remote sensing of structurally

complex semi-natural vegetation—an example from heathland. *Int. J. Remote Sens.* 8:31–42.
Warren, P.L., and Huchinson, C.F. (1984). Indicators of rangeland change and their potential for remote sensing. *J. Arid Envir.* 7:107–126.
Watt, A.S. (1981). A comparison of grazed and ungrazed grassland in East Anglian Breckland. *J. Ecol.* 69:509–536.
Weaver, R.E. (1987). Spectral separation of moorland vegetation in airborne Thematic Mapper data. *Int. J. Remote Sens.* 8:43–55.
Weismiller, R.A., Kristof, S.J., Scholtz, D.K., Anuta, P.E., and Momin, S.A. (1977). Change detection in coastal zone environments. *Photogramm. Eng. Remote Sens.* 43:1533–1539.
West, D.C., Shugart, H.H., and Botkin, D.B. (eds.) (1981). *Forest Succession Concepts and Application*. Springer-Verlag, NY.
Wickware, G.M., and Howarth, P.J. (1981). Change detection in the Peace-Athabasca Delta using digital Landsat data. *Remote Sens. Envir.* 11:9–25.
Williams, K., Hobbs, R.J., and Hamburg, S.P. (1987). Invasion of annual grassland in northern California by *Baccharis pilularis* ssp. *consanguinea*. *Oecologia* (Berlin) 72:461–465.
Yates, H., Strong, A., McGinnis, D., Jr., and Tarpley, D. (1986). Terrestrial observations from NOAA operational satellites. *Science* 231:463–470.
Yool, S.R., Star, J.L., Estes, J.E., Botkin, D.B., Eckhardt, D.W. and Davis, F.W. (1986). Performance analysis of image processing algorithms for classification of natural vegetation in the mountains of southern California. *Int. J. Remote Sens.* 7:683–702.

11. Remote Sensing of Landscape Processes

Geoff Pickup

Landforms, through their effects on climate, hydrology, soils, and vegetation, determine much of the spatial variability in biosphere functioning. Landform characteristics are the product of geological structure coupled with tectonic and climatic history and, at the larger spatial scales, change only slowly (Table 11.1). They are, therefore, usually treated as constant over the time intervals during which ecosystems experience and respond to change. At smaller spatial scales, the pace of change accelerates and the surficial characteristics of a landform may be modified quite rapidly. This produces a mosaic of surfaces that are either gaining or losing sediment or remaining stable. These surfaces have different ages, sediment characteristics, and disturbance regimes. They may also respond to shifts in climate or other controlling variables at rates similar to those of the ecosystems that occupy them. Landscape and ecosystem processes may then become linked either directly or through feedback mechanisms.

The most important of the linkages between landscape and ecosystem occurs via the soil. Even small changes in landform behavior can involve significant modification or redistribution of the soil layer. This affects the pattern of infiltration, moisture storage and runoff, as well as the distribution of nutrients. Significant changes in vegetation may then take place that feed back to the landform process, intensifying or moderating it.

Changes in the surficial characteristics of landforms will be an important consequence of the man-induced global climatic shifts that have been fore-

Table 11.1. Classification of Geomorphic Features by Scale [Modified after Baker (1986).]

Spatial Scale (km^2)	Characteristic Units	Creation Time (Years)	Persistence Time (Years)
10^7	Continents, ocean basins	10^7 to 10^9	10^8 to 10^9
10^6	Physiographic provinces, shields	10^7	10^8
10^4	Medium-scale tectonic units (sedimentary basins, mountain massifs, domal uplifts)	10^6 to 10^7	10^7 to 10^8
10^3	Smaller tectonic units (fault blocks, troughs, sedimentary subbasins, individual mountain ranges)	10^6 to 10^7	10^7
10^2 to 10^3	Large scale erosion/depositional units (deltas, major valleys, piedmonts)	10^5 to 10^6	10^6
10 to 10^2	Medium-scale depositional units or landforms (ridges, terraces, dune fields)	10^3 to 10^4	10^4 to 10^5
10^{-1} to 10	Larger geomorphic process units (hill slopes, floodplains)	10 to 10^3	10^3 to 10^5
10^{-5} to 10^{-1}	Medium-scale geomorphic process units (gullies, fans, areas of sheet erosion, stream channels)	10^{-1} to 10^2	10 to 10^3

cast (e.g., Allison and Peck, 1987; Wasson and Clark, 1987). The rate and extent of these changes will not be uniform so their effect on ecosystem processes will be variable both spatially and temporally. There will also be significant economic consequences as snowpack size and duration shift, as patterns of soil erosion change, as new areas are affected by silting, as river flood plains become more or less prone to inundation, and as slopes in areas with increased rainfall become less stable.

This chapter descibes how remote sensing technology may be used to describe, explain, and predict some of the geomorphic changes that may affect biological activity at a time scale of tens to hundreds of years. It begins with a consideration of the different types of remotely sensed data and how those data can be related to landscape properties. Few of these properties can be measured directly but remotely sensed data can often provide useful surrogate information on both properties and processes. Deriving these surrogates, however, may require consideration of spatial or temporal variability in the data rather than the single-scene approach now common. Landscape change may be sporadic and spatially discontinuous, making it difficult to understand. The chapter continues with a discussion of some of the models of change that are widely accepted by geomorphologists. These are equilibrium-based and describe the relationships between landforms and the fluctuating climatic conditions that control their properties. Once an understanding of landscape change is available, it may become possible to predict future change. The chapter concludes with a discussion of progress in predicting soil erosion, which is currently one of the most active areas of research attempting to harness remote sensing to the prediction of landscape change. Consideration is also given to the problem of scale because, while most our knowledge of these processes is point based, predictions of future change need to be at the regional level to be useful. Remote sensing is gradually overcoming the first barrier to regional prediction, namely, that of data. The next barrier is the development of process models that operate at the regional scale.

Remote Sensing of Landscape Properties

The Remote Sensing Process

Remote sensing involves measurement in the electromagnetic spectrum using instruments on satellites or aircraft to characterize the land, atmosphere, or ocean. The most common form of remote sensing is aerial photography, whereby information on reflected light in the visible range of the spectrum is stored on film. It is also possible to acquire information in the near-infrared (NIR) using special types of film.

Another common form of remote sensing uses multispectral scanners mounted on satellites that operate in the visible and NIR range and collect information in a series of spectral windows. Multispectral scanners with a 57 by 79 m resolution have been in use since 1972 on the Landsat series of satellites. Higher-resolution data (30 by 30 m) extending into the middle infrared (MIR) are becoming available from the Landsat Thematic Mapper (TM) while the (SPOT) HRV offers a 20 by 20 m resolution in spectral windows similar to those of the Landsat Multi-Spectral Scanner (MSS) and 10 m in panchromatic mode.

Multispectral scanners do not provide information on landscape characteristics per se. However, because reflectance in different ranges in the spectrum differs with such factors as soil and rock color and mineralogy or vegetation type and phenology, terrestrial radiance is frequently a surrogate measure of landscape properties. Ground features may then be characterized and discriminated by their combined response in several spectral windows. The relationship between spectral response and ground characteristics is made complex, however, by the fact that the individual pixels contain a mixture of surfaces. Relating a spectral "signature" to a landform characteristic, therefore, usually requires ground verification or skilled interpretation.

Remote sensing is not restricted to the visible, NIR, and MIR regions of the spectrum. Landsat TM and a number of meteorological satellites also measure the thermal energy emitted by the earth's surface, which is a function of surface temperature. Thermal data have been used in many applications, including estimation of soil moisture, geological mapping, and identification of landforms buried by thin desert sand sheets. Their principal use, however, is meteorological.

So far, only limited use has been made of microwave sensors. These respond to the dielectric properties of the earth's surface, which are strongly related to soil moisture content (e.g., Owe et al., 1988). Active microwave (radar) and passive microwave instruments have been used on both satellites and aircraft. At present, the available passive microwave data mostly come from low-resolution meteorological satellites and are more suitable for climatological use than for landform studies. Higher-resolution microwave sensors proposed for satellites of the 1990s (e.g., NASA, 1987a) may change this and, when used with suitable hydrologic models, could greatly improve estimates of runoff for the prediction of erosion at the drainage basin scale.

Active microwave data in the form of synthetic aperture radar (SAR) will become more significant in the 1990s with the launch of the ERS-1 and Radarsat satellites. Multiband SAR shows great potential for characterizing both terrain and vegetation through its roughness properties (NASA, 1987b). Radar altimeters perform best over water where there is less variability in backscatter, but may also provide topographic data over land (NASA, 1987c). Indeed, successful topographic mapping has already been carried out using stereo radar data acquired from aircraft (Mercer and Kirby, 1987).

There is growing interest in the use of aircraft-mounted scanners for high-resolution remote sensing. At present, however, much of the work is experimental and this facility is not widely available. At the simplest and cheapest level, video cameras may be used as sensors in the visible and NIR range of the spectrum with the collected information stored on videotape for subsequent digitizing (Everitt et al., 1987). Thermal and MIR cameras are also becoming available although sensor lag is a problem. At a more sophisticated level, scanners and high-resolution spectrometers are in use. These instruments offer a range of spectral windows that may be changed at will. Laser altimeters mounted in aircraft can also provide high-resolution information on elevation. Aircraft charter and instrument rental are, however, sufficiently expensive to deny these facilities to many.

Determining Landscape Properties

Landforms are routinely mapped from air photographs, producing subjective results that vary with the skill of the interpreter and the type and quality of the photography. Remotely sensed data from satellites can be processed digitally to produce repeatable results quickly over very large areas. There is, however, still a subjective element in that the interpreter has to identify or provide information on the spectral characteristics of landscapes. Satellite-based data also have the disadvantage of being collected at a much lower spatial resolution than conventional air photography and so some features may be obscured. Remote sensing satellites provide three types of information that can be used to assess landscape behavior:

1. The radiance or emittance of the earth's surface on a pixel-by-pixel basis.
2. The spatial variability of radiance or emittance from which spatial patterns may be detected.
3. Patterns of change through time if information is acquired from several passes of the satellite.

Comparatively little use is made of spatial and temporal variability because of the added cost and complexity of processing. This is unfortunate because spatial and temporal data contain information that greatly increase the potential of remote sensing in landform studies (e.g., Pickup and Chewings, 1988a).

Information on landscape properties derived by remote sensing may be used in three ways:

- To describe the surface properties of landforms.
- To derive information on variables such as slope or vegetation cover that influence the rate at which landform processes operate.
- To describe the state of a landscape in terms of some evolutionary process such as the extent of erosion.

A range of landform surface properties can be potentially determined from remotely sensed data. For example, many minerals have distinctive spectral signatures, although the ability to recognize and differentiate between those signatures is closely related to the number of spectral bands available (e.g., Evans, 1988). It is also possible to identify differences in surface roughness using synthetic aperture radar data (NASA, 1987b). Sometimes, surface properties provide a clear unambiguous signature indicating dominance of a particular characteristic. More often, however, composite signatures occur, suggesting the presence of more complex structures. It may then become necessary to disentangle particular landscape properties using statistical techniques. The surface characteristics are normally used to infer the operation of a set of geomorphic processes. Occasionally, however, the frequent coverage provided by remote sensing satellites makes it possible to observe those processes in action, particularly if they operate at a large scale. These opportunities are rare because when things are happening in fluvial environments, for example, the landscape is usually obscured by cloud. This problem is now being overcome because sensors in the microwave region have the ability to penetrate cloud.

Estimation of vegetation cover from satellite data is a common operation (e.g., Foran, 1988) but the derivation of slope has been restricted, for practical purposes, by the coarse resolution of satellite-mounted sensors and lack of stereo cover. SPOT HRV data overcome this problem and recent tests indicate that contour lines with a 20-m interval can be produced routinely (Rodriguez et al., 1988). There is even potential for the automatic production of digital terrain models by correlation with accuracies in the 10-m class.

Statistical measures describing landscape characteristics abound in geomorphology but have not been widely adapted to remote sensing data. Differences in terrain attributed to geology have been described by texture measures (e.g., Weska et al., 1976). Erosion severity in arid areas has been linked to scene variance and spatial autocorrelation (Frank, 1984; Pickup and Chewings, 1988a). The field of fractals as landscape descriptors also offers exciting possibilities (e.g., Mark and Aronson, 1984).

Process Domains

Most studies of landform behavior that use remotely sensed data involve the transformation, enhancement, and classification of imagery into landscape units. Visual interpretation and manual subdivision frequently produce better results in this process than digital methods because landscape units show a great deal of internal diversity. This diversity is rarely described successfully by texture measures because it varies in alignment and occurs at a range of spatial frequencies.

One approach to landscape classification is to divide an area into process domains (Millington et al., 1987). These are discrete areas or types of landform with similar characteristics that share a common history and whose behavior is shaped by a particular set of processes or events of a certain frequency. The concept is useful because it allows geomorphologists to break up complex landscapes into simpler units for which behavioral models can be derived.

An example of the process domain approach is to divide a river valley into channels, floodplains, and hill slopes. River channels change relatively quickly and, in some climates, are adjusted to events of moderate magnitude and frequency (Wolman and Miller, 1960; Wolman and Gerson, 1978). Floodplains change more slowly and may take thousands of years to adjust because the volume of sediment required to change them significantly is greater than the short-term supply. Some floodplains may respond only to catastrophic events (Nanson and Erskine, 1988) and may remain out of equilibrium with the current climate for very long periods. Hill slopes may not change significantly for thousands of years. A landscape thus might be thought of as an assemblage of landforms adjusted to events of different frequency and displaying different amounts of lag in their response to environmental change. Those parts of the landscape that, by virtue of their location, can be modified only by rare events will display strong lag effects and may never fully adjust to climatic change. Other parts of the landscape that act as conduits for water and sediment will respond to change more quickly and, while they can be modified radically by large events, on average will be more closely adjusted to the moderate ones. These will show the effects of climatic change first.

The division of landscapes into a set of process domains is usually more complex than this and raises the question of scale. Process geomorphologists have traditionally measured landform behavior for short periods at a few points in the landscape or at many points in restricted locations because of the logistic difficulties of acquiring data for larger areas. Their models and outlook are biased toward short-term microscale processes and they lack the synoptic view. There are, however, meso- and macroscale processes at work, and at these scales, the results of point-based and localized studies frequently provide more noise than signal when it comes to determining what is going on in the landscape. They have also made it difficult for geomorphologists to recognize, comprehend, and accept the effects of macroscale processes.

Remote sensing is particularly useful in providing information on large-scale processes and allowing geomorphologists to "think big." Indeed, where geomorphologists have been unable to carry out process studies directly and have had to rely on remote sensing alone for their data, landforms tend to be explained in terms of processes operating at the regional level. A case in point is Baker's (1983) use of cataclysmic flooding to explain Martian channel systems up to 100 km wide and 2,000 km in length. Terrestrial macroprocesses have also left their traces. The Lake Missoula floods in the Columbia River basin in the late Pleistocene involved discharges of about 21 million $m^3 s^{-1}$ and flow depths of 100 m (Baker, 1978) and their effects even extended to the abyssal sea floor. The magnitude of these events and the landforms they created can be appreciated only from satellite images. Less spectacular but still of regional importance are the huge changes that occurred in the channels of the Riverine Plain in southeastern Australia (Schumm, 1968) with the onset and retreat of glaciation and its associated effects on discharge. The impact of enormous floods on arid landscapes has also been detected on the floodplains and alluvial fans of central Australia.

Remote sensing allows geomorphologists to change scales either by data resampling or by the use of sensors with differing resolution. The behavior of a whole system often is marked by features that are not apparent when only a part of it is studied. These features may show the effects of catastrophic events or of more gradual underlying trends produced by tectonic activity or underlying geology. Changes in scale allow a hierarchy of process domains to be identified, with the effects of mesoscale processes embedded in and influenced by events at the macroscale. Few geomorphologists have taken advantage of these features but once they do, some unexpected landscape-generating events and processes are likely to be revealed.

Remote Sensing of Landscape Change

Conceptual Models of Landscape Behavior

Landscapes are made up of assemblages of individual landforms. Thus when a landscape responds to a change in a controlling variable such as climate, the result is the combined response of a whole array of individual landforms. Landscape-level changes therefore can be highly complex because the rate of change varies across the landscape and individual landforms do not always respond in the same way. There may also be lags in landform response to the extent that landform adjustment to short-term climatic change is rarely completed across the whole landscape. These factors make it difficult when records are short to distinguish between landform changes that are fluctuations about a stable mean condition adjusted to a particular climate and those that represent long-term change. It is useful, therefore, to begin by considering the variability associated with equilibrium conditions before considering what happens when a controlling variable changes. It is also important to understand the associated temporal and spatial patterns of landscape response since these may sometimes be detected by remote sensing techniques.

Equilibrium is frequently treated as a static condition in which the properties of a system remain constant as long as the controlling variable does not change. Few landforms exhibit this property and four types of equilibrium behavior (Figure 11.1) are commonly recognized at a given time scale (Chorley and Kennedy, 1971). Steady-state equilibrium is found in those parts of the landscape where the long-term supply of sediment is equal to the rate of removal. However, because both supply and removal are episodic and related to individual events such as storms or floods whose magnitude and frequency can vary, there will be significant fluctuations about the mean condition. Dynamic equilibrium exists where the long-term sediment supply is smaller than the potential rate of removal and the deficit is made up by erosion, a condition often occurring in the headwaters of a stream. Alternatively, it may be the condition in sediment sinks such as deltas where sediment supply is greater than removal. Thus fluctuations occur about a mean condition representing episodic change, but there is also a trend. A third type of behavior is metastable equilibrium. Here, the landform may occupy one of several states, all of which are equally stable. Normally, landform characteristics fluctuate about one of these states but, if an event large enough to push behavior across the threshold condition

11. Remote Sensing of Landscape Processes

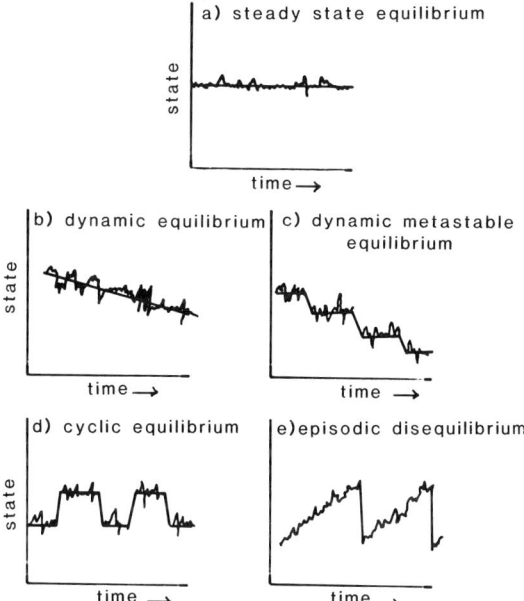

Figure 11.1. Some types of equilibrium behavior in geomorphic systems. Each graph shows behavior through time at a point. There will also be distinct patterns of spatial variability. [From Chorley and Kennedy (1971) and Nanson and Erskine (1988). Reproduced with the permission of Academic Press.]

between states occurs, stability will be restored when the landform occupies one of the alternative states. Gravel-bed rivers sometimes show this tendency, some reaches being braided whereas others only have a single channel. Dynamic metastable equilibrium exists when a landform is capable of shifting between alternative states but the long-term mean of each state displays a trend.

When there is a step change, as opposed to a short-term fluctuation, in one of the controlling variables, landforms may respond fully, partially, or not at all. If the landform responds fully, then once initial lag effects have been overcome, it will move to a new equilibrium condition. Partial adjustment takes place when there is a strong lag effect or when the change in the controlling variable is not of sufficient duration to allow the new equilibrium to be fully established before the controlling variable changes yet again. An example of partial adjustment is shown in Figure 11.2 for a river channel assuming a static equilibrium response model (i.e., no fluctuations about the equilibrium condition). The horizontal dotted lines represent the equilibrium bed elevations in a channel cross section at a number of constant flows. The solid line represents the actual position of the bed over time. At any given flow, the bed is moving toward the equilibrium bed

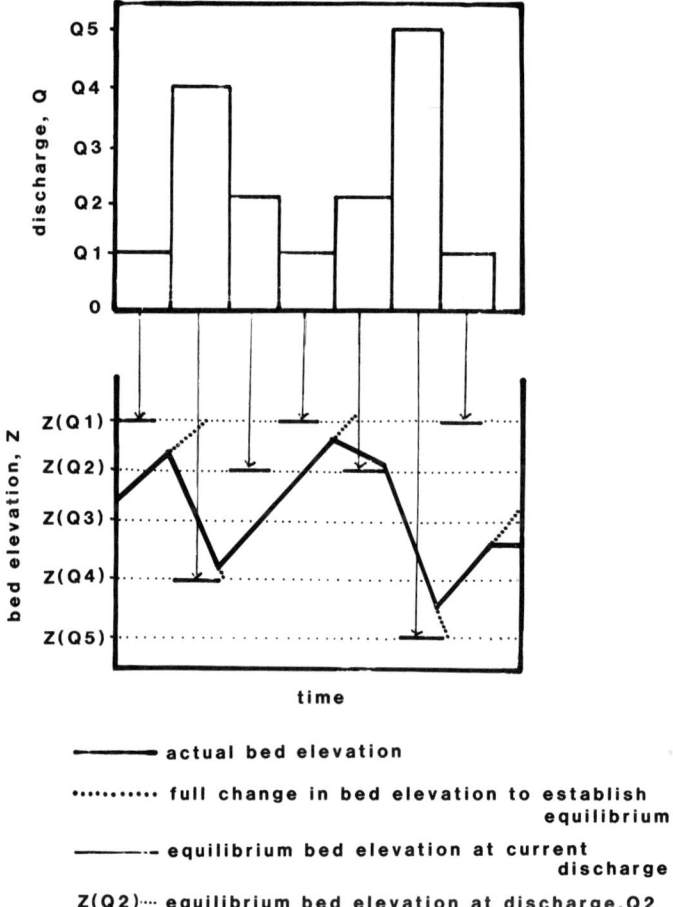

Figure 11.2. Geomorphic response to fluctuations in a controlling variable such as climate (see text for explanation).

elevation for that flow. This trend is illustrated for several different flows by the heavy dotted lines. The equilibrium bed elevation may never be reached because the flow changes and the bed is continuously seeking a new equilibrium position.

The no-change condition is seen when the shift in climate is not of sufficient duration to overcome lag effects or great enough to produce events which exceed the erosion threshold of a particular landform. This happens quite frequently with short-term fluctuations in climate even though they are substantial. It can also occur in those parts of the landscape that are shaped only by the largest event such as great floods. Because of the rarity of these events, they may not be experienced within the time frame of short-term climatic change. An example might be the 1,000-year return period flood, which has a low probability of occurrence during a climatic fluctuation taking place over, say, a 50-year period.

Unless a sequence of catastrophic events is involved, the effects of climatic change on landforms may take decades, or even centuries, to emerge. Even then, because of lags in landform response, only the most active parts of the landscape may change sufficiently to allow detection. An example of this process comes from southeastern Australia, where short-term climatic variability is the norm. Examination of flood records suggests that, since 1799, there have been three separate periods when large floods were relatively frequent (Figure 11.3) and two intervening periods when flood flows were much smaller (Riley, 1981). Warner (1987b) has described these periods as flood-dominated regimes (FDRs) and drought-dominated regimes (DDRs). The differences between FDRs and DDRs last three to five decades and are such that floods of a given frequency vary in magnitude by factors of two to four. The last of these shifts has been studied in

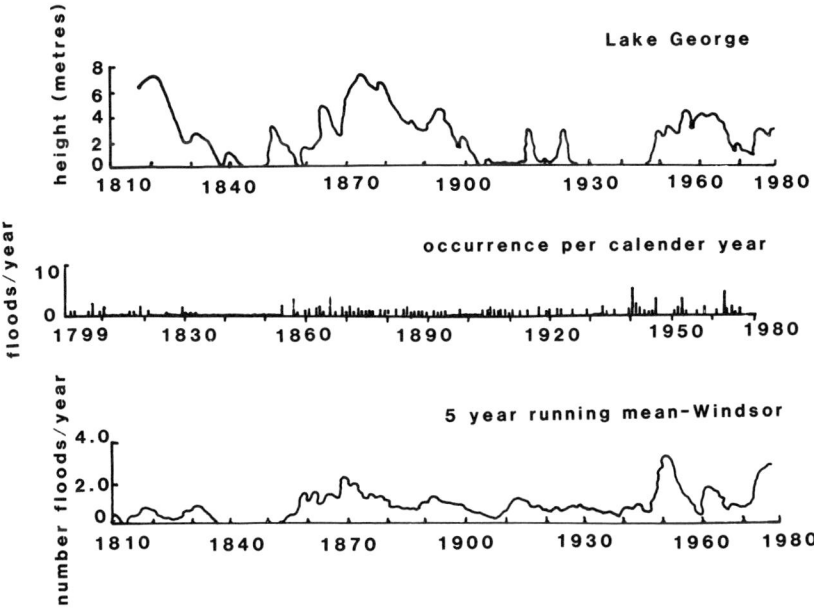

Figure 11.3. Short-term climatic fluctuations in eastern Australia as shown by stage heights in Lake George and the flood record of the Hawkesbury River at Windsor, N.S.W. Lake levels smooth variations in rainfall and show a series of wet and dry periods. Flood occurrences illustrate increasing frequency of large events as average rainfall becomes greater. [From Riley (1981). Reproduced with permission of the American Water Resources Association.]

some detail (Pickup, 1976; Erskine and Bell, 1982; Warner, 1987a) and a consistent pattern emerges.

The response to successive FDRs and DDRs is closely related to both vegetation response and flood magnitude. In the dry period, the loss of cover from hill slopes results in a substantial increase in sediment delivery to channels. Flows are not sufficient to remove this material which is quickly vegetated and stabilized, gradually choking the smaller channels. Larger channels respond by deposition on the banks, which reduces their size and their capacity to transport bedload (Nanson and Erskine, 1988). In the wet period, hill slopes revegetate, partly cutting off the supply of sediment, while increased flow in both major and minor channels scours out the accumulated sediment and enlarges them. There are, however, spatial variations in the pattern of response that are still not properly understood (Warner, 1987a).

Approaches to Change Detection from Remotely Sensed Data

Detection of change usually begins with an attempt to identify differences between sets of imagery acquired at different times. However, the existence of a difference does not always mean that there has been a change as there are many other sources of variation in imagery. This section describes some of the reasons for this variation and identifies ways of dealing with it. It also examines some of the limitations of remotely sensed data.

The most valuable source of information on landscape change comes from aerial photography. In Australia, this photography spans periods of 30 to 50 years and has been acquired for both routine mapping and special projects. It is particularly important because the period involved coincides with change from a DDR to an FDR in many areas.

Comparing aerial photography acquired at different times poses special problems which can make it difficult to distinguish real from apparent change. Differences in sun angle at the time of acquisition, film type, photographic processing, and photograph age combine to produce complex relationships between true color or grayness and the tone of the image. These relationships can vary between the edge and the center of each photograph. Considerable skill therefore is required in visual air photo interpretation and changes in spatial pattern or texture are used as much as brightness for change detection.

The advent of low-cost digitizing equipment in the form of scanners, video cameras, and frame grabbers will greatly improve air photo interpretation. Spatially variable image correction procedures can be applied and brightness levels can be adjusted by histogram equalization or by modifying the range of pixel values between light and dark target areas. Color images may be converted to monochrome either by the use of monochrome cameras or by adding red, green, and blue signals together.

The ability to identify changes in landform behavior from satellite-based remotely sensed data is limited by the lack of archived data and the spectral and spatial resolution of the sensor involved. Landsat MSS data have been available since 1972 but the spatial resolution of the MSS is low, data are coarsely quantized, and the signal-to-noise ratio is relatively low. Landscape changes thus must be substantial and occur over large areas to be distinguishable. SPOT HRV and Landsat TM data have a higher resolution and accuracy but have only been available for a short time and so are still of limited value for detecting change.

Change detection involves subtracting one data set from another to determine what differences exist. If the two data sets are not comparable, which is frequently the case, two main types of radiometric correction may be needed. First, if the data are acquired at different times of the year, the amount of incoming, and thus reflected, solar radiation will not be the same. This problem can be partly overcome by the normal cosine correction procedure. This does not allow for differences in the amount of shadow associated with changes in solar position and the effects of the atmosphere. Such differences may be partly compensated for by band ratioing or by recalibrating radiance data based on stable target areas but they can never be fully removed. The answer is to acquire data at or close to the same time each year, thereby ensuring the same scene illumination. A second correction procedure is needed when comparing data collected by different satellites. Since 1972, there have been five satellites in the Landsat series and each MSS has had slightly different response characteristics. A good way of standardizing among the different Landsats is to use the physical radiance values rather than the cailbrated data normally available (Robinove, 1982). It is also important to carry out the normal radiometric correction procedures of destriping and, if necessary, coherent noise removal (Landsats 4) so that observed changes over time are real and not just artifacts of sensor noise.

The need for geometric correction arises because there are small variations in the alignment of each satellite over time and differences in the spatial distortions produced by each MSS. There are also substantial differences between the orbits of Landsats 1, 2, and 3 and those of Landsat 4 and 5. It is, therefore, rarely possible to achieve a perfect pixel for pixel overlay even with data acquired from the same satellite. Rectification procedures that allow data from one image to be resampled to fit over another can reduce spatial errors to less than one pixel but can never remove them entirely. It is therefore best when subtracting one rectified image from another to use some form of spatial averaging to reduce local variability resulting from imperfect overlaying. Fully rectified imagery based on map coordinates can now be bought, which makes monitoring much simpler.

Even when standardized radiometric data are used, there may still be difficulties in separating real and apparent landscape change. These arise because changes in vegetation cover and greenness may vary from season to season. In the arid zone, for example, vegetation cover may routinely vary by 50% or more depending on rainfall. There may also be substantial differences in the spatial pattern of change from one rainfall event to another (Foran, 1988). Thus it is necessary to have data for a long enough period to filter out short-term variability when detecting true landscape change.

Some authors have used short-cut methods to change detection that involve principal components analysis of sequential images (e.g., Byrne et al., 1980; Millington et al., 1987). When this analysis is undertaken, gross changes attributable to differences in incoming radiation and atmospheric effects are contained in component one. Minor differences associated with changes in the land surface, and so on, are identified with the higher-order components. This technique is only approximate and it is customary to reject values below a certain threshold in the higher-order components. Information is also lost as to the band in which the change has occurred. It may be, therefore, unsuitable for detecting the subtle variations often associated with landscape change.

In spite of the difficulties involved in standardizing remotely sensed information for change detection, a number of studies have successfully used Landsat MSS data to identify changes in landform characteristics. Most of these successes have come from areas where change is relatively rapid or where soil and sediment reflectance is not confounded by the effects of vegetation. In the Nile delta, for example, Klemas and Abdel-Kader (1982) have found that it is possible to monitor coastal erosion, sediment transport, and changes to lakes using unenhanced Landsat MSS products. The effects of vegetation are not always a problem. In arid central Australia, Pickup and Chewings (1988a) have identified erosion and deposition at a number of sites using MSS data sets acquired ten to 12 years apart. However, they used changes in vegetation cover and greenness associated with the redistribution of water and sediment to indicate erosion and deposition rather than detecting it directly. Their approach could not be used in areas where the relationship between plant cover and erosion status was not strong.

Landsat TM probably offers the highest potential for detecting landscape change once a long-term archived data set accumulates. This potential arises because of the larger number of spectral bands and higher spatial resolution that make it possible to detect relatively localized changes in surface mineralogy. Once again, however, short-term changes in vegetation cover can make it difficult to distinguish longer-term changes resulting from erosion and deposition. For example, Millington et al. (1987) have shown that TM data can be used to detect change in north African playa basins but the best results come from the unvegetated playas themselves. Activity in neighboring rivers and on alluvial fans is less easy to identify and even more difficult to separate from seasonal vegetation differences.

Using Landscape Process Models with Remotely Sensed Data

Models and Their Limitations

Ideally, a distributed model of sedimentation processes in a landscape should describe four elements of behavior:

1. Spatial and temporal variations in the sediment transport capacity of the system.
2. Spatial and temporal variations in the supply of sediment that is available for transport.
3. The pattern of sediment redistribution that results where transport capacity and available load differ.
4. The change in sediment supply or transport capacity at a point when sediment is eroded or deposited and system geometry or other characteristics are affected.

None of the currently available models meet this ideal. Most have been developed to predict sediment yield at a single point such as a drainage basin outlet (e.g. Moore, 1984) so they do not provide adequate descriptions of behavior across the landscape. Instead, the variability within a basin is represented by splitting subareas into a series of lumped parameter compartment models linked by water and sediment routing procedures. These compartment models involve a great deal of spatial averaging and frequently represent spatially variable activity as a point process. This approach is necessary because the data required for modeling may not be available at all locations within a drainage basin, some processes may not be amenable to measurement at the landscape scale, and the information required to test model behavior at many locations simultaneously cannot be collected. The result is that even the best models contain empirical relationships or parameters that have to be calibrated by comparing observed and modeled behavior using optimization techniques (e.g., Borah et al., 1981).

Some progress is now being made in the development of distributed process models on small drainage basins that provide detailed information on the spatial variability of sediment transport rates. These models operate by calculating surface runoff from rainfall on a highly localized basis and then estimating the sediment transport rate from runoff, local slope, and sediment properties.

At the drainage basin scale, the response of runoff to rainfall is highly dependent on the location and size of saturated source areas that act as runoff-producing zones (Moore et al., 1986). These areas expand and contract as the catchment is progressively wetted or dries up. Variations in wetness within a basin are closely related to local slope, the hydraulic properties of the soil profile, and the degree of convergence of the local topography. O'Loughlin (1986) has used these characteristics to produce a three-dimensional saturation zone model, which, when coupled with a digital terrain model, is capable of describing the location of the runoff producing zones. O'Loughlin's model has since been coupled with a sediment transport model that allows for the divergence or convergence of topography to produce estimates of erosion and deposition in three-dimensional terrain (Moore and Burch, 1986a). This model has not been tested since it assumes that the rate of sediment transport is determined by transport capacity whereas in practice, it is often more closely related to supply. The patterns of erosion and deposition produced by the model are, however, realistic in appearance.

Three-dimensional erosion models offer potential for the prediction of landscape change but their use is limited by lack of data on terrain characteristics and for model testing. Remote sensing may provide some of these data and a number of groups are now attempting to couple watershed runoff and sediment models with data from satellites (e.g., Peterson et al., 1987). Areas currently in use or under development include the determination of catchment characteristics and sediment availability from MSS, TM, or HRV data. Areas with future potential are rainfall-runoff models, which use soil moisture data from microwave sensors, and the mapping of topography using stereo imagery and radar altimeters. The remainder of this section concentrates on the possibility of deriving information on sediment availability from remotely sensed data.

Sediment Availability

The supply of sediment within a drainage basin is normally much less than the amount that erosional processes such as wind, rain splash, and runoff are capable of transporting. The imbalance occurs because the surface is protected by vegetation and the individual soil particles are bound up as soil aggregates. Erosive forces must exceed a threshold to remove vegetation or break up aggregates before transport can begin. This only happens in a few locations and over limited periods—hence the ability to accumulate a soil layer even though soil-forming processes operate very slowly.

Changes in sediment supply arise when there is a change in vegetation cover or in the rate at which soil aggregates are broken down. Many factors can produce these changes but they frequently occur in response to short- to long-term fluctuations in rainfall and land-use practices. Remote sensing has proved to be very effective in monitoring some of these changes, particularly when they happen rapidly and involve large areas (e.g., Graetz et al., 1986). Color Plate 3 provides an example of this in which vegetation cover changes in response to both rainfall and grazing activity. The upper image shows an area of arid central Australia in 1980 in which vegetation cover varies substantially across the landscape. The lower image is for 1983 and shows the effects of different land-use practices. Cattle concentrate their activities around water points to which they must return at regular intervals to drink. The high level of activity in these areas results in a reduction in vegetation cover. One paddock at the lower right is being used heavily, and the grazing effect has extended beyond the water points to affect most of the paddock. The red areas therefore might be used to indicate zones with high sediment availability. These images show that remote sensing can be used to provide information on both spatial and temporal variability. The question is then whether existing predictive models are capable of handling that information.

Physically based procedures for predicting changes in sediment in supply in response to changes in vegetation cover or land-use practice are not available so it is normal to develop empirical relationships based on plot experiments that are extrapolated to similar areas. In the United States, this procedure is well developed for crop lands and soil erosion potential is routinely estimated using Wischmeier and Smith's (1978) Universal Soil Loss Equation (USLE):

$$A = R \cdot K \cdot L \cdot S \cdot C \cdot P$$

where A is potential soil erosion (tons/acre/year), R is a precipitation factor, K is a soil erodibility factor, L is a slope length factor, S is a slope steepness factor, C is a cropping practices factor, and P is a management practices factor.

When the database is sufficient to provide estimates of these factors, remotely sensed data and classification procedures have been used to identify soil, cover, and land-use types (e.g., Niemann et al., 1987). Appropriate values of K, C, and P may then be assigned to particular areas. The USLE, being empirical, is not suitable for many areas because the data to use it are not available. Physically based procedures are under development (e.g., Moss et al., 1979, 1980; Moore and Burch, 1986b) but they have only been applied to a few soil types, making it still necessary to rely on erosion plot data in many areas.

Estimation of sediment supply from small plots is an acceptable procedure in agricultural areas where the landscape is divided into small, relatively homogeneous fields in which vegetation or crop type and management practice are uniform. Natural landscapes are more diverse, making plot-based measurements difficult to extrapolate to larger areas. Thus, while attempts have been made to estimate sediment yield using the USLE, they have not been particularly successful unless an empirical correction factor known as the sediment delivery ratio is used. New methods are needed to estimate the amount and spatial distribution of sediment supply in natural landscapes and to determine the processes that influence it.

Little work has been done on the distribution of sediment supply because of the difficulties of measurement. Sediment yield is normally measured by sampling the suspended load in streamflow at a gauging station at the basin outlet. These data make it possible to compare basins in terms of average vegetation cover, average slope, and so on, but do not allow for spatial variability. They also result from the interaction of sediment supply and transport capacity rather than sediment supply alone.

An area of research that may help in deriving the regional pattern of sediment supply in grazing lands is based on animal distribution models. Grazing animals remove vegetation cover and break up soil aggregates, making the soil more liable to erosion. These animals do not use an area uniformly but instead concentrate their activity in particular areas. In the arid zone, for example, cattle concentrate on water points and more palatable vegetation communities (e.g., Hodder and Low, 1978) whereas sheep distribution is also affected by proximity to camping sites (Stafford Smith, 1988). Some progress has been made in modeling the distribution of animal activities (Senft et al., 1983) and recently Pickup and Chewings (1988b) have been able to derive a suitable model from remotely sensed data. Their approach uses a convection–diffusion analogy to determine the average distribution of animals in a particular vegetation type coupled with a stochastic process to generate local variability. The parameters of the model have to be fitted from observed cattle distributions, but the change in Landsat band 5 above a particular threshold over a period of time can be a useful surrogate since it is closely related to percent vegetation cover (Foran and Pickup, 1984). The threshold level is set by plotting a curve showing changes in band 5 against distance from water. This curve shows a decline away from water owing to grazing effects but it eventually stabilizes at a level that represents the amount of change that can be related to natural conditions. Once the grazing distribution is known, the extent of trampling may be generated from it by routing animals along the shortest path to water and counting the number passing through each grid cell (Color Plate 4).

Modeling Landscape Change Directly from Remotely Sensed Data

Conceptual Basis

Distributed process models that predict sediment transport, erosion, and deposition from the pattern of water flow in an area may be suitable for small drainage basins but have achieved little success in predicting change at the landscape scale. The problem arises because errors in the estimation of flow, sediment supply, or transport rate that inevitably arise at particular locations propagate through the model. This affects behavior at other locations, compounding the original error. Real trends in erosion and deposition are then submerged by model-generated errors. These errors may be very large. All models use sediment transport equations of dubious reliability. Where these equations have been tested, calculated sediment transport rates may deviate from observed rates by factors of three or more (e.g., Alonso et al., 1981). Errors in both the timing and the magnitude of the flow rates used to calculate sediment transport may also be seen. No data are available describing these errors throughout a catchment but there are many examples showing model performance at a few points. In virtually every case, even after the best models have been calibrated against observed catchment behaviour, errors remain, especially at peak flow (e.g., Moore et al., 1986).

An alternative approach is to model the patterns of the sediment flows alone using observed sediment movement in the landscape. This method is particularly suited to remotely sensed data from arid areas where differences in vegetation cover are closely linked to patterns of erosion and deposition (e.g., Pickup and Nelson, 1984) and may be detected using Landsat MSS. It may also be suitable for use with Landsat TM whose finer resolution and additional spectral bands make it possible to identify soil surface changes associated with erosion and deposition. This is especially true in flat areas where erosion and deposition redistribute soil as a thin veneer over a relatively large area. In steeper areas, changes tend to be more confined to smaller areas such as channels and involve more vertical loss or accumulation of sediment, which is less easy to detect.

The sediment flow approach is based on the idea that, at the regional scale, sediment is not entrained and transported smoothly down a system or in time as most models suggest. Instead, it is supplied discontinuously from a number of points or distributed sources and transported intermittently, often spending long periods in storage before remobilization. There is, however, some order. Each time a sediment transport event occurs, eroded material moves in scour–transport–fill (STF) sequence (Figure 11.4) with erosion and sediment entrainment at the upper end of the sequence, deposition at the lower end, and a transfer zone in between in which sediment transport capacity and sediment supply are roughly balanced. These STF sequences are intrinsic to the process of sediment transport from a restricted source area in unsteady flow. Water travels faster than most of the sediment it carries so the flood wave tends to overtake the sediment wave, resulting in deposition. Sediment transport may also decrease downslope in arid areas as runoff is lost by infiltration. If deposits are large enough or become stabilized by vegetation, they can reduce flow velocity, producing further deposition. The STF sequence then becomes fixed or self-enhancing (Pickup, 1988a). Alternatively, it may not be stabilized at all and be destroyed in the next flood event.

The STF sequences can develop at a variety of spatial scales over a period of time. When flows are small or the rate of erosion is low, they are likely to occur as low-amplitude, high-spatial-frequency features (Pickup and Chewings, 1988a). This produces a buffered landscape in which sediment is unlikely to travel far before it is trapped by one of the vegetated deposition areas even when a large flow occurs. If large flows occur frequently or erosion rates are high, the STF sequences will be much larger and sediment will travel longer distances before deposition occurs.

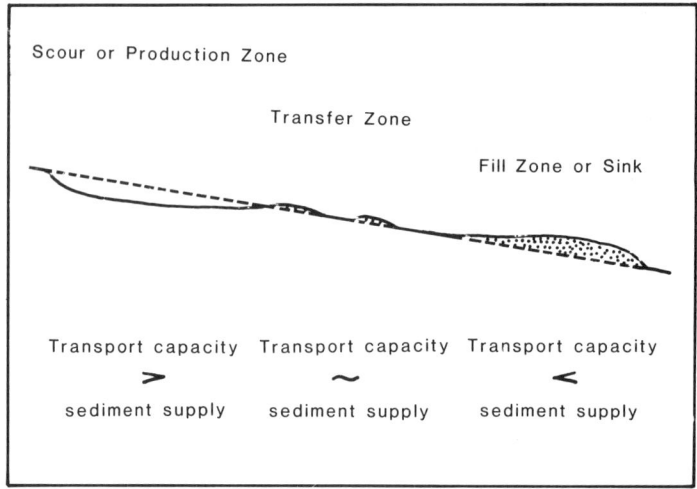

Figure 11.4. A scour–transport–fill sequence. [From Pickup (1988a). Reproduced with the permission of Academic Press.]

Over time, a distinctive pattern of STF sequences may develop in a landscape. Some will take place in response to topographic variations as sediment is shed from convex slopes and accumulates in concavities. Others may develop because certain areas consistently lose soil as a result of natural erodibility or the presence of an erosion-producing activity. Not all STF sequences will be active simultaneously. Some may result from very large events and may only operate when those events recur. Others, particularly the smaller ones, may be active during high-frequency events, only to be partly destroyed or nonoperational during the extremes. Whatever their behavior, they describe and integrate continuing processes of sediment redistribution in an area and form a basis for predicting future development.

The STF sequence concept has proved particularly useful in developing models of landscape behavior for the floodplains, alluvial fans, and alluvial footslopes of arid central Australia. These landscapes may be divided into inactive areas, where little change is occurring, and active areas, where there is obvious recent erosion and deposition. Active areas consist of a mosaic of overlaid and interlocking erosion cells of various sizes that are the two-dimensional equivalent of a set of STF sequences. Each cell consists of a source or scour zone, a transfer zone, and a sink, although some cells may have been captured by others. It is also possible for small cells to be partly subsumed by larger ones. As erosion intensifies, the mosaic changes its characteristics. Source zones expand, transfer zones may be converted into source zones, and sinks expand as more material accumulates. The largest erosion cells expand at the expense of the smaller ones until the landscape consists almost entirely of large cells.

Model Implementation

STF sequences have been directly modeled from Landsat MSS data in flat arid landscapes characterized by erosion cell behavior. This is possible because erosion results in a decrease in the greenness and amount of vegetation cover whereas deposition increases them. Many vegetation indices are sensitive to such changes, including the stability index of Pickup and Nelson (1984), which filters out the effects of differences in soil and vegetation color. It also expresses the state of the landscape as a single variable with erosion at one end of the scale, stability in the middle, and deposition at the other end. This allows mapping of the main STF sequences. A change in the behavior of a sequence may then be represented by changing its geometry to allow for elongation, contraction, breakup into smaller sequences, or variation in intensity.

Modifying the geometry of a single STF sequence is a simple process, as Figure 11.5 indicates. It is more difficult in the case of erosion cell mosaics where the landscape is made up of a complex array of two-dimensional overlaid sequences of different sizes. Some success has been achieved, however, using stochastically based image texture generation procedures (Pickup and Chewings, 1988a).

A model of erosion cell mosaic behavior can be constructed by repro-

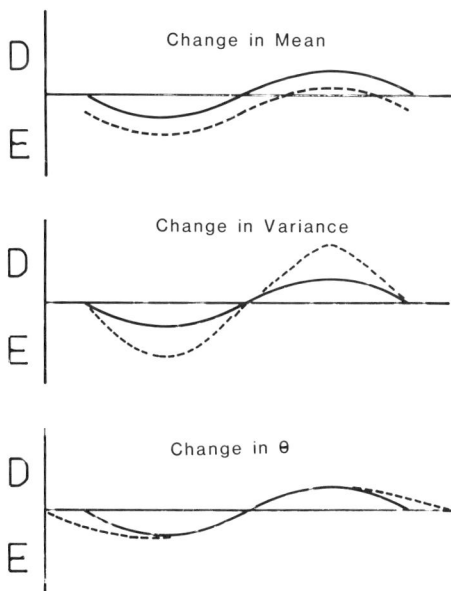

Figure 11.5. Schematic diagram showing how changes in a simple STF sequence may be represented by shifts in mean, variance, and the θ parameter, which controls the level of spatial interaction. E denotes erosion, whereas D is deposition. [From Pickup (1988a). Reproduced with the permission of Academic Press.]

ducing four types of change in the soil stability index values for an area. First, the mean may change. Second, the number of pixels in stable, eroded, and depositional states may change, affecting variance. Third, the individual areas of erosion and deposition may spread, contract, or be broken up into smaller patches, producing changes in the spatial autocorrelation function of the stability index. Fourth, new erosion cells may develop. There is, however, a great deal of continuity in erosion cell mosaic behavior so the development of new cells is limited. This makes it possible to forecast change in the soil stability index values by modifying existing patterns in the landscape. Modeling of changes in spatial dependency and in the frequency distribution of soil stability index values is all that is needed to produce an adequate forecast.

The model used in the forecasting procedure is a simultaneous autoregressive (SAR) process and may be written as:

$$Y(s) = \sum_{r \in N} \theta_r Y(s+r) + \sqrt{\rho w(s)}$$

where $Y(s)$ is the soil stability index value at location s; s refers to a pair of x,y grid coordinates; r indicates the coordinates of a member of the neighborhood set, N, around location s; θ_r are the model weighting parameters; ρ is the variance of the noise series; and $w(s)$ is a sequence of random variates with zero mean and unit variance. Methods for fitting the model are described by Kashyap and Chellappa (1983).

A version of this model that appears to describe erosion cell mosaic behavior well is the first-order, eight-neighborhood version (Pickup and Chewings, 1988):

$$Y_{ij} = \theta_1 y_{i-1,j} + \theta_2 y_{i+1,j} + \theta_3 y_{i,j-1} + \theta_4 y_{i,j+1}$$
$$+ \theta_5 y_{i-1,j-1} + \theta_6 y_{i-1,j+1} + \theta_7 y_{i+,j+1}$$
$$+ \theta_8 y_{i+1,j-1} + \sqrt{\rho w_{ij}} + \bar{Y}$$

where i and j are grid coordinates and y values are deviations about the mean. The model behaves like a two-dimensional diffusion process and operates by reducing or enhancing high-frequency spatial variability. It can represent three of the four types of change needed to model different states of an eroding or recovering system. Changes in the overall system state are represented by shifts in the mean. Changes in the variance of the stability index and the spread or breakup of patches within the mosaic are handled by the θ and ρ parameters. The $w(s)$ series contains information on the location of erosion and deposition and is known as the underlying pattern series.

The modeling procedure relies on the continuity over time and space that exists in erosion cell behavior and uses the landscape to forecast itself. The underlying pattern series represents the latent pattern of erosion and deposition in an area while the mean, variance, and θ parameters describe the extent of erosion. As the landscape becomes more eroded, a variance is imposed, scaling the underlying pattern series. This scaling is represented by the ρ parameter. Spatial dependency is also imposed as an erosion cell mosaic develops by intensifying very weak features in the underlying pattern series. This structure is controlled by the θ parameters.

Modeling landscape behavior as erosion intensifies uses the assumption that a landscape Y in a particular state of erosion can be related to an underlying pattern series, w, by a linear filter, A_1 (Pickup, 1988b) such that:

$$Y_1 = A_1 w$$

The same landscape in a more eroded state, Y_2 is related to w by

$$Y_2 = A_2 w$$

Given estimates of A_1 and A_2, and assuming that w does not change over time, the more eroded state may be forecast as:

$$\hat{Y}_2 = A_2 \hat{W} = A_2 A_1^{-1} Y_1$$

where the symbol ^ indicates an estimate. The linear filters, A_1 and A_2 are SAR models obtained using the fitting procedures described above. The estimate of A_1 is derived by fitting an appropriate model to Y_1 and $\hat{W}$ is calculated by inverse filtering. The A_2 model is an unknown but can be obtained by using landscapes that are similar to Y_1 but are in a more eroded condition as prototypes. Once such a prototype is identified, A_2 is estimated by the normal SAR model-fitting procedure. Color Plate 5 shows an example of the modeling process.

The approach appears simple but tests show that it is capable of producing reasonable forecasts of change for large areas with very complex patterns of erosion and deposition (Pickup and Chewings, 1988a). It seems to work because the underlying pattern series contains most of the information on the likely pattern of soil movement. Also, because the model parameters used to generate a more eroded condition are derived from a similar or neighboring area, they contain the averaged response of the landscape, bypassing the need for detailed simulation over a sequence of time and distance steps that process-based models require.

The advantage of the method is its simplicity. Because model parameter estimates are derived from a prototype, they already embody the response of a particular landscape to change. The prototype therefore must be selected with care, or an unrealistic forecast could result. The problems of the method are that it uses a single average model for a whole landscape when several different ones could be more appropriate. It is also incapable of forecasting erosion patterns that are not already latent in the landscape, such as the erosion zones around new stock watering points. The ability to introduce new erosion patterns remains a problem although some progress is now being made by inserting structures or trends into images using the grazing distribution models described in a previous section.

Conclusions

The enormous increase in the quantity, resolution, and type of remotely sensed data from both satellite and aircraft-mounted sensors planned for the 1990s poses new challenges in the study of landscape processes. Geomorphologists have relied on air photographs ever since they have been available but have been slow in taking up the use of satellite data. This problem is being rectified as low cost personal-computer-based image-processing facilities make the basic hardware and software available and as the speed and capacity of these machines increase. The difficulty now is that models and concepts lag behind data availability.

Most attempts to model landscape processes using remotely sensed data are likely to use existing so-called physically based models. While the performance of these models can be expected to improve with better data, their limitations will become increasingly apparent and a limit will be quickly reached. This will occur because such models rely on point-based relationships and attempt to reproduce large-scale system behavior as the sum of a whole set of localized behaviours. Synergistic effects are neglected and the problems of extrapolating from the point to the region with equations that are not fully accurate even at the point will multiply. New models thus are required that will describe the behavior of whole systems, incorporating processes that operate at a variety of scales.

It is difficult to predict where these models will come from. One possibility is the field of image modeling where the classical statistical approaches are now giving way to methods based on scale-dependent processes (e.g., Lovejoy and Schertzer, 1985). Whatever the source, some radical changes are needed in our approach to landscape behavior if the potential offered by remotely sensed data is to be fully realized.

Acknowledgments

The author is grateful to Richard Hobbs, Steve Running, and Dean Graetz for their comments on this chapter.

References

Allison, G.B., and Peck, A.J. (1987). Man-induced hydrologic change in the Australian environment. Bureau of Mineral Resources, Geology and Geophys. Rep. 282, pp. 35–37.

Alonso, C.V., Neibling, W.H., and Foster, G.R. (1981). Estimating sediment transport capacity in watershed modelling, *Trans. ASAE* 24:1211–1220, 1226.

Baker, V.R. (1978). Palaeohydraulics and hydrodynamics of scabland floods. pp. 59–79. In V.R. Baker and D. Nummedal (eds.), *The Channeled Scabland*. Nat. Aeron. and Space Admin. Washington, DC.

Baker, V.R. (1983). Large-scale fluvial palaeohydrology. pp. 453–478. In K.J. Gregory (ed.), *Background to Palaeohydrology*. Wiley, Chichester, England.

Baker, V.R. (1986). Introduction regional landforms analysis. pp. 1–26. In N.M. Short and R.W. Blair (eds.), *Geomorphology from Space—A Global Overview of Regional Landforms*. Spec. Pub. 486, Nat. Aeron. and Space Admin., Washington, DC.

Borah, D.K., Alonso, C.V., Prasad, S.N. (1981). Single event model for routing water and sediment on small catchments. Appendix 1. *Stream Channel Stability*. U.S. Dept. Agric. Sedimentation Lab., Oxford, MS.

Byrne, G.F., Crapper, P.F., and Mayo, J.K. (1980). Monitoring land cover by principle components analysis of multitemporal Landsat data. *Remote Sens. Envir.* 10:175–184.

Chorley, R.J., and Kennedy, B.A. (1971). *Physical Geography: A Systems Approach*. Prentice-Hall, London, England.

Erskine, W.D., and Bell, F.C. (1982). Rainfall, floods and river channel changes in the upper Hunter. *Aust. Geog. Stud.* 20:183–196.

Evans, D. (1988). Multi-sensor classification of sedimentary rocks. *Remote Sens. Envir.* 25:129–144.

Everitt, J.H., Escobar, E., and Nixon, P.R. (1987). Near real-time video systems for rangeland assessment. *Remote Sens. Envir.* 23:291–311.

Foran, B.D. (1988). Detection of yearly cover change with Landsat MSS on pastoral landscapes in central Australia. *Remote Sens. Envir.* 23:333–350.

Foran, B.D., and Pickup, G. (1984). Relationships of aircraft radiometric measurements to bare ground on semi-desert landscapes in central Australia. *Aust. Range. J.* 6:59–68.

Frank, T.D. (1984). The effect of change in vegetation cover and erosion patterns on albedo and texture of Landsat images in a semi-arid environment. *Ann. Assoc. Amer. Geogr.* 74:393–407.

Graetz, R.D., Pech, R.P., Gentle, M.R., and O'Callaghan, J.F. (1986). The application of Landsat image data to rangeland assessment and monitoring: The development and demonstration of a land image-based resource information system. *J. Arid Evir.* 10:53–80.

Hodder, R.M., and Low, W.A. (1978). Distribution of free ranging cattle at three sites in the Alice Springs District, Central Australia. *Aust. Rang. J.* 1:95–102.

Kashyap, R.L., and Chellappa, R. (1983). Estimation and choice of neighbours in spatial interaction models of images. *IEEE Trans. Infor. Theory* 29:60–72.

Klemas, V., and Abdel-Kader, A.M.F. (1982). Remote sensing of coastal processes with emphasis on the Nile Delta. *Procs. Int. Symp. on Remote Sensing of Environment, Cairo, Egypt* 1:381–415. Envir. Res. Inst, Michigan, Ann

Lovejoy, S., and Schertzer, D. (1985). Scale and dimension dependence in remote sensing: analysis and simulation. *Proc. Workshop on Image Understanding in Remote Sensing*. 10 July 1985, University College, London, 3(1)–3(23).

Mark, D.M., and Aronson, P.B. (1984). Scale dependent fractal dimensions of topographic surfaces: An empirical investigation with applications in geomorphology and computer mapping. *Math. Geol.* 16:671–683.

Mercer, J.B., and Kirby, M.E. (1987). Topographic mapping using STAR-1 radar data. *Geocarto* 3:39–42.

Millington, A.C., Jones, A.R., Quarmby, N., and Townshend, J.R.G. (1987). Remote sensing of sediment transfer processes in playa basins. pp. 369–381. In L. Frostick and I. Reid (eds.), *Desert Sediments, Ancient and Modern*. Spec. Pub. 35, Geol. Soc., London, England.

Moore, I.D., and Burch, G.J. (1986a). Modelling erosion and deposition: Topographic effects. *Trans. ASAE* 29:1624–1630, 1640.

Moore, I.D., and Burch, G.J. (1986b). Sediment transport capacity of sheet and rill flow: Application of units stream power theory. *Water Resources Res.* 22:1350–1360.

Moore, I.D., Mackay, S.M., Wallbrink, P.J., Burch, G.J., and O'Loughlin, E.M. (1986). Hydrologic characteristics and modelling of a small forested catchment in southeastern New South Wales. Pre-logging condition. *J. Hydrol.* 83:307–335.

Moore, R.J. (1984). A dynamic model of basin sediment yield. *Water Resources Res.* 20:89–103.

Moss, A.J., Walker, P.H., and Hutka, J. (1979). Raindrop-stimulated transportation in shallow water flows: an experimental study. *Sediment. Geol.* 22:165–184.

Moss, A.J., Walker, P.H., and Hutka, J. (1980). Movement of loose, sandy detritus by shallow water flows: an experimental study. *Sediment. Geol.* 25:43–66.

Nanson, G.C., and Erskine, W. (1988). Episodic changes of channels and floodplains on coastal rivers in New South Wales. In R.F. Warner (Ed.), *Australian Fluvial Geomorphology*. Academic Press, Sydney, Australia.

NASA. (1987a). HMMR—High Resolution Multifrequency Microwave Radiometer. *Instrument Panel Report. Earth Observing System IIe*. Nat. Aeron. and Space Admin., Washington, DC.

NASA. (1987b). SAR—Synthetic Aperture Radar. *Instrument Panel Report. Earth Observing System IIf*. Nat. Aeron. and Space Admin., Washington, DC.

NASA. (1987c). Altmetric system. *Instrument Panel Report. Earth Observing System IIh*. Nat. Aeron. and Space Admin., Washington, DC.

Niemann, B.J., Sullivan, J.G., Ventura, S.J., and Chrisman, N.R. (1987). Results of the Dane County Land Records Project. *Photogramm. Eng. Remote Sens.* 53:1371–1378.

O'Loughlin, E.M. (1986). Predictions of surface saturation zones in natural catchments by topographic analysis. *Water Resources Res.* 22:794–804.

Owe, M., Chang, A., and Golus, R.E. (1988). Estimating surface soil moisture from satellite microwave instruments and a satellite-derived vegetation index. *Remote Sens. Eviron.* 24:331–345.

Peterson, G.W., Connors, K.F., Miller, D.A., Day, R.L., and Gardner, T.W. (1987). Aircraft and satellite remote sensing of desert soils and landscapes. *Remote Sens. Environ.* 23:253–271.

Pickup, G. (1976). Geomorphic effects of changes in runoff Cumberland Basin NSW. *Aust. Geogr.* 13:188–193.

Pickup, G. (1988a). Modelling arid zone soil erosion at the regional scale. In R.F. Warner (ed.), *Australian Fluvial Geomorphology*. Academic Press, Sydney, Australia,

Pickup, G. (1988b). Hydrology and sediment models. In: M.G. Anderson (Ed.), *Modelling Geomorphic Systems*. Wiley, New York, pp. 153–215.

Pickup, G., and Chewings, V.H. (1988a). Forecasting patterns of soil erosion in arid lands from Landsat MSS data. *Int. J. Remote Sens.* 9:69–84

Pickup, G., and Chewings, V.H. (1988b). Estimating the distribution of grazing and patterns of cattle movement in a large arid zone paddock: An approach using animal distribution models and Landsat imagery. *Int. J. Remote Sens.* 9:1469–1490.

Pickup, G., and Nelson, D.J. (1984). Use of Landsat radiance parameters to distinguish soil erosion, stability and deposition in arid central Australia. *Remote Sens. Environ.* 16:195–209.

Riley, S.J. (1981). The relative influence of dams and secular climatic change on downstream flooding, Australia. *Water Resources Bull.* 17:361–366.

Robinove, C.J. (1982). Computation with physical values from Landsat digital data. *Photogramm. Eng. Remote Sens.* 48:781–784.

Rodriquez, V., Gigord, P., de Gaujac, A.C., and Munier, P. (1988). Evaluation of the stereoscopic accuracy of the SPOT satellite. *Photogramm. Eng. Remote Sens.* 54:217–221.

Schumm, S.A. (1968). River adjustment to altered hydrologic regimen. Prof. Paper, 598 U.S. Geological Survey. Washington, DC.

Senft, R.L., Rittenhouse, L.R., and Woodmansee, R.G. (1983). The use of regression models to predict spatial patterns of cattle behaviour. *J. Range Manag.* 36:553–557.

Stafford Smith, M. (1989). Modelling: Three approaches to predicting how herbivore impact is distributed in rangelands. New Mexico State Univ. Agric. Exp. Stn. Res. Rep. 628.

Warner, R.F. (1987a). The impacts of alternating flood- and drought-dominated regimes on channel morphology at Penrith, New South Wales, Australia. Sci. Pub. 168, Int. Assoc. Hydrol., pp. 327–338.

Warner, R.F. (1987b). Spatial adjustments to temporal variations in flood regime in some Australian rivers. pp. 15–40. In K. Richards (ed.), *River Channels: Environment and Process.* Blackwell, Oxford, England.

Wasson, R.J., and Clark, R.L. (1987). The quaternary in Australia—past, present and future. Rep. 282, Bur. Mineral Resources, Geol. and Geophys., pp. 29–34.

Weska, J.S., Dyer, C.R., and Rosenfeld, A. (1976). A comparative study of texture measures for terrain classifications. *IEEE Trans. Sys. Man, Cybernet.* SMC-6: 269–285.

Wischmeier, W.H., and Smith, D.D. (1978). *Predicting Rainfall Erosion Losses—A guide to Conservation Planning.* Handbook 537, U.S. Dept. Agric., Washington, DC.

Wolman, M.G., and Gerson, R. (1978). Relative scales of time and effectiveness of climate in watershed geomorphology. *Earth Surf. Proc.* 3:189–208.

Wolman, M.G., and Miller, J.P. (1960). Magnitude and frequency of forces in geomorphic processes. *J. Geol.* 68:54–74.

12. Synoptic-Scale Hydrological and Biogeochemical Cycles in the Amazon River Basin: A Modeling and Remote Sensing Perspective

Jeffrey E. Richey, John B. Adams, and
Reynaldo L. Victoria

Deforestation of tropical river basins is one of the primary variables in global change scenarios. The overall hydrological problem for these basins could be summarized as: "How would extensive land-use change modify the routing of water and its chemical load from precipitation input through the drainage system back to the atmosphere and to the ocean?" Studies of catchment behavior for the Amazon and for other great world rivers not only are important regionally, but would be compatible in scale with general circulation models (Dooge, 1982) for extrapolation of local changes to global effects.

The application of remote sensing to river basin processes would seem to have particular potential in these difficult environments. Intensive field campaigns, such as the First ISLSCP (International Satellite Land Surface Climatology Program) Field Experiment (Sellers et al., this volume) and the Hydrologic–Atmospheric Pilot Experiments (Andre et al., 1986), have coordinated ground measurements, aircraft, and remote sensor experiments to quantify water and energy fluxes for relatively uniform and accessible environments. Schultz (1988) has reviewed the application of remote sensing techniques to specific problems in hydrology. However, there has been little progress to date on studies of large-scale catchment behavior for tropical basins. Problems of scale and vast area, sparseness of records, logistic realities and difficulty of access for ground measurements, dense vegetation and frequent heavy cloud cover, and lack of suitable scientific constructs constrain not only remote sensing but any integrative analysis as well.

Given the rapidity of land-use change, it is essential to launch basin-level studies of hydrological and biogeochemical properties in the tropics immediately. Such an analysis must optimize the resources, knowledge, and logistic support currently available. In this chapter, we develop a research strategy combining field- and remotely sensed data with modeling for determining the fluxes of water and elements in the Amazon River basin on synoptic scales (catchment and regional space scales of 100 to 500 km and time scales of 10 to 15 days). We believe that existing information, in particular remote sensing data, are sufficient to implement the strategy now. Though the concept is developed for the Amazon basin, it is intended to be transferrable to other large basins.

An Analytical Perspective for Large Drainage Basins

The Amazon basin is the largest contiguous stand of tropical rain forest and savannah in the world, 6 million square kilometers, consisting of several major physiographic zones, each with characteristic geology, soils, landscape, climate, and vegetation (Figure 12.1). The rain forest is being replaced with degraded scrub at the rate of 20,000 km²/yr (1987 estimate, A. Setzer personal communication).

It is convenient to think of basin-scale biogeochemical and water cycles as a combined problem in regional water balances and subsequent downstream routing (Figure 12.2). Then the balance of water at any site and time can described by

$$R = P - ET \pm SM \qquad (1)$$

where P is precipitation, R is the effective runoff, ET is actual evapotranspiration, and SM is change in soil moisture and groundwater storage. The research problem is to evaluate Equation (1). Issues that must be addressed include:

1. What are the distributions and their variability of the components of the hydrological and biogeochemical cycles in the different sectors of the drainage basin across space and time scales?
2. What are the surface properties (vegetation, soils, topography) in specific regions, and how will they change over time?
3. How do properties of soil, vegetation, and topography influence the processes controlling water and element fluxes? Any ability to "predict" change requires a physical, not just statistical, understanding of routing properties.
4. What are the consequent exchanges of water and energy between the land surface and the atmosphere?

12. Synoptic-Scale Cycles in the Amazon 251

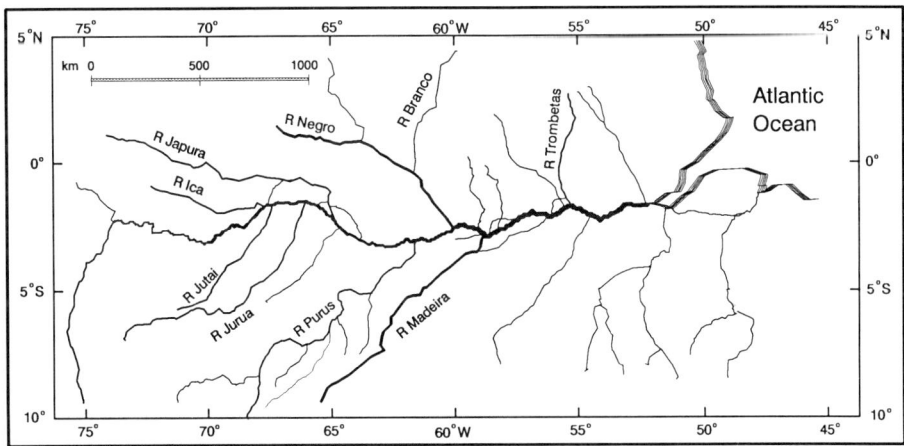

Figure 12.1. The Amazon basin (note the location of major tributaries for later reference).

$$R = P - ET \pm SM$$

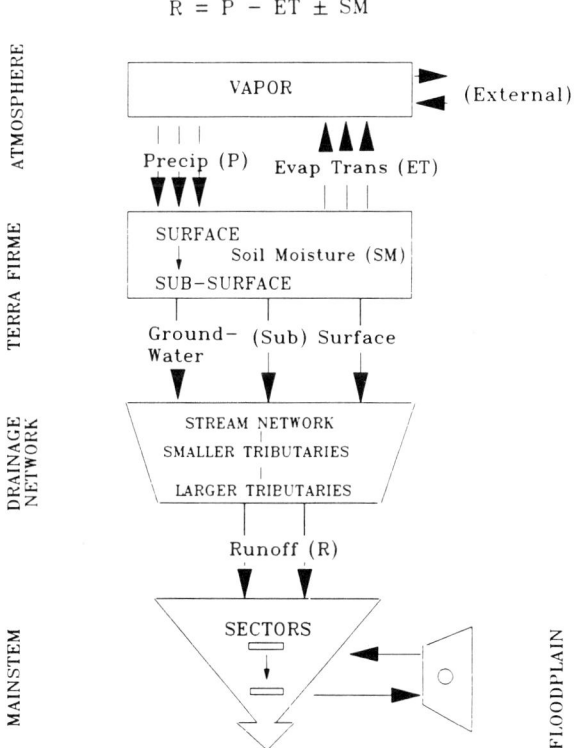

Figure 12.2. A synoptic river basin perspective, where runoff (R) is equal to precipitation (P) minus evapotranspiration (ET) plus or minus soil moisture (SM).

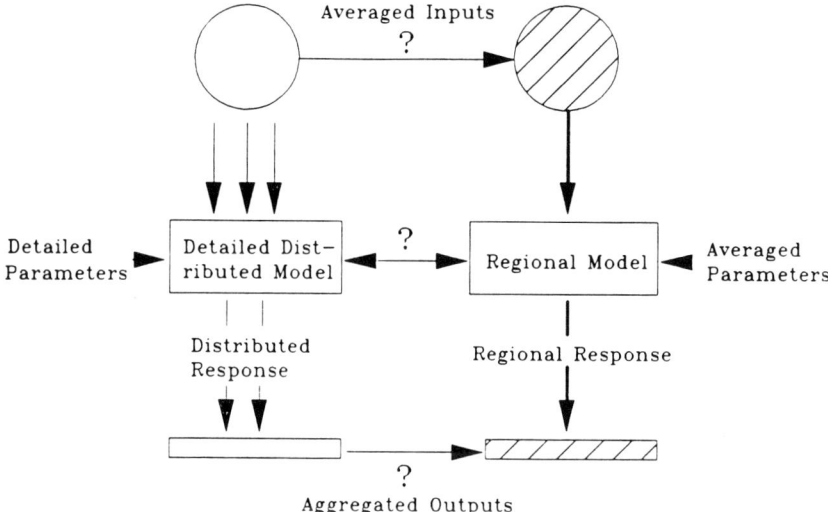

Figure 12.3. Schematic of hierarchy of scales in hydrological responses between detailed local sites and aggregated regional responses. [Adapted from *Patterns to Processes*, NASA Eos Science Steering Committee Report, Vol. II, p.73.]

A Regional Water Balance Perspective

The first problem in solving Equation (1) for a large drainage basin such as the Amazon is to establish the hierarchy of time and space scales at which the processes that control the fluxes of interest are operable (Figure 12.3). Hydrological and ecological modelers have traditionally focused most of their attention on Equation (1) at the very small to moderate spatial scales of square meters to several hectares. In theory, water flow at a site can be described in terms of physical properties based on the equations of fluid mechanics representing the conservation of mass and linear momentum. Application of a strict physical approach to modeling, even at these scales, is difficult because the mechanisms driving the hydrological process are incompletely understood, and may be difficult to express mathematically. Systems models, which are conceptual to the extent that their structure is based on the aggregate physical pathways describing water movement in a catchment (e.g, Crawford and Linsley, 1966; Johanson et al., 1984), are more common in practice. However, the actual transfer functions representing these aggregate pathways are primarily based on a black-box approach that is fine-tuned for specific applications.

The integrated effect of small-scale cycling may or may not influence cycling at larger temporal or spatial scales. The question of how spatially to average the hydrological parameters of mesoscale areas when their component parameters are spatially variable and not well characterized at a smaller scale is one that has drawn the attention of many investigators (e.g., Eagleson, 1978; Brutsaert et al., 1985; Klemes, 1983). Heterogeneity of the environment with regard to overland flow variables, catchment characteristics, measurement problems, and differences in response between smaller and larger catchments makes it difficult to extrapolate from one site to larger areas (Pilgrim et al., 1982; Beven, 1983; Dooge, 1982).

Early efforts in very-large-scale, or global, hydrology emphasized the estimation of global water balances (e.g., Budyko and Drozdov, 1953). The advent of General Circulation Models (GCMs) has offered the possibility of introducing the dynamics of the forcing function (precipitation) into global-scale hydrological models. Key interfaces between surface and atmospheric hydrological processes in GCMs occur on the catchment and regional space scales of 100 to 500 km and time scales of 10 to 15 days. Most applications of GCMs have treated the rainfall-runoff transformation at only the crudest level, although a new generation of models has a more realistic treatment of the transformations.

Application to the Amazon: The Database

The second problem in defining Equation (1) for a particular drainage basin is actually to assemble appropriate data and then interpret the data. For the Amazon, ongoing projects and the literature represent important sources of data. Research on the hydrology and meteorology of the Amazon has been conducted since the mid-1970s by the Instituto Nacional de Pesquisas da Amazônia (INPA, Manaus), the Centro de Energia Nuclear na Agricultura (CENA, Piracicaba), and the Instituto de Pesquisas Espaciais (INPE, São José dos Campos).

Brazil is the second largest user of Landsat outside of the United States; INPE maintains a receiving station in Cuiaba. INPE has had Multi-Spectral Scanner (MSS) data over the Amazon available since the early 1970s and thematic mapper (TM) data since 1983, and is starting to collect Systeme Probatoire d'Observation de la Terre (SPOT) data. Advanced Very High Resolution Radiometer (AVHRR) data are available over much of the basin, as are data from the Geostationary Operational Environmental Satellites (GOES). Radambrasil (1972) completed a side-looking radar survey of geomorphology, vegetation, and soils.

As the "backbone" for large-scale hydrological analyses, the Departmento Nacional de Aguas e Energia Elétrica (DNAEE, Brasilia) maintains a gauging network of 600 precipitation, river stage, and meteorological (solar radiance, surface temperature, humidity, wind) stations throughout the basin. In the Madeira subbasin, these gauges are typically less than 200 km apart. Daily records of at least a ten-year duration are available for most stations, and longer records exist for specific stations. In addition, the ministry has compiled annual and monthly averages for the ten-year period.

CAMREX (Carbon in the Amazon River Experiment), a joint project of the University of Washington (Seattle), INPA, and CENA, has extensive data on river discharge and chemistry, and will be expanding this work in the area of main stem–floodplain exchanges. The International Atomic Energy Agency (IAEA) Amazônia I project of INPA, CENA, and INPE emphasizes local basin and regional water routing tied to potential effects of deforestation. It is currently compiling data on soil types and associated nutrient and hydrological properties, and is establishing a basin-wide network for sampling the ^{18}O of water vapor, precipitation, and runoff.

Actual site-specific studies of hydrological processes in the Amazon are few. Several "model basin" studies have attempted to quantify the evapotranspiration and the fraction of rainfall intercepted by the forest canopy of the terre firme Amazon forest by water balance methods (Franken et al., 1982; Leopoldo et al., 1982). INPA, through the IAEA program, is starting a field-scale experiment for characterizing runoff processes at both a natural and a deforested site. Determining the effects of precipitation intensity on the type of runoff process, on infiltration, and on resulting soil moisture at both sites is planned. Water stress–foliage relations are being investigated across a gradient from cleared land into forest as part of this project.

Annual evapotranspiration (ET) has been calculated by energy budget methods (Villa Nova et al. 1976), and has been measured more precisely at a natural forest site near Manaus by a combined team of INPE, INPA, and the Institute of Hydrology (U.K.) (Shuttlesworth et al., 1984). From such field experiments, runoff type and rates as a function of precipitation and vegetative cover for use in a mesoscale representation could be determined for specific sites. The National Aeronautics and Space Administration (NASA)/INPE Amazon Boundary Layer Experiments 2A and 2B provided linkages between local regional climatology and gas fluxes.

Application to the Amazon

From the information available for the Amazon, we can begin to solve Equation (1). Condensational energy release from convective precipitation is of sufficient magnitude to influence global weather and climatic patterns. Atmospheric water vapor over the Amazon basin comes primarily from the Atlantic, with water vapor from the Pacific excluded by the Andes (Salati et al., 1979; Lettau et al., 1979). Approximately 50% of rainfall is recycled via ET from the forest (Villa Nova et al., 1976; Salati et al., 1979; Lettau et al., 1979). Water vapor export to the south has been suggested by Oliveira (1986) and James and Anderson (1984), and we have found ^{18}O of water vapor in Brasilia characteristic of Amazon vapor (R. Victoria, unpublished data). Such a transport would appear to be very important in determining the weather of the central part of the continent; if so, deforestation could have a significant impact on climate to the south of the Amazon itself.

A pronounced difference in wet and dry seasons between the northern and southern sides of the basin is caused by the slow seasonal migration of the continental convective bands over tropical South America. The most important synoptic scale (rain-producing) systems in the Amazon may be sea breeze lines, which sometimes propagate inland all the way to the Andes. The secondary maximum of precipitation in the southern Amazon probably can be accounted for by the interaction of cold fronts with convective precipitation. More than 3,500 mm yr^{-1} of precipitation falls in the northwest lowlands, decreasing to less than 2,000 mm yr^{-1} in the extreme northeastern and southern parts of the basin, and then increasing to over 3,000 mm yr^{-1} along the coast from the Guaianas to Para, and up to 7,000 mm yr^{-1} at sites on the east side of the Andes (Salati et al., 1979). South of the equator there is a distinct dry period from June to August, whereas north of the equator the dry period lasts from January to March.

The mean precipitation of 2,500 mm yr^{-1} results in an average discharge of about 200,000 m^3 s^{-1}, or 20% of the world's runoff to the oceans. The most striking feature in the discharge regime of the main stem of the Amazon River in comparison with other, smaller rivers is not only the magnitude of the flux, but also its relative uniformity over the annual cycle; Figure 12.4 (Richey et al., 1990a). The predominant variability at interannual time scales in the Amazon hydrograph is coupled, at least in part, to variations in the El Nino–Southern Oscillation cycle (Richey et al., 1989). Current knowledge of the biogeochemistry of the river system comes primarily from CAMREX (Richey et al., 1980, 1986, 1990b; Devol et al., 1987; Hedges et al., 1986). The biogeochemistry of carbon in the main stem carries signals of drainage basin processes operative at several different time and space scales. Variations in discharge and concentrations of dissolved and particulate materials in transport occur on a relatively damped and predictable basis, with changes taking place over several weeks and tens to hundreds of kilometers. Concentrations of particulate organic carbon (POC) range from 6 mg liter^{-1} upriver to 2 mg liter^{-1} downriver in the main stem and from 6 mg liter^{-1} in the Rio Madeira to less than 1 mg liter^{-1} in the Rio Negro. Dissolved organic carbon (DOC) averages 4 to 6 mg liter^{-1} in the main stem and up to 12 mg liter^{-1} in the Rio Negro. Upriver dissolved inorganic carbon concentrations of about 1,200 μM are diluted by tributaries and floodplain drainage to 600 μM at the site most downriver. This intermediate signal, however, also appears to be affected by processes operating on very different scales. The chemical compositions of the bulk organic matter in transport appear to be established on a decade basis or longer over large areas of the Andean and tributary basins (and perhaps on the floodplain). The maximum residence time of fine POC is less than 600 years and of DOC less than 150 years, whereas coarse POC is essentially contemporary. Conversely, the level and dynamics of in situ oxidation indicate that there is also a "fast-dynamic" signal, where the relevant scale parameters are hours to days over distances of

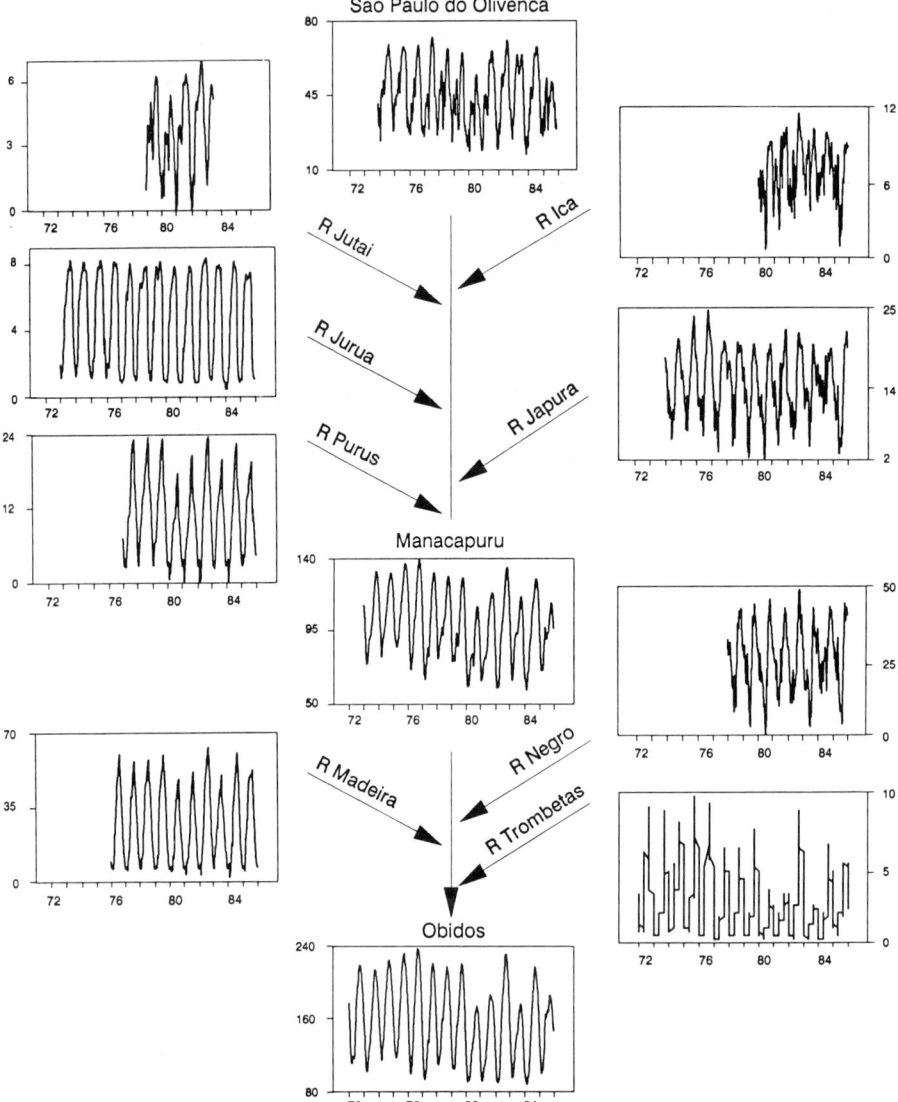

Figure 12.4. Discharge (10^3 m^3 s^{-1}) regime of the Amazon River main stem and major tributaries, 1972–1985. [Richey et al. (1990a).]

several kilometers or less. Evasion of CO_2, invasion of O_2, and in situ oxidation were of comparable magnitude, 3 to 8 μmol m^{-2} s^{-1}. Export of total organic carbon (TOC) from the Amazon is 36.1 Tg yr^{-1}.

A River Basin Analytical Strategy

The above review of the information available for the Amazon and consideration of the components of Eq. 1 indicate that it is indeed feasible to continue. To account for scaling and the realities of the data currently and potentially available, we have chosen the strategy of defining regional, aggregated water and energy balances concurrent with modeling of more detailed site-specific processes (Figure 12.5). The objective is to calculate

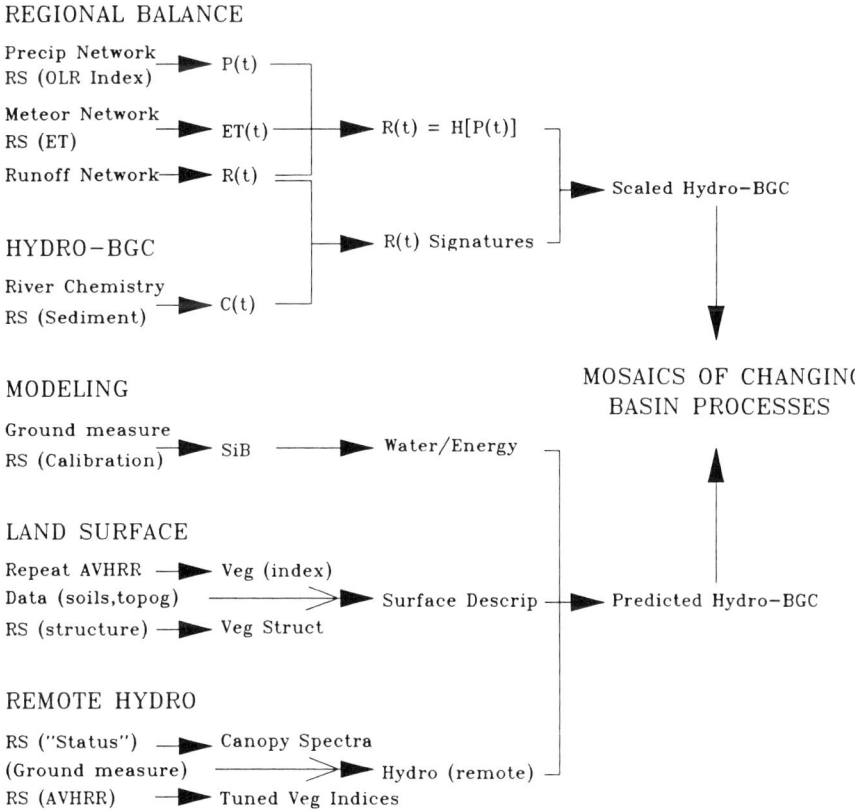

Figure 12.5. Overall research strategy, combining regional balances, linkages between hydrology (hydro) and biogeochemistry (BGC), local balances derived from modeling and experiments, description of the land surface, and scaling of hydrological processes using remote sensing (RS).

the time- and space-averaged balances of water and energy as a function of vegetation, soils, and topography for subdrainages of the basin.

The aggregated empirical approach provides boundary conditions on extrapolations from more detailed studies and modeling at local sites. Remote sensing furnishes the basis for extending local observations to regional scales. The critical test of the essentially independent approaches will be to determine their convergence. With water fluxes described, it will be possible to link hydrologic and biogeochemical cycles. Results should be expressed on the basis of pixel resolution, based on some understanding of subpixel processes, both from hydrological and spectral perspectives.

The Regional Model: $R(t) = H[P(t)]$

To provide overall constraints on water and energy fluxes using data that are realistic to obtain, the plan starts at the regional scale, with river discharge (runoff) and then precipitation at the major tributary level (10^5 to 10^6 km^2). The relation between precipitation inputs and runoff outputs can be examined generally via:

$$R(t) = H[P(t)] \qquad (2)$$

where H represents the operator(s) transforming inputs to outputs. Then ET can be constrained annually as the difference between P and R, and examined in more spatial and temporal detail via regional applications by energy calculations. If the distributions of P, R, H, and ET could be done with sufficient precision, insight into the regional distributions of soil moisture would be possible.

Regional Runoff

River stage records, the data from which discharge is calculated, are among the most complete and the most accurate data available for remote basins. Hence calculation of R provides the strongest constraints on the overall budget analysis (e.g., Figure 12.4). The flow patterns of water in river channels can then be described with routing models, which account for inputs from the upstream reach, primary gauged tributaries, local ungauged tributaries and channels, exchanges with the floodplain, and exports to the next downriver reach (Richey et al., 1990a).

An important requirement in describing the discharge regime for a large river is to account for the time sequence of the inundation of floodplains. This information is important not only for water routing but also for estimation of the extent of biogenic gas fluxes; floodplains are an important source of methane to the troposphere. Remote sensing has direct applications to this problem. We have mapped floodplain inundation patterns for different water levels using brightness differences between MSS images taken of the same locale over time (J. Adams, unpublished data). I. Fung (unpublished data) has found seasonal differences in polarization of microwave data over the entire floodplain that correlate to river stage. These types of data, combined with the mass balance routing models, may make it possible to calculate the storage capacity of the main channel and floodplain at different discharges.

Regional Precipitation: Synoptic Analyses

Precipitation patterns can be determined by analyses of the data from rain-gauge networks. Together with the analyses of these data, other synoptic analyses in the atmosphere can be done by making use of the operational daily analyses from, for example, the European Center for Medium Range Weather Forecast (ECMWF) and radiosonde data.

Because of the well-known sparseness of precipitation data in these regions, it is necessary to improve the methods for obtaining such information from satellite data. Relation of satellite observations to the rain-gauge network can be considered through the use of the proxy precipitation records derived from measurements of outgoing long-wave radiation (OLR) obtained from GOES or AVHRR. Through comparison of these data with the precipitation network data, it might be possible to improve current algorithms; e.g., Richard and Arkin (1981). The Tropical Rainfall Mission (TRMM) could provide valuable data.

H: The Rainfall-Runoff Operator

The lag between precipitation and observed stream discharge, as represented by H [Equation (2)] is a result of the passage of water through vegetation and the soil column to the atmosphere and to the drainage network and the routing of that water through the stream channels to the measuring points. Some indication of the lag between precipitation and discharge may suggest the dominant runoff processes occurring in the basin, if the distance between gauging stations is close enough.

Several possibilities exist for examining aggregated rainfall-runoff responses. The lag between peak precipitation and maximum discharge can be calculated empirically for station pairs in the same basin that represent reference topography and canopy cover. In the Amazon, the lag between the observation of a water mass tracer, $\delta^{18}O$, in precipitation and its subsequent appearance in a river channel is of the order of four to eight weeks and can be traced (Mortatti et al., 1986; Richey et al., 1990a; Goncalves, 1979; Clarke, 1982).

The possible use of geomorphic information to determine the most critical processes governing the rainfall-runoff response of large rivers deserves attention. In particular, stream order and subbasin shape may be used to characterize both the type of runoff occurring in a subbasin and the lag time between maximum precipitation intensity and peak flow in upstream tributaries (where runoff peaks are not substantially attenuated by the channel). Drainage network structure in remote areas is readily mapped by remote sensing.

Geomorphic unit hydrograph (GUH) theory (Rodriguez-Iturbe and Valdes, 1979), which is based primarily on drainage network structure, is potentially applicable to this problem. Although the rainfall-runoff component of GUH models is essentially probabilistic (that is, the most common application has been to predict flood frequency distributions rather than the characteristics of a particular runoff hydrograph), GUH theory has shown some promise for estimating runoff response for large, complex catchments where runoff is governed primarily by the channel network. It has been applied with some apparent success to the Mamon River in central Venezuela (Rodriguez-Iturbe et al., 1982; Valdes et al., 1979). One advantage of the GUH approach is that there are few or no site-specific parameters to be estimated from prior rainfall-runoff records.

Regional Evapotranspiration and Soil Moisture

Evapotranspiration is a much more stable process than precipitation (e.g., it is less variable over time, and variations arise over a longer time period). Methods for determining evaporation locally at a point using micrometeorologic data are available, but have not been useful or practical for application at the basin scale, Most rainfall-runoff models treat evaporation as a function of potential evaporation, which, in turn, is based on a bulk transfer coefficient and a reduction factor. These methods of calculating evaporation depend on a knowledge of soil moisture that can vary greatly over relatively small distances. A mesoscale representation of this process, therefore, must incorporate information about the spatial variability of soil moisture, as well as its mean value.

Regional potential ET can be calculated using the climatological data (insolation, humidity, temperature, wind speed) available from the meteorological network employing the Monteith (1973) method adapted to a tropical forest (Villa Nova et. al., 1976). Based on detailed eddy correlation methods, Shuttlesworth et al. (1988) have modified the empirical (bulk) formulation fitted for the micrometeorological data and have shown that this model yields results within 15 to 20% of the actual fluxes, and describes the seasonal cycle. Given the variability in the base climatological data, it is not always possible to calculate ET even on a monthly basis for all years. In some cases where data are limited, it may be necessary to calculate the monthly or annual means, and to derive the distribution of these averages from theoretical considerations (e.g., Eagleson, 1978).

Fixed station data are not extensive enough to characterize the spatial variability of *ET* in the same way that the precipitation data can be used to characterize the spatial variability of the precipitation process. A significant problem is to determine the actual mesoscale distribution of the energy terms from which ET is calculated as a function of vegetation type. Remote sensing of the radiances corresponding to these energy terms may be the best way to do this. Within the likely evolution of the capability for detecting these terms over remote basins, however, the challenge is large. For example, at the present time it is not possible to measure soil moisture under dense tropical forest canopies using remote sensing. The problems of remote sensing of ET and soil moisture are reviewed elsewhere in this volume.

Modeling: From Local to Regional

A better understanding of the small-scale processes and their relation to vegetative cover represented in the lag phenomenon discussed above is necessary before the runoff process can be parameterized at the mesoscale.

Surface Hydrology in General Circulation Models

Driven by the need to improve the representation of the land surface processes in GCMs, models of these interactions have been developed recently which take into account morphological and physiological characteristics of the vegetation as well as physical characteristics of the soil. These include the Biosphere-Atmosphere Transfer Scheme (BATS) of Dickinson and Henderson-Sellers (1988) and the Simple Biosphere Model (SiB) of Sellers et al. (1986).

The transfer of sensible heat and water vapor between the canopy/ground and the atmosphere is modeled using the "resistance analogue" concept (Monteith, 1973). However, the characteristics of each major type of vegetation/soil are taken explicitly into consideration in the calculations, while the morphological properties of the vegetation are taken into account in the calculations of the transfer of momentum between the canopy and atmosphere (the drag exerted by the vegetation on the atmosphere) as well as in the calculations of surface albedo. Of paramount importance for the transfer of water vapor from the soil to the atmosphere is the stomatal resistance that, by and large, controls this transfer. Another important factor in the transfer of water vapor from the canopy is the evaporation of precipitation water intercepted in the canopy (interception loss). SiB and BATS explicitly calculate this loss, as well as a simplified formulation of the soil water balance, including runoff. Given the properties of vegetation and soils and soil moisture, plus wind speed, air temperature and humidity at a reference level above the canopy, visible and near-infrared (IR) incoming solar radiation, and precipitation, these models can be used to calculate the fluxes of vapor more accurately than models that use simple energy balance formulations.

Model Application to the Amazon

The BATS model has been used to begin to assess the effects of Amazon deforestation on the atmosphere (Dickinson and Henderson-Sellers, 1988). Application of the SiB to the Amazon is in its initial phase (P. Sellers and C. Nobre, unpublished data). For application of the SiB or BATS-type models to the Amazon, the models have to be calibrated with measurements of air temperature and humidity above the canopy, precipitation and downward fluxes of solar and near-IR radiation, and wind speed for typical ecosystems (terra firme forests, igapo forest, campina forest, floodplain vegetation, and vegetation of areas under use such as grass).

How a physical model developed for a particular set of conditions can be used to scale to larger areas, until ultimately direct comparisons with the independently calculated regional water balances can be made, is a challenge. Site-specific calibrations, which in themselves require field work of almost heroic effort, still do not address the problem of parameterization over larger areas. The long-term challenge for remote sensing is to be able to measure the parameters required to calibrate these models directly.

Once mean properties of vegetation and soil are known for a typical catchment area in the Amazon, climatological series of the precipitation and atmospheric parameters could be input into one-dimensional versions of these models to calculate the time evolution of ET, runoff, and soil moisture. Given a historical sequence of input data, these models can be integrated in time to give the variation in ET, soil moisture, and runoff.

Sensitivity analyses could then be used to determine which basin-scale processes control the large-scale dynamics. Categories of variables to be aggregated include spatial invariant variables (solar radiation), surface-condition-dependent variables (infiltration, reflected radiation), and surface-dependent time-invariant variables (soil properties) (Jackson, 1985). These processes would be singled out for more detailed modeling and, to the extent possible and necessary, enhanced data collection. Studies of the sensitivity of the model calculations can be done for different vegetation/soil types of the Amazon and for different scenarios of land use, given the same input data.

From Local to Regional Hydrology with Remote Sensing

A central issue in our research scheme is how to scale from the regional water balances (which can be defined unambiguously) down to spatial and temporal scales at which a knowledge of actual processes, as represented by the physical-scale model, can be represented. Superimposed on the hydrological and chemical problem is the scaling inherent in going from remote detection of properties on the scale of the 1-km pixel, which is practical for large areas and is characteristic of AVHRR (or Eos-era

MODIS and HMMR), down to pixels on scales of 10 to 100 m (MSS, TM, SPOT, and Eos-era HIRIS, or SAR).

Given the nature of the canopy, it would be difficult to separate out individual terms of the water budget for detection. Rather, a more productive approach may be to ask the question, "Is there a characteristic spectral signature from the canopy indicative of the hydrological condition (as determined by ground measurements) of a site?"

We are currently investigating this problem using Landsat MSS and TM data. Our preliminary results indicate that by using a spectral mixing model that accounts for subpixel scale heterogeneity, we are able to measure the fraction of the forest canopy that has lost green leaves. This defoliated fraction of the forest increases in area during the dry season in the terra firme and increases in the floodplain when the forest is flooded. In the future, the 10-nm spectral resolution of HIRIS will allow measurement of subtle shifts in the depths and widths of water and chlorophyll absorption bands in green biomass. During dry seasons, it may be possible to detect changes in these absorption bands as a function of water stress by calibrating to ground stations. Leaf area index (LAI) would also be expected to decrease during dry seasons, thereby decreasing ET. Improved measures of LAI from Landsat and HIRIS will be made possible by using a spectral mixing model that accounts for the fractions of green vegetation, stem and branch biomass, soil, and shade (Adams et al., 1989). As the amount of deforestation increased, it would be increasingly important to account for soil exposed at the subpixel scale.

New results from the NASA AVIRIS aircraft images suggest that it may be possible to make direct measurements of water vapor in and over vegetation canopies using narrow bandpasses centered in the weak water bands (e.g., at 960 nm). This suggests the possibility of using HIRIS in conjunction with thermal data (TIGER/ITER) to image the water vapor concentration associated with Amazon vegetation. By calibrating such images at ground stations, it may be possible to understand better the spatial and seasonal changes in ET at the 30-m pixel scale and to extrapolate to the MODIS scale.

Subpixel mixing at the 1-km scale is not well understood at present. It is also scene dependent. If spectral mixing at the 30-m scale using TM and HIRIS can be understood in terms of processes observed at ground stations, the same spectral end-members could be used with MODIS data on a regional scale. We are currently studying scaling from Landsat TM to AVHRR for the Amazon; however, the wavelengths of AVHRR limit the use of spectral mixing models.

In this fashion, a spectral model can be developed to relate hydrological properties to canopy—and thus site—water status. Then the hydrological status for larger areas could be predicted by applying the model to a detailed vegetation description for that area (as represented in a Geographic

Information System model). Results would be compared with the scaled-process hydrology calculations resulting from the comparison of the regional balances with the scaled physical model.

Coupling of Hydrological and Biogeochemical Cycles

The Riverine Perspective

Runoff is important not only as an output from land but also as the transport medium for sediment and nutrients. The concentration of major ions in the tributaries, and subsequently in the main stem, is strongly influenced by the weathering regimes in the respective catchments. Superimposed on these patterns are the effects of nutrient cycling by the vegetation. Outflow from a tributary thus represents the net result of the hydrological and biogeochemical transformations occurring in its drainage basin. The main channel of the Amazon then represents an integration of all that took place upstream in the respective catchments.

Results to date of CAMREX show characteristic differences in the patterns of discharge, sediment and nutrient concentrations, and geochemical tracers among the different tributaries. To link observed patterns of river chemistry to terrestrial processes, we can start by considering the relative fluxes of inorganic and organic components from the respective tributaries. For example, to compare drainage basins of different sizes, Richey and Ribeiro (1987) computed the area-normalized flux of water and nitrate for the major tributaries. Clearly, such analyses must take into consideration differences in nutrient supply via geochemical weathering (*sensu* Stallard and Edmond, 1983); the challenge is separating the biological from purely geochemical signals.

The organic composition associated with total suspended sediments remains relatively constant for a particular site (Hedges et al., 1986; Richey et al., 1990b). The ability to be able to determine basin-wide sediment concentrations yields critical information not only on sediment transport/erosion but also on links to biogeochemical cycles.

We have used MSS images of the Amazon to map changes in the near-surface turbidity in channel and floodplain environments (see Color Plate 6). Field work near Manaus during similar flow conditions has verified the general turbidity patterns observed in the MSS image (J. Richey and L. Mertes, unpublished data). The mixing model approach used (Adams et al., 1989) can be employed to calibrate and estimate the actual changes in surface turbidity across the floodplain. The calibration of the MSS Landsat scenes can use semiempirical methods to calculate the surface turbidity visible in the images. Non-linear regressions between Landsat signals and near-surface turbidity measurements can have correlations as high as 98%.

The modeling and MSS-based determination of sediments could be expanded to all major tributaries. Model development could be expanded to relate sediment changes to topography and landuse change. The application of TM, and ultimately HIRIS, data to finer-scale determinations of sediment concentration, and possibly direct assessment of chemical parameters (e.g., highly colored humic materials), could be explored.

The Terrestrial Perspective

Research currently being developed elsewhere that emphasizes the direct application of remote sensing to such biospheric processes as decomposition, primary production, and biogenic gas exchange could be incorporated. This information is summarized elsewhere in this volume.

Define Basin Structure: GIS Model Setup

The long-term intent is to be able to track the time rate of change of vegetation in the basin at several levels of resolution, including (1) the presence/absence of primary forest; (2) the relative distribution of primary forest, regrowth, savannah, and scrub; and (3) the canopy architecture associated with the different vegetation communities. Then the objective is to relate properties of the hydrological and biogeochemical cycles to these properties.

Given the multisource, complex data requirements and modeling interfaces needed, a Geographic Information System (GIS)-based model is a logical technology to express such a model. The "region" represented by each grid is defined for progressively finer resolution by the drainage network, starting at the major tributary basin level (10^5 to 10^6 km^2) and proceeding to local basins (10^3 km^2).

For initial setup, existing data sets for topography (RADAM), vegetation cover, and soils can be used. Vegetation maps from composite AVHRR images, tuned with Landsat, represent the type of feasible work required to update the model description. The vegetation description could be upgraded by assembling composite AVHRR images over the entire basin. Increased resolution of canopy structure could be investigated by application of mixing models with Landsat TM combined with field measurements (J. Adams, unpublished data).

Presentation

Potential output products of this research plan might include:

1. Mass balance summaries of hydrological and biogeochemical parameters by grid area of the Amazon basin, including precipitation, ET, river discharge, meteorological variables, biomass, soils, topography, and riverborn sediments and elements.
2. Basic data going into the mass balance calculations.
3. Color-coded mosaics of these data assembled for local regions, and ultimately for the basin.

The requirements of this research plan include access to a series of three-dimensional fields derived from model output, geographic information systems, and near-simultaneous observations of the earth from the current and next generation of satellites; that is, global data sets. Virtually all such data, particularly satellite images, must be processed using specialized computer facilities designed for computations on large arrays, the results of which are presented in image form. These images may be maps of various types, or photographic images where tones or colors are keyed to specific parameters.

Few laboratories in the country are equipped to process large satellite images or mosaics of images. For example, most university laboratories cannot store, process, and read out a full Landsat TM scene without first subdividing the data and reassembling the images. It is even more difficult to assemble multiple scenes or other types of data covering large areas without adequate equipment and facilities.

The research approach described here relies on a multidisciplinary approach to describing drainage-basin-scale hydrological and biogeochemical cycles in the Amazon basin. As stated above, we believe that the expertise and technology currently exist to make significant progress on a problem of global importance.

Acknowledgments

The evolution of the concepts represented in this chapter benefitted in particular from conversations with C. Nobre, A. Setzer, L. Molion, and R. Cunha (INPE), E. Salati, L. Martinelli. and J. Mortatti (CENA); M.N.G. Ribeiro, W. Franken, A. Marques, and B. Forsberg (INPA); P. Vose (IAEA), P. Sellers and C. Tucker (NASA GSFC), and T. Dunne (UW).

This research was supported by the National Science Foundation under grant BSR-8107522, and the International Atomic Energy Agency BRA/0/010-08. Contribution number 30 of the CAMREX project and number 1818 of the School of Oceanography, University of Washington.

References

Adams, J.B., Smith, M., and Gillespie, A. (1989). Simple models for complex natural surfaces: a strategy for the hyperspectral era. *IEEE Geosci. Rem. Sen.* 1:16–21.

Andre, J.C., Goutorbe, J.P., and Perrier, A. (1986). HAPEX-MOBILHY: A hydrologic atmospheric experiment for the study of water budget and evaporation flux at the climatic scale. *Bull Amer. Meteorol. Soc.* 67:138.

Beven, K. (1983). Surface water hydrology—runoff generation and basin structure. *Rev. Geophys. Space Phys.* 21:721–730.

Brutsaert, W., Alley, W.M., Chappell, C.F., Georgakakos, K.P., Gupta, V.K., Jackson, T.J., Milly, P.C.D., and Sellers, P. (1985). Hydrological and meteorological experimentation at the mesoscale. *Eos* 66(34):601–602.

Budyko, M.I., and Drozdov, O.A. (1953). Characteristics of the moisture circulation in the atmosphere. *Izv. Akad. Nauk SSSR Ser. Geogr. Geofiz.* 4:5–14.

Clarke, R.T. (1982). Isotopic methods in hydrological studies of the Amazon forest. Int. Atomic Energy Agency IAEA BRA-78-006.
Crawford, N.H., and Linsley, R.K. (1966). Digital simulation in hydrology: Stanford Watershed Model IV. Tech. Rep. No. 39, Stanford Univ., Stanford, CA.
Devol, A.H., Richey, J.E., Quay, P., and Martinelli, L. (1987). Dissolved gases and air-water exchange in the Amazon River. *Limnol. Oceanogr.* 32:235–248.
Dickinson, R.E., and Henderson-Sellers, A. (1988). Modelling tropical deforestation: a study of GCM land-surface parameterizations. *Quart. J. Roy. Meteorol. Soc.* 114:439–462.
Dooge, J.C.I. (1982). Parameterization of hydrologic processes. In P. Eagleson (ed.), *Land Surface Processes in Atmospheric General Circulation Models*. Cambridge Univ. Press, Cambridge, England.
Eagleson, P. (1978). Climate, soil, and vegetation. 1. Introduction to water balance dynamics. *Water Resource Res.* 14:705–712.
Franken, W., Leopoldo, P.R., Matsui, E., and Ribeiro, M. (1982). Interceptacao das precipitacoes em foresta amazonica de terra firme. *Suppl. Acta Amazonica* 12(3):15–22.
Goncalves, A.R.L. (1979). Determinacao do tempo de residencia da agua de chuva em algumas bacias hidrograficas atraves de valores de isotopos estaveis. Master's thesis, Univ. Sao Paulo, Sao Paulo, Brazil.
Hedges, J.I., Clark, W., Quay. P.D., Richey, J.E., Devol, A.H., and Ribeiro, N. (1986). Composition and fluxes of organic matter in the Amazon River. *Limnol. Oceanogr.* 31:717–738.
Hedges, J.I., Ertel, J.R., Quay, P.D., Grootes, P.M., Richey, J.E., Devol, A.H., Farwell, G.W., Schmidt, F.H., and Salati, E. (1986). Organic carbon-14 composition in the Amazon River system. *Science* 231:1129–1131.
Jackson, R.D. (1985). Evaluating evapotranspiration at local and regional scales. *Proc. IEEE* 73:1086.
James, X., and Anderson, Y. (1984). The seasonal mean flow and distribution of large-scale weather systems in the southern hemisphere: The effects of moisture transports. *Quart. J. Roy. Meteorol. Soc.* 110:943–966.
Johanson, R.C., Imhoff, J.C., Kittle, J.L., and Donigian, A.S. (1984). *Hydrologic Simulation Program—FORTRAN (HSPF) Users Manual*. Release 8.0, Rep. EPA-600/3-84-006, Envir. Protection Agency, Washington, DC.
Klemes, V. (1983). Conceptualization and scale in hydrology. In: Scale Problems in Hydrology. *J. Hydrol.* (Spec. issue) 65:1–23.
Leopoldo, P.R., Franken, W., Matsui, E., and Salati, E. (1982). Estimativa de evapotranspiracao de floresta amazonica de terra firme. *Suppl. Acta Amazonica* 12(3):23–28.
Lettau, H.K., Lettau, and Molion, L.C.B. (1979). Amazonia's hydrologic cycle and the role of atmospheric recycling in assessing deforestation effects. *Mo. Weather Rev.* 107:227–237.
Monteith, J.L. (1973). *Principles of Environmental Physics*. American Elsevier, NY.
Mortatti, J., Salati, E., Victoria, R., and Ribeiro, M. (1986). Analysis of the isotopic behavior of hydrogen and oxygen of water in the main channel of the River Solimoes/Amazon. *Mitt. Geol. Palaont. Inst.* 58:259–266.
Oliveira, A. (1986). Interacoes entre os sistemas frontais na America do Sul e a conveccao na Amazonia. MSc., 4008-TDL/239, Inst. de Pesquisas Espaciais, Sao Jose dos Campos, Brazil.
Penman, H.L. (1948). Natural evaporation from open water, baresoil and grass. *Proc. Roy. Soc. A.* 193:120–145.
Pilgrim, D.H., Cordery, I., and Baron, B.C. (1982). Effects of catchment size on runoff relationships. *J. Hydrol.* 58:205–221.

Radambrasil. (1972). Mosaico semi-contralado de radar, escala 1:250,000. Ministerio das Minas e Energia Dep., Nat. Producao Mineral, Rio de Janeiro, Brazil.

Richard, F., and Arkin, P. (1981). On the relationship between satellite-observed cloud cover and precipitation. *Mo. Weather Rev.* 109:1081-1093.

Richey, J.E., Brock, J.T., Naiman, R.J., Wissmar, R.C., and Stallard, R.F. (1980). Organic carbon: oxidation and transport in the Amazon River. *Science* 207:1348-1351.

Richey, J.E., Devol, A.H., Hedges, J.I., Forsberg, B., Victoria, R., and Ribeiro M.N.G. (1990a). Distributions and fluxes of carbon in the Amazon River. *Limnol. Oceanogr.*, in press.

Richey, J.E., Meade, R.H., Salati, E., Devol, A.H., Nordin, C.F., and dos Santos, U. (1986). Water discharge and suspended sediment concentrations in the Amazon River: 1982-1984. *Water Resources Res.* 22:756-764.

Richey, J.E., Mertes, L.A.K., Oliveira, E., Forsberg, B., Dunne, T., Victoria, R., and Tancredi, A. (1989). Sources and routing of the Amazon River floodwave. *Global Biogeochem. Cycles*, in press.

Richey, J.E., Nobre, C., and Desser, C. (1990b). Amazon River discharge and climate variability: 1903-1985. *Science*, in press.

Richey, J.E., and Ribeiro, M.N.G. (1987). Element cycling in the Amazon Basin: A riverine perspective. pp. 245-250. In R. Dickinson (ed.), *The Geophysiology of Amazonia.* Wiley, NY.

Rodriguez-Iturbe, I., Sanabria, M.G., and Caamano, G. (1982). On the climatic dependence of the IUH: A rainfall-runoff analysis of the Nash Model and geomorphoclimatic theory. *Water Resources Res.* 18(4):887-903.

Rodriguez-Iturbe, I., and Valdes, J.B. (1979). The geomorphologic structure of hydrologic response. *Water Resources Res.* 15(6):1409-1420.

Salati, E., Dall'Olio, A., and Matsui, E. (1979). Recycling of water in the Amazon Basin: An isotopic study. *Water Resources Res.* 15:1250-1258.

Schultz, G.A. (1988). Remote sensing in hydrology. *J. Hydrol.* 100:239-265.

Sellers, P.J., Hall, F.G., Asrar, G., Strebel, D.E., and Murphy, R.E. (1988). The first ISLSCP field experiment (FIFE). *Bull. Amer. Meteorol. Soc.* 69:22-27.

Sellers, P.J., Mintz, Y., Sud, Y.C., and Dalcher, A. (1986). A simple biosphere model (SiB) for use within general circulation models. *J. Atmos. Sci.* 43:505-531.

Shuttlesworth, W.J., Gash, J.H.C., Lloyd, C., Moore, C.J., Roberts, J., A. Marques Filfo, deO., Fisch, G., Silva Filho, V. de P., Ribeiro, M.N.G., Molion, L.C.B., de A Sa, L.D., Nobre, C.A., Cabral, O.M.R., Patel, S.R., and de Moraes, J.C. (1984). Eddy correlation measurements of energy partition for Amazonian forest. *Quart. J.Roy. Meteorol. Soc.* 110:1143-1162.

Stallard, R., and Edmond, J.M. (1983). Geochemistry of the Amazon: 2. The influence of geology and weathering environment on the dissolved load. *J. Geophys. Res.* 88:9671-9688.

Valdes, J.B., Fiallo, Y., and Rodriguez-Iturbe, I. (1979). A rainfall-runoff analysis of the geomorphologic IUH. *Water Resources Res.* 15(6):1421.

Villa Nova, N., Salati, E., and Matsui, E. (1976). Estimativa de evapotranspiracao na Bacia Amazonica. *Acta Amazonica* 6(2):215-228.

13. Remote Sensing of Marine Photosynthesis

John S. Parslow and Graham P. Harris

The oceans occupy 70% of the earth's surface, and estimates of their contribution to global photosynthesis range from 10% to 50% (Perry, 1986) or 20 to 50×10^{15} g C y^{-1} (McCarthy, 1984; Martin et al., 1987). Photosynthesis in the oceans is of interest as the basis of marine food chains (Ryther, 1969) and for its role in global biogeochemical cycles (e.g., Sundquist, 1985). The nature and dynamics of marine producers differ markedly from those of their terrestrial counterparts. The pool of living plant carbon in the oceans is small (about 0.5 to 5.0 gC m^{-2}) and consists principally of microscopic unicellular organisms that turn over rapidly, on time scales of the order of days (Harris, 1980a). These turnover rates are thought to be controlled primarily by light and nutrient limitation. Because of the low biomass concentrations, estimates of photosynthetic rates have been based primarily on the measurement of rates of incorporation of ^{14}C-labeled isotopes in incubations (Harris, 1984). While there has been a long-standing discussion of the interpretation of these data (Peterson 1980; Harris, 1984), debate has recently intensified with the development of alternative methodologies (e.g., Shulenberger and Reid, 1981; Jenkins and Goldman, 1985). Evidence of the importance of extremely small cells (picoplankton) (Johnson and Sieburth, 1979, 1982; Platt and Li, 1986) and of incubation artifacts such as metal contamination (Fitzwater et al., 1982) has cast further doubt on the large historical set of marine primary production estimates.

Biological oceanographers also face an extremely difficult sampling problem. The rapid turnover rates of phytoplankton mean that biomass and photosynthetic rates can change completely over periods of a few days (Harris, 1980a). Spatial patchiness and advection further complicate the interpretation of point samples (Harris and Griffiths, 1987). There is always a temptation to extrapolate from point samples to surrounding regions, especially in the open ocean where there may be no obvious sign of horizontal spatial structure. Remote sensing has transformed our picture of the oceans by directly revealing mesoscale variation in sea-surface temperature and chlorophyll.

Thus in trying to resolve methodological inconsistencies, biological oceanographers have to contend with an inadequate sampling of spatial and temporal variation in biomass and productivity (Harris and Griffiths, 1987). It is not clear to what extent discrepancies between point samples represent real variation, nor is it obvious how to compare point samples with estimates (e.g., oxygen utilization and particle traps) that integrate over space and/or time, but may involve other assumptions concerning horizontal and vertical transport. Given these problems, the variation in global production estimates is not surprising. Recent evidence suggests that fluxes of direct significance to climate change may be even more dependent on intermittent events, and subject to greater uncertainty as a result of sampling problems, than has previously been suspected. Remote sensing is vital to resolving the spatial and temporal variation, but extrapolating from remotely sensed variables to fluxes will always depend on linking remote measurements to estimates of in situ processes. The challenges involved in designing a sampling program to improve this link remain considerable, as this chapter will discuss.

Marine photosynthesis and climate change are potentially causally linked at a range of scales. At scales of a few years to tens of years, changes in ocean and atmospheric circulation associated with climate change can be expected to change the spatial and temporal distribution of marine photosynthesis, with implications for marine food chains and commercial fisheries. At the same time, it has been suggested that marine photosynthesis and the burial of carbon in marine sediments may play a major role in long-term (scales of thousands of years) fluctuations in atmospheric CO_2 and global temperatures (McElroy, 1983). Remote sensing can play a monitoring role, allowing us to detect changes in the distribution of phytoplankton biomass that are not apparent in point samples. It can be used to improve estimates of fluxes among key pools, including atmospheric CO_2. Finally, it may be used to test models that couple physical, chemical, and biological processes at a regional or global level.

This chapter begins with a discussion of the remotely sensed parameters that may be relevant to marine photosynthesis. It then reviews simple models that link parameters that may be observed remotely (predominantly surface pigment concentrations) to in situ photosynthetic rates. Finally, the link between photosynthesis, other flux estimates, and climate change is discussed.

Remote Sensing

This section will concentrate on the remote sensing of the oceans in the visible region of the spectrum to obtain estimates of phytoplankton pigment concentrations and other constituents. Other variables, such as solar irradiance at the surface, sea-surface temperature, wind speed, sea-surface height, and surface roughness, can be estimated from satellite. These may directly affect photosynthetic rates, or may force or reflect physical processes controlling mixing and advection. This chapter will not discuss the measurement techniques for these variables, but will refer to their application later.

The reflectance spectrum of the oceans is now understood quite well in terms of the scattering and absorption of light by various constituents within the water column. Pure water has an absorption minimum in the blue, strong absorption in the red, and increased scattering at short wavelengths, so that the upwelling radiance from very clear oligotrophic water is heavily dominated by wavelengths of less then 500 nm. Chlorophyll, with its strong absorption peak in the blue (440 nm) and absorption minimum in the green (550 nm), tends to reduce the ratio of blue to green light, and this is the basis of current techniques for estimating chlorophyll content. The spectrum, and blue–green ratios in particular, are also affected by other constituents such as dissolved organics (yellow substance) and nonphototrophic particulates. The empirical relationships currently used to estimate pigment concentrations rely on the fact that these other constituents tend to covary with chlorophyll, at least in open ocean (so-called Case I) waters (Gordon and Morel, 1983).

The upwelling radiance in the red tends to increase with chlorophyll content, as the increased backscattering from particulates outweighs the increased absorption. Upwelling radiance spectra also contain a contribution near 685 nm owing to passive solar fluorescence (Gower and Borstad, 1981; Topliss, 1985). Although this peak is very small in magnitude, it is more tightly coupled to living phytoplankton than other spectral features. It may also reveal aspects of the physiological state of phytoplankton in the surface layer (Topliss and Platt, 1986), a possibility that will be discussed later.

To date, experience in the remote sensing of phytoplankton from space has come primarily from the Coastal Zone Color Scanner (CZCS) carried by Nimbus-7. This experimental sensor operated from 1978 to 1986 and measured upwelling radiance in four narrow visible bands at 443, 520, 550, and 670 nm, one broad near-infrared (IR) band, and one thermal IR band, with a surface resolution of about 0.8 km. There is now a large body of literature on the processing of CZCS data to obtain estimates of phytoplankton pigment concentration and the diffuse attenuation coefficient (e.g., Gordon and Morel, 1983; Sturm, 1983; Robinson, 1985). This processing occurs in three stages:

1. The use of calibration data to convert raw counts to radiance.
2. The removal of atmospheric effects (primarily aerosol and Rayleigh scattering) to obtain water leaving radiance.
3. The use of empirical relationships to estimate pigment (chlorophyll-a plus phaeopigment) concentrations from blue–green radiance or reflectance ratios.

This is not the place for an exhaustive review but some important lessons with implications for future sensors have been learned. Calibration is particularly important (Gordon, 1987), as most of the signal (80 to 90%) must be removed as theoretically calculated Rayleigh radiation. A slow drift in the sensitivity of CZCS has had to be corrected by assuming that the reflectance of certain oligotrophic ocean areas is constant in some statistical sense (Gordon et al., 1983). Setting aside questions of accuracy, this is hardly a desirable basis for monitoring response to climate change.

The removal of aerosol scattering is particularly difficult and currently depends on assumptions about the shape of reflectance spectra and assumptions of horizontal homogeneity in aerosol scattering spectra (Gordon and Clark, 1980a; Gordon and Morel, 1983). Plans for future sensors include improved spectral resolution and extensions into the near-IR. It is hoped that this will allow aerosol correction over a greater variety of water types.

The restriction of accurate estimates of pigment concentration to Case I waters may not be of particular significance to studies of global oceanic production. However, the near-shore waters may be the dominant sites of some important processes, such as carbon burial in sediments (Walsh, 1984). Given better spectral resolution, it may be possible to use more sophisticated reflectance models and to derive more robust concentration estimates for a number of independently varying constituents (Gordon et al., 1988). These models should have greater applicability in near-shore waters.

The ocean reflectance depends primarily on optical properties within one extinction depth (Gordon and Clark, 1980b), and pigment estimates from reflectance correspond roughly to this layer. Chlorophyll and phaeopigment both contribute to absorption, and best empirical correlations have been obtained with their sum, which is generally referred to as c_K (mg m^{-3}). Typical estimated errors in Case I waters are of the order of 20% to 40%, although discrepancies between empirical relationships from different regions may exceed this level (Gordon and Morel, 1983). An empirical relationship has also been used to estimate the diffuse attenuation coefficient K from band ratios (Austin and Petzold, 1981).

Pigment Concentration and Photosynthesis

Statistical Approaches

Given that the basic remotely sensed variable is pigment concentration c_K in the surface waters (i.e., depth $z < K^{-1}$), the problem is to relate this to photosynthetic rate and other fluxes or biomass variables of interest. In point samples from a particular depth, the ratio of photosynthetic rate to chlorophyll content, known as the assimilation number or photosynthetic capacity (Harris, 1978), is commonly measured using ^{14}C incorporation and extracted chlorophyll-a. However, the variable of interest here is the depth-integrated primary production P_T (mg C m^{-2} d^{-1}), determined by vertical profiles of phytoplankton biomass and assimilation number, which depend in turn on associated vertical profiles of physical and chemical conditions. Additional assumptions and approximations, based on historical experience and theoretical development (Harris, 1978), are required to make this connection between c_K and P_T.

The simplest approach is to look for a direct empirical relationship between P_T and c_K. By combining data from cruises ranging from the Arctic through the subtropical gyres to the Peruvian upwelling, Eppley et al. (1985) obtained a log-log plot showing a reasonably linear relationship over several orders of magnitude, with $P_T = 1{,}000\ c_K^{0.5}$ (mg C m^{-2} d^{-1}). The scatter within and between cruises about this line corresponds to an uncertainty of about one order of magnitude. According to this relationship, the ratio P_T/c_K declines as c_K increases. An explanation for this behavior will become apparent shortly. Eppley et al. found a slight seasonal effect, and strong regional differences in the ratio P_T/c_K, but these seem likely to be due primarily to differences in c_K, given the strong nonlinearity.

A sequence of papers has attempted to analyze the factors controlling P_T in the southern Californian bight (Smith and Eppley, 1982; Smith et al., 1982; Eppley et al., 1985). Early results (Smith and Eppley, 1982) showed that 65% of the variance in P_T could be explained by a regression of $\log(P_T)$ on a temperature anomaly and day length. Smith et al. (1982) found that the same variables explained the same percentage of the variance in P_T/c_K. In the data considered by Eppley et al. (1985), a regression of $\log(P_T/c_K)$ on $\log(c_K)$, temperature anomaly and day length yielded an r^2 of 0.56. Regression r^2 values do not necessarily give a good indication of the prediction errors. Eppley et al. (1985) did test the predictive value of their regression on several cruises excluded from the regression calculations. The results were fairly sobering: the regression line explained only 18% of the variance for these cruises, although errors were dominated by one anomalous cruise.

One might conjecture that at least part of the variability in the relationship between P_T and c_K is attributable to variability in the ratio of surface biomass to depth-averaged biomass. Smith and Baker (1978) investigated the relationship between the average chlorophyll concentration $<c>$ (mg m^{-3}) in the euphotic zone (above the 1% light depth) and surface chlorophyll c_K, using data from a variety of locations. A log-log regression yielded

$$<c> = 0.95 \, c_K^{0.788}, \quad \text{with a } r^2 \text{ of } 0.91$$

Brown et al. (1985) found a similar trend using data obtained in the North Atlantic, with

$$<c> = 1.10 \, c_K^{0.89}$$

Note that $<c>$ tends to exceed c_K when the latter is low, whereas the situation is reversed when c_K is high. In oligotrophic areas where c_K is small, there is commonly a pronounced subsurface chlorophyll maximum (e.g., Venrick et al., 1973), whereas in eutrophic waters with high biomass, maximum concentrations are found close to the surface.

Light Dependence and Depth Integration

The nonlinearity in the relationship between $<c>$ and c_K is too weak to explain the strong nonlinearity in P_T/c_K. The latter is more readily explained by a consideration of the dependence of photosynthesis on light intensity I (E m^{-2} h^{-1}) (E = Einstein). For a given phytoplankton culture exposed to different light intensities, photosynthetic rates normally increase linearly at low light intensities, saturate at higher light intensities, and possibly decrease at still higher intensities (for reviews see Harris, 1978, 1980b; Kirk, 1983). The portion of the curve that does not involve photoinhibition can be described by any two of three parameters: the initial slope α (mg C mg Chla^{-1} E^{-1} m^2) (Chla = Chlorophyll-a), the maximum assimilation number P_{max} (mg C mg Chla^{-1} h^{-1}), and I_K (E m^{-2} h^{-1}), which equals P_{max}/α. Various mathematical forms have been proposed for the P versus I curve: a good review can be found in Jassby and Platt (1976).

Given the depth dependence of Chla, light intensity, and appropriate P versus I parameters, it is possible in principle to calculate P_T (mg C m^{-2} h^{-1}) by integrating over depth. There have been many papers on this theme, including Talling (1957), Steele (1962), Harris (1978), and Bannister (1979). The integration can be performed analytically, and simple formulae for P_T obtained, under the simplifying assumptions that the chlorophyll-a concentration c, the diffuse attenuation coefficient K, and the P versus I parameters are constant over the euphotic zone. The light intensity at depth z is then $I_z = I_0 e^{-K \cdot z}$, where I_0 is surface irradiance. For example, using Smith's (1936) formula:

$$P = \frac{P_{\max} I}{(I_K^2 + I^2)^{0.5}}$$

integration gives:

$$P_T = \frac{<c> P_{\max} \ln(I^* + (1 + I^{*2})^{0.5})}{K}$$

where $I^* = I_0/I_K$ (Platt, 1986). Arguing that I^* is large, Talling (1957) approximated this by:

$$P_T = <c> P_{\max} \ln(2 I^*)/K$$

Alternatively, using Steele's (1962) formula,

$$P = \alpha I e^{-I/eI_K}$$

which includes photoinhibition, one obtains

$$P_T = \frac{<c> P_{\max} e (1 - e^{-I^*/e})}{K}$$

An additional complication is provided by the diurnal cycle in I_0. We are primarily interested here in daily areal production P_T in mg C m^{-2} d^{-1}. We could use a P versus I curve from daily incubations, so that I and I_0 represent daily average irradiance, and multiply the above formulas by day length. Alternatively, we could use an "instantaneous" P versus I curve and integrate the expressions over time, given $I_0(t)$ (Talling, 1970). There are potential problems in both approaches. Diurnal changes in photosynthetic parameters have long been documented (e.g, Yentsch and Ryther, 1957a, Harris, 1978), so that a single short-term P versus I curve may not be applicable throughout the day. On the other hand, if cells in situ are mixed through a range of depths, they may not be exposed to very high light intensities for prolonged periods, and saturation or photoinhibition in long-term incubations at shallow depths may be misleading (Harris and Piccinin, 1977; Marra, 1978).

The K Effect

There is one common feature of these simple models that is worth comment: P_T is inversely proportional to the diffuse attenuation coefficient K. This in itself can account for a major part of the nonlinear relationship between P_T and c_K described by Eppley et al. (1985). Phytoplankton are responsible for a significant but varying proportion of light attenuation. If they accounted for a fixed proportion (K proportional to $<c>$), then P_T would be independent of $<c>$. In practice, the relationship between K and c is affected by the absorption and scattering attributable to pure water and to constituents that vary in a nonlinear way with pigment concentration (Smith and Baker, 1978). An old relationship attributable to Riley (1956) is:

$$K = 0.04 + 0.0088<c> + 0.054<c>^{2/3}$$

Using this relationship, a plot of $\log[<c>/K]$ versus $\log[<c>]$ is only mildly nonlinear for $<c>$ between 0.1 and 10 mg m^{-3}, and has an average slope of 0.57. If one includes the relationship between $<c>$ and c_K found by Brown et al. (1985), the formulas predict that P_T should be proportional to $c_K^{0.51}$ over a similar range. These simple models explain Eppley et al.'s (1985) empirical regression surprisingly well. (Note that the power law will break down, and integrated production will saturate, at very high values of $<c>$; see Takahashi and Parsons, 1972).

Two Approximations

The point of using the simple models is not to derive the empirical regression, but to explain some of the scatter about it. Their usefulness will depend on the extent to which variations in I_0, α, and $P_{\max}$ can be predicted. Two approximations have been proposed to reduce these requirements. Talling (1957) and others (e.g., Harris et al., 1980) have argued that $I^* = I_0/I_K$ tends to have a constant value around 5 and that P_T is relatively insensitive to changes in I^*. All three formulas then agree that $P_T \approx <c> P_{\max} 2.3/K$, and only one parameter need be estimated, assuming $<c>$ and K can be remotely sensed. In contrast, Platt (1986) has suggested that saturation should be ignored, leading to the approximation $P_T \approx \alpha I_0/K$. The error involved in ignoring saturation, when diurnal variation in I_0 is taken into account, is estimated to be about 50% when $I^* = 5$. (Both arguments calculate I^* with respect to maximum or noon values of I_0.) The second approximation requires that I_0 be known, but if I_0 can be calculated from remotely sensed cloud cover (Gautier et al., 1980; Bishop and Marra, 1984), then only one parameter must be estimated.

Talling's approximation ($P_T = 2.3 P_{\max} <c>/K$) was used by Harris et al. (1980) to consider Great Lakes data, and also by Eppley et al. (1987) for data in the eastern tropical Pacific. Harris et al. (1980) plotted P_T versus $P_{\max}<c>/K$ and observed slopes ranging from 1.75 to 2.47 over three years and two locations. The variation in P_T about a line of slope 2.3 was about ±50%. Values of $P_{\max}$ in this eutrophic environment were low, ranging from <0.5 to about 3.0 mg C mg Chla^{-1} h^{-1}, depended rather weakly on temperature, and showed a stronger dependence on changes in the ratio of euphotic zone depth z_e to mixing zone depth z_m. However, variations in P_{max} for a given euphotic zone depth exceeded 1 mg C mg Chla^{-1} h^{-1}. Eppley et al. (1987) regressed P_T/c_K on P_{max}/K using data collected in the eastern tropical Pacific in 1967 to 1968. The regression line had a slope of 2.3, and explained 71% of the variance.

These studies suggest that the P_{max} approximation is useful in the field, when P_{max} is measured using incubations. In order to use this approximation for remotely sensed data, we must be able to predict P_{max}. This is the most commonly measured photosynthetic parameter, and there is a very large set of accumulated field and laboratory measurements showing a wide range of values. A recent review (Kelly, 1989) reports values ranging from 0.2 to 20 mg C mg Chla^{-1} h^{-1}. Recent studies have tended to report higher values: the review by Parsons and Takahashi (1973) reported only one study with values exceeding 8 mg C mg Chla^{-1} h^{-1}, whereas eight of the papers cited in Kelly's review reported values exceeding this figure. Kelly notes that values reported for terrestrial plants rarely exceed 4 mg C mg Chla^{-1} h^{-1} and questions the high values being reported for phytoplankton. Possible methodological problems are discussed in the following.

Why should P_{max} vary so widely? Physiological changes in P_{max} are known to result from adaptation to different light and nutrient regimes. A primary mechanism of adaptation in phytoplankton involves changes in the C:Chla ratio, θ (Jorgansen, 1969; Laws and Bannister, 1980). It is the specific growth rate μ(h^{-1}), which is important to phytoplankton population dynamics; this is related to assimilation number by:

$$\mu = \frac{P(I)}{\theta} - r$$

where r is the carbon-specific respiration rate. Within certain constraints, changes in I may be compensated for by changes in θ, especially if the growth rate is controlled by other limiting factors, such as nutrient supply. Whether P_{max} will also change depends on the rate-limiting step for P_{max}, and the nature of changes in chlorophyll per cell. For example, if the rate-limiting step lies in the electron transport chain, and a decrease in θ is brought about by increases in the photosynthetic unit (PSU) number, P_{max} will not change (Prezelin, 1981). However, if the rate-limiting step lies in the dark reaction, and enzyme concentrations can vary independently of chlorophyll content, large variations in P_{max} can result.

Platt (1986) has tested the alternative linear approximation directly, using eight data sets from different regions of the North Atlantic, and regressing P_T/c_T on I_0. Here, c_T is the total depth-integrated chlorophyll (mg m^{-2}), which is related to $<c>$ by $c_T = <c>z_e$, where z_e is the depth of the euphotic zone. By taking z_e as the 1% light depth or 4.6/K, one obtains $c_T = <c>4.6/K$. The linear approximation then becomes:

$$P_T = \frac{c_T \alpha I_0}{4.6}$$

or, in Platt's formulation,

$$\frac{P_T}{c_T} = \psi I_0,$$

with $\psi = \alpha/4.6$. The regressions using the North Atlantic data were generally significant, explaining 60% to 86% of the variance, with values of ψ ranging from 0.31 to 0.66 (mg C mg Chla^{-1} E^{-1} m^2). These correspond to values of α ranging from 1.4 to 3.0 (same units). In the Bedford Basin, directly measured values of α, divided by 4.6, exceeded the calculated value of ψ by more than two times. Platt (1986) attributed the major part of this discrepancy to the error involved in ignoring light saturation, as observed values of I^* were about 5. Platt et al. (1988) consider this approximation further, examining errors associated with the assumption of vertical homogeneity as well as the effects of light saturation.

The linear approximation also appears to be potentially useful, but we must again consider our capacity to predict α. A survey of some recently published results from both field and laboratory studies yielded values of α ranging from 0.3 to 15, with most in the range 1 to 10 mg C mg Chla^{-1} E^{-1} m^2. There is evidence of phylogenetic and physiological variation in α, which is often analyzed as the product of two parameters: an absorption coefficient k_c (m^2 mg Chla^{-1}) and a maximum quantum efficiency ϕ_m (mol C fixed (E absorbed)$^{-1}$). Measured values of k_c range between 0.005 and 0.021 (Harris, 1978; Kirk, 1983). Variations in the ratio of ancillary pigments to Chla, in the spectrum of incident light, and in the packaging of chlorophyll within the cell, can all affect k_c (Harris, 1978, Falkowski et al., 1985; Morel et al., 1987; Sathyendranath et al., 1987). The theoretical maximum for ϕ_m is 0.1 to 0.125, but observed values range from 0.02 to 0.11 (e.g., Falkowski et al., 1985; Cleveland and Perry, 1987; Sathyendranath et al., 1987). There appears to be a tendency for stressed cells to have low values of ϕ_m, suggesting that entire photosynthetic units may be inactivated. Taking $k_c = 0.02$ m^2 mg Chla^{-1} and $\phi_m = 0.10$ gives an upper bound for α of 20 mg C mg Chla^{-1} E^{-1} m^2, which exceeds the observed values by some margin.

The reported values of α vary far more widely than the relatively restricted range of ψ values reported by Platt (1986). It is not clear whether α is more conservative in the ocean than laboratory studies suggest, or whether there are compensating effects owing to depth integration, or whether the data set analyzed by Platt is completely representative. It is also worth remembering that the ψ values reported by Platt are themselves averages over cruises or regional data sets. The extent of variations in ψ among individual stations was not reported.

Topliss and Platt (1986) have reported an inverse correlation between α and the fluorescence efficiency η (E emitted/E absorbed) in observations from the North Atlantic. They have suggested that it may be possible to measure η remotely by measuring the fluorescence peak in upwelling radiation near 685 nm, and hence to predict some of the variation in α. The inverse relationship between α and η presumably reflects an inverse relationship between ϕ_m and η, which suggests an alternative approach to determining P_T remotely. The reflectance spectrum that is actually measured is affected by phytoplankton in so far as they absorb and scatter light. The current empirical formulas for estimating c_K do not discriminate scattering and absorption by phytoplankton and other constituents. If this becomes possible with better spectral resolution and more sophisticated reflectance models (e.g., Gordon et al., 1988), remote sensing may provide estimates of $k_c \cdot c_K$ directly. It may then be more useful to write the linear approximation as

$$P_T = \frac{\phi_m (k_c\, c_K) I_0}{K}$$

at least to the extent that ϕ_m is more predictable than α. If Topliss and Platt's (1986) results can be shown to apply widely, it may be possible to use the fluorescence peak to estimate ϕ_m.

Adaptation to Environmental Fluctuations

A steady-state continuous culture in the laboratory under fixed conditions of light and nutrient availability may exhibit a reproducible P versus I curve with fixed values of α and $P_{\max}$. One can study a number of such cultures and observe the dependence of these parameters on light and nutrient conditions under steady-state conditions. However, natural populations may never experience steady-state conditions of this kind (Harris, 1980a). Vertical mixing, diurnal variations in light intensity, and episodic inputs of nutrients across themoclines or from zooplankton excretion all provide perturbations on a variety of time scales (Harris, 1980a, 1986). Laboratory studies suggest that perturbations may produce larger changes in photosynthetic parameters than a range of steady-state conditions. In chemostats and turbidostats at a wide range of nutrient-limited and light-limited growth rates, changes in $P_{\max}$ and α at steady-state can be relatively small (Parslow and Harrison, unpublished data). However, depression of $P_{\max}$ (and α) is widely observed on a diurnal basis in natural populations (Yentsch and Ryther, 1957), and in batch cultures entering senescence (e.g., Cleveland and Perry, 1987). A reasonable interpretation is that extreme fluctuations in $P_{\max}$ and α arise from imbalances between different rate-limiting steps, resulting from a failure or lag in physiological adaptation.

In a variable environment, the P versus I parameters become variables that change on time scales ranging from minutes to days or weeks. At short time scales, the exposure of cells to high light intensities for a few minutes may yield very different $P_{\max}$ values than exposure over periods of hours or days (Harris, 1973; Harris and Piccinin, 1977; Marra, 1978). Cells are also faced by diurnal changes in light intensity, and day-to-day changes in I_0, mixing rates, and nutrient supply rates (Harris, 1980a). From this point of view, the prediction of P_T requires a dynamic model of adaptation at appropriate time scales as well as knowledge of the environmental driving variables.

The choice between the two approximate models discussed above is partly related to adaptation time scales. The $P_{\max}$ model assumes that physiological or community adaptation leads to values of I_0/I_K around 5. If this adaptation occurs on time scales that are short compared with day-to-day fluctuations in I_0 and z_m, then we would expect little fluctuation in I_0/I_K. If I_0/I_K equaled 5 exactly, the $P_{\max}$ and α formulas would disagree by exactly a factor of 2.17, which represents the error incurred in the α formula in neglecting light saturation. This could be corrected for by dividing by 2.17; that is,

$$P_T = \frac{\alpha I_0 c}{2.17 K}.$$

(This correction would approximately remove the discrepancy between $\alpha/4.6$ and μ reported by Platt (1986) for the Bedford Basin.) At the other extreme, if adaptation is slow compared with day-to-day fluctuations in I_0, the $P_{\max}$ model will fail to explain the significant short-term variation in P_T that will be predicted by the corrected α model. In practice, adaptation time scales are typically of the order of one to several days (Harris et al., 1983) so that the real situation is likely to lie somewhere between these two extremes. It is conceivable that regional differences in the success of the models may reflect differences in the amount of short-term variance in I_0, K, and the depth of the mixed layer z_m. It is also worth remembering that some data sets may be biased toward high I_0.

Given phylogenetic variation in photosynthetic parameters (Ryther, 1956; Harris, 1986), changes in phytoplankton community composition in response to environmental perturbation may be of equal or greater importance in determining primary production rates (Harris et al., 1983; Harris, 1986). The important driving variables are fluctuations in I_0, K, and z_m, which together determine the recent light history of the populations in surface waters. These parameters are linked through the "critical depth" concept of Sverdrup (1953) because respiration continues throughout the mixed layer and photosynthetic gains must exceed respiratory losses if the populations are to grow and survive. There is evidence of compensatory changes in I_K and $P_{\max}$ in populations in surface waters (Harris, 1978) in response to changes in I_0 and z_m.

Nutrient Limitation

Phytoplankton growth is thought to be nutrient limited over large areas of the ocean, including the extensive subtropical gyres. In these oligotrophic waters, chlorophyll concentrations are very low and this limits P_T, although somewhat countered by the correspondingly low values of K. The P versus I models treat light dependence explicitly and nutrient dependence implicitly, in so far as it affects the parameters P_{max} and α. Early field data suggested that phytoplankton in these waters grew slowly and had low P_{max} values. More recent measurements have produced higher P_{max} values and claims that growth rates are relatively high (McCarthy and Goldman, 1979).

There are a number of serious methodological problems in oligotrophic waters. Incubation artifacts such as metal contamination (Fitzwater et al., 1982) and damage to fragile cells during collection or filtering (Harris et al., 1989) can lead to underestimates of photosynthetic rates. There is a problem of separating phototrophs and heterotrophs with overlapping size ranges, exacerbated by high apparent dark uptake rates of ^{14}C (Harris et al., 1989). While the subtropical gyres historically were regarded as constant environments, there has been increasing attention to perturbations on time scales from seconds to months. The extent to which field data from different times and/or locations can or should be compared is increasingly questioned.

The capacity of remote sensing to resolve some of these sampling problems is discussed in the following. However, until the methodological problems for captive water bodies are overcome, the resulting uncertainties about photosynthetic rates must extend to remote sensing estimates.

Marine Photosynthesis, Carbon Flux, and Climate Change

So far, we have discussed the problem of estimating depth-integrated primary production, P_T. Climate modelers are primarily interested in marine primary production as a way of transferring carbon out of the surface layers of the ocean, where it exchanges relatively freely with the atmosphere, into the deep ocean or sediments. Several carbon fluxes can be considered, including transfer out of the euphotic zone, transfer into deep water, and burial in sediments. All are important, but on different time scales, ranging from the seasonal to the interglacial.

Much of the carbon fixed within the euphotic zone is consumed and respired there, and the flux of organic carbon out of the zone can be thought of as net community production. This flux has been assumed to consist primarily of particulate detrital material, and it can then be measured directly using particle traps (e.g., Karl and Knauer, 1984; Martin et al., 1987; Pace et al., 1987). An alternative approach has been based on an assumption of nitrogen balance within the euphotic zone. Phytoplankton take up ammonia and urea, supplied by recycling of organic nitrogen within the euphotic zone and nitrate that has been mixed into the euphotic zone from below. The flux of particulate organic carbon out of the euphotic zone involves an associated flux of particulate organic nitrogen that must, in turn, be replaced by inorganic nitrogen (nitrate) from below. The fraction F of photosynthesis based on nitrate uptake has been called "new production" and assumed to equal the fraction exported (Eppley and Peterson, 1979).

Incubations using the stable isotope ^{15}N can be utilized to estimate F in the field (Dugdale and Goering, 1967). These field data suggest a systematic variation in F with total primary production, summarized by Eppley and Peterson (1979) in the empirical relation:

$$F = \begin{cases} 0.0025\, P_T & P_T < 200 \text{ mg C m}^{-2}\text{ d}^{-1} \\ 0.5 & P_T > 200 \text{ mg C m}^{-2}\text{ d}^{-1} \end{cases}$$

This relationship is consistent with a classical view of oligotrophic communities as tightly coupled assemblages of small phototrophs and heterotrophs with efficient recycling of carbon and nitrogen. These communities are contrasted with eutrophic or bloom communities consisting of large producers and consumers, weakly coupled in time, that produce a large flux of organic material in the form of senescent cells and fecal pellets.

From a remote sensing point of view, this empirical relation between F and P_T seems ideal, allowing an immediate jump from photosynthesis to carbon flux. However, the classical view of new production in oligotrophic systems has recently been challenged. A seasonal oxygen maximum that accumulates within the euphotic zone has been argued to represent net carbon fixation of as much as 50 g C m^{-2} y^{-1} (Jenkins and Goldman, 1985). This should equal the vertical organic carbon flux, if organic carbon does not accumulate. This interpretation is supported by apparent oxygen utilization rates at greater depths (Jenkins, 1982). As these waters were historically believed to support total primary production rates of less than 50 g C m^{-2} y^{-1}, the oxygen data and interpretation have aroused some debate (e.g., Platt and Harrison, 1986; Reid and Schulenberger, 1986). Even given higher recent estimates of total primary production of 100 to 150 g C m^{-2} y^{-1}, they still represent F values of 30 to 50%, compared with historical estimates of about 10% for these waters (Eppley and Peterson, 1979).

These discrepancies may be explained by new mixing models that predict higher average nitrate fluxes across the pycnocline as a result of intense mixing events that are intermittent in space and time (Klein and Coste, 1984). According to this explanation, the major part of the carbon flux out of oligotrophic surface waters is due to occasional blooms, with decoupled grazing and low recycling, which have been undersampled by traditional field programs. The classical relation between F and P_T may still hold locally, but cannot be applied to spatial and temporal averages. Alternatively, it has been suggested that classical results have paid insufficient attention to the vertical structure of the water column, and in particular to the deeper portion of the euphotic zone, where new production may be proportionally much higher (Altabet, 1988).

Remote sensing may contribute to improved estimates of carbon flux by providing improved spatial and temporal resolution. However, this resolution seems likely to be limited to approximately 1 km and one day in the open ocean under ideal conditions; compositing to avoid cloud may hide or blur the events of interest (Denman and Abbott, 1988). The horizontal and temporal dimensions of the postulated intermittent mixing events are not clear. Moreover, a mixing event can be detected remotely only by some surface expression, either in chlorophyll or a physical parameter. It is entirely possible that the nutrient input and biological response are restricted to subsurface layers. Research is required to establish how much remote sensing can contribute to carbon flux measurements, especially since fluxes of carbon to the deep ocean or sediments are likely to be even more strongly dominated by episodic events.

Effects of Climate Change on Marine Photosynthesis

One would not expect any direct effect of increased CO_2 levels on marine photosynthesis: dissolved inorganic carbon (DIC) in the oceans is thought to be saturating. The response will depend instead on indirect effects on atmospheric and oceanic physics. Some (highly speculative) scenarios may be useful. The Pacific equatorial upwelling represents an important zone of total and new production, contributing as much as 20% or more of global new production according to Chavez and Barber (1987). Production in this zone is already known to be climate sensitive, being severely diminished in El Nino years. Climate changes could lead to a long-term increase or reduction in production.

One would expect the spring bloom in high-latitude regions to represent a significant source of new production. The lack of a spring bloom in the subarctic Pacific (Heinrich, 1962) has been the subject of continuing interest. Present interest focuses on the role of microzooplankton grazers (Frost, 1987) and possible iron limitation (Martin and Fitzwater, 1988). However, the importance of the relatively shallow winter halocline remains unclear (Evans and Parslow, 1985). It seems possible that changes in ocean circulation could produce a spring bloom in this region, or result in the failure of the spring bloom in the North Atlantic.

The subtropical convergence in the southern hemisphere has been implicated as an important CO_2 sink (Pearman and Hyson, 1986). It appears to be a highly productive region (Harris et al., 1987, 1988), and CZCS images suggest a permanent band of elevated chlorophyll (G. Feldman, personal communication). The satellite images also show a strong north–south seasonal movement of the high-chlorophyll zone. It is not clear what the effects of large-scale climate change might be, but fluctuations in zonal westerlies south of Australia, associated with the southern oscillation index (SOI), are known to produce strong interannual variations in the position of the front and in recruitment to commercial fisheries (Harris et al., 1988).

In most cases, monitoring of surface chlorophyll may be sufficient to detect regional effects of the kind discussed here. It is true in marine systems, as elsewhere, that apparent effects of climate change can only be judged significant against an adequate baseline picture of spatial and seasonal patterns and 'normal' levels of interannual variability. The historical CZCS data, and the data provided by the ocean color satellites of the 1990s, can make a vital contribution to this baseline.

Conclusions

It appears to be increasingly accepted that incubation of point samples at discrete stations cannot be used as a basis for reliable and accurate estimates of regional fluxes. If important processes in the ocean are intermittent and patchy, the kind of spatial and temporal coverage offered by satellites is essential. However, satellites cannot be used to measure fluxes such as photosynthetic rates directly. We are then faced with the need to relate the variables that can be measured, especially surface pigment concentrations c_K, to fluxes of interest, such as P_T. This review has shown that, while some of the variability in the ratio P_T/c_K can be explained simply by variation in K, there is a large residual variance that reflects physiological and phylogenetic variation, driven by fluctuations in the physical and chemical environment. This residual variance may be reduced somewhat by the use of regional or seasonal parameters, but a large part is due to fluctuations on the same spatial and temporal scales as c_K or P_T itself.

It should be possible to use other remotely sensed data such as sea-surface temperature, solar irradiance, surface roughness, and wind speed to help predict P_T/c_K (and other flux ratios). This is essentially a modeling problem, whether it is tackled using simple empirical models or more complex realistic models. Both physical and biological processes will have to be addressed. As the process models must be based on data acquired from ship-based studies, sampling strategies must be designed using remote sensing to sample as full a range of environmental variation as possible. The difficulties involved should not be underestimated: a large number of intensive regional research programs will be required.

The amount of effort and the type of measurements required will depend on the time and space scales at which predictive ability is sought. Experience has shown the problems of bias in scaling up from sparse data: estimates of seasonal or annual production are better based on such techniques as oxygen balance and sediment traps which average over space or time. As in any modeling program that aims at prediction, success will depend on constant and conservative appraisal of the assumptions and uncertainties associated with theoretical and empirical components. Many of the empirical studies cited have emphasized the proportion of the variance explained. For prediction, it is more important to know the magnitude of the residual variance.

It is interesting to compare the problems of estimating biological fluxes in marine and terrestrial ecosystems. The problems of scaling up from physiological studies of individual leaves, through a wide range of spatial scales and biological structures, to the scales appropriate to remote sensing or climate modeling are more obvious in terrestrial systems. The problem of relating remotely sensed data to plant biomass may be more difficult in terrestrial systems, but terrestrial structures tend to change more slowly, allowing some separation in time scales between structure and function. In the oceans, biomass changes rapidly and the obvious appeal of remote sensing is its ability to resolve these changes. However, as emphasized earlier, environmental forcing changes the relationship between biomass and fluxes on similar space and time scales.

There is a lesson for marine scientists in the terrestrial use of biophysical models of water and energy exchange to link remotely sensed data to those parameters (e.g., vapor pressure deficit) that directly control terrestrial photosynthesis. In the ocean, a three-dimensional mesoscale model of upper-layer physics, driven by remotely sensed insolation, wind stress, and sea surface height and tested by comparing predictions of sea-surface temperature with observations, may provide the vertical structure and nutrient fluxes needed to determine photosynthetic parameters. A combined biological and physical model could be tested by comparing both sea-surface temperature and surface pigment with remote observations.

The estimation of fluxes would be just a component, rather than the ultimate goal, of models of the type just discussed. In fact, we should be wary of any attempt to reduce biological oceanography or remote sensing to the estimation of particular regional or global fluxes. The scenarios discussed suggest possible changes in spatial and temporal biomass and flux distributions resulting from climate change. The problem is to detect such changes against a noisy background of fluctuations on a range of time and space scales. There is a clear need to use existing archives (e.g., CZCS) to establish baseline patterns of short-term, seasonal and interannual variability in biomass (and a need to develop statistical techniques to deal with such large, patchy, high-dimensional data sets).

Static estimates of fluxes are also of little use in the prediction of climate change, which is currently based on general circulation models with very coarse spatial resolution. These models require process submodels to predict fluxes of energy, carbon dioxide, and so on in response to climatic driving variables. In the oceans, this will require coupled atmosphere–

ocean circulation models with the biology included. One way forward may be to develop the diagnostic, mesoscale models discussed above, and then integrate these with global, prognostic climate models.

References

Altabet, M.A. (1988). Variations in nitrogen isotopic composition between sinking and suspended particles: implications for nitrogen cycling and particle transformation in the open ocean. *Deep-Sea Res.* 35:535–554.

Austin, R.W., and Petzold, T.J. (1981). The determination of the diffuse attenuation coefficient of sea water using the Coastal Zone Color Scanner. pp. 239–256. In J.F.R. Gower (ed.), *Oceanography from Space*. Plenum Press, NY.

Bannister, T.T. (1979). Quantitative description of steady-state, nutrient saturated algal growth, including nutrient saturation. *Limnol. Oceanogr.* 24:76–96.

Bishop, J.K.B., and Marra, J. (1984). Variations in primary production and particulate carbon flux through the base of the euphotic zone at the site of the Sediment Trap Intercomparison Experiment (Panama Basin). *J. Mar. Res.* 42:189–206.

Brown, O.B., Evans, R.H., Brown, J.W., Gordon, H.R., Smith, R.C., and Baker, K.S. (1985). Phytoplankton blooming off the U.S. East Coast. A satellite description. *Science* 229:163–167.

Chavez, F.P., and Barber, R.T. (1987). An estimate of new production in the equatorial Pacific. *Deep-Sea Res.* 34:1229–1243.

Cleveland, J.S., and Perry, M.J. (1987). Quantum yield, relative specific absorption and fluorescence in nitrogen-limited *Chaetoceros gracilis*. *Mar. Biol.* 94:489–498.

Denman, K.L., and Abbott, M.R. (1988). Time evolution of surface chlorophyll patterns from cross-spectrum analysis of satellite color images. *J. Geophys. Res.* 93:6789–6798.

Dugdale, R.C., and Goering, J.J. (1967). Uptake of new and regenerated forms of nitrogen in primary productivity. *Limnol. Oceanogr.* 12:196–206.

Eppley, R.W., and Peterson, B.J. (1979). Particulate organic matter flux and planktonic new production in the deep ocean. *Nature* 282:677–680.

Eppley, R.W., Stewart, E., Abbott, M.R., and Heyman, U. (1985). Estimating ocean primary production from satellite chlorophyll: Introduction to regional differences and statistics for the southern Californian bight. *J. Plankton Res.* 7:57–70.

Eppley, R.W., Stewart, E., Abbott, M.R., and Owen, R.W. (1987). Estimating ocean production from satellite-derived chlorophyll: Insights from the Eastropac data set. *Oceanologica Acta, Proc. Int. Symp. Equatorial Vertical Motion*, Paris, May 6–10, 1985, pp. 109–113.

Evans, G.T., and Parslow, J.S., (1985). A model of annual plankton cycles. *Biol. Oceanogr.* 3:327–347.

Falkowski, P.G., Dubinsky, Z., and Wyman, K. (1985). Growth-irradiance relationships in phytoplankton. *Limnol. Oceanogr.* 30:311–321.

Fitzwater, S.E., Knauer, G.A., and Martin, J.H. (1982). Metal contamination and its effect on primary production measurements. *Limnol. Oceanogr.* 27:544–551.

Frost, B.W. (1987). Grazing control of phytoplankton stock in the open subarctic Pacific Ocean: A model assessing the role of mesozooplankton, particularly the large calanoid copepods Neocalanus spp. *Mar. Ecol. Prog. Ser.* 39:49–68.

Gautier, C., Diak, G., and Masse, S. (1980). A simple physical model to estimate

incident solar irradiance at the surface from GOES satellite data. *J. Appl. Meteorol.* 19:1005-1012.
Gordon, H.R. (1987). Calibration requirements and methodology for remote sensors viewing the ocean in the visible. *Remote Sens. Envir.* 22:103-126.
Gordon, H.R., Brown, J.W., Brown, O.B., Evans, R.H., and Clark, D.K. (1983). Nimbus-7 Coastal Zone Color Scanner: Reduction of its radiometric sensitivity with time. *Appl. Optics* 22:3929-3931.
Gordon, H.R., Brown, O.B., Evans, R.H., Brown, J.W., Smith, R.C., Baker, K.S., and Clark, D.K. (1988). A semi-analytic radiance model of ocean color. *J. Geophys. Res.* 93:10909-10924.
Gordon, H.R., and Clark, D.K. (1980a). Atmospheric effects in the remote sensing of phytoplankton pigment. *Boundary-Layer Met.* 18:299-313.
Gordon, H.R., and Clark, D.K. (1980b). Remote sensing optical properties of a stratified ocean: An improved interpretation. *Appl. Optics* 19:3428-3430.
Gordon, H.R., and Morel, A. (1983). *Remote Assessment of Ocean Color for Interpretation of Satellite Visible Imagery: A review.* Springer-Verlag, NY.
Gower, J.F.R., and Borstad, G.A. (1981). Use of the in vivo fluorescence line at 685 nm for remote sensing surveys of surface chlorophyll-a. pp 329-338. In J.F.R. Gower (ed.), *Oceanography from Space.* Plenum Press, NY.
Harris, G.P. (1973). Diel and annual cycles of net phytoplankton photosynthesis in Lake Ontario. *J. Fish. Res. Bd. Canad.* 30:1779-1787.
Harris, G.P. (1978). Photosynthesis, productivity and growth: The physiological ecology of phytoplankton. *Ergeb. Limnol. (beih. Arch. Hydrobiol.)* 10:1-171.
Harris, G.P. (1980a). Temporal and spatial scales in phytoplankton ecology. Mechanisms, methods, models and management. *Canad. J. Fish. Aq. Sci.* 37:877-900.
Harris, G.P. (1980b). The measurement of photosynthesis in natural populations of phytoplankton. pp. 129-187. In I. Morris (ed.), *The Physiological Ecology of Phytoplankton.* Blackwell, Oxford, England.
Harris, G.P. (1984). Phytoplankton productivity and growth measurements: Past, present and future. *J. Plankton Res.* 6:699-713.
Harris, G.P. (1986). *Phytoplankton Ecology: Structure, Function and Fluctuation.* Chapman and Hall, London, England.
Harris, G.P., Davies, P., Nunez, M., and Meyers, G. (1988). Interannual variability in climate and fisheries in Tasmania. *Nature* (London). 333:754-757.
Harris, G.P., Ganf, G.G., and Thomas, D.P. (1987). Productivity, growth rates and cell size distributions of phytoplankton in the SW Tasman Sea: Implications for carbon metabolism in the euphotic zone. *J. Plankton Res.* 9:1003-1030.
Harris, G.P., and Griffiths, F.B. (1987). On means and variances in aquatic food chains and recruitment to the fisheries. *Freshwater Biol.* 17:381-386.
Harris, G.P., Griffiths, F.B., and Thomas, D.P. (1989). Light and dark uptake and loss of ^{14}C: Methodological problems with productivity measurements in oceanic waters. *Hydrobiologia* 173:95-105.
Harris, G.P., Haffner, G.D., and Piccinin, B.B. (1980). Physical variability and phytoplankton communities. II. Primary productivity by phytoplankton in a physically variable environment. *Arch. Hydrobiol.* 88:393-425.
Harris, G.P., and Piccinin, B.B. (1977). Photosynthesis by natural phytoplankton populations. *Arch. Hydrobiol.* 80:405-457.
Harris, G.P., Piccinin, B.B., and Van Pyn, J. (1983). Physical variability and phytoplankton communities: V. Cell size, niche diversification and the role of competition. *Arch. Hydrobiol.* 98:215-239.
Heinrich, A.K. (1962). The life histories of plankton animals and seasonal cycles of plankton communities in the oceans. *J. Cons. Int. Explor. Mer.* 27:15-24.

Jassby, A.T. and Platt, T. (1976). Mathematical formulation of the relationship between photosynthesis and light for phytoplankton. *Limnol. Oceanogr.* 21:540–547.

Jenkins, W.J. (1982). Oxygen utilization rates in the North Atlantic subtropical gyre and primary production in oligotrophic systems. *Nature* (London) 300:246–248.

Jenkins, W.J., and Goldman, J.C. (1985). Seasonal oxygen cycling and primary production in the Sargasso Sea. *J. Mar. Res.* 43:465–491.

Johnson, P.W., and Sieburth, J. McN. (1979). Chroococcoid cyanobacteria in the sea: A ubiquitous and diverse phototrophic biomass. *Limnol. Oceanogr.* 24:928–935.

Johnson, P.W., and Sieburth, J. McN. (1982). In-situ morphology and occurrence of eucaryotic phototrophs of bacterial size in the picoplankton of estuarine and oceanic waters. *J. Phycol.* 18:318–327.

Jorgansen, E.G. (1969). The adaptation of plankton algae IV Light adaptation in different algal species. *Physiol. plantarum* 22:1307–1315.

Karl, D.M., and Knauer, G.A. (1984). Vertical distribution, transport and exchange of carbon in the northeast Pacific Ocean: Evidence for multiple zones of biological activity. *Deep-Sea Res.* 31:221–243.

Kelly, G.J. (1989). A comparison of marine photosynthesis with terrestrial photosynthesis: A biochemical perspective. *Oceanogr. Mar. Biol. Ann. Rev.* 27, in press.

Kirk, J.T.O. (1983). *Light and Photosynthesis in Aquatic Environments*. Cambridge Univ. Press, Cambridge, England.

Klein, P., and Coste, B. (1984). Effects of wind-stress variability on nutrient transport into the mixed layer. *Deep-Sea Res.* 31:21–37.

Laws, E.A., and Bannister, T.T. (1980). Nutrient and light-limited growth of *Thalassiosira fluviatilis* in continuous culture with implications for phytoplankton growth in the ocean. *Limnol. Oceanogr.* 25:457–473.

Marra, J. (1978). Phytoplankton photosynthetic response to vertical movement in a mixed layer. *Mar. Biol.* 46:203–208.

Martin, J.H., and Fitzwater, S.E. (1988). Iron deficiency limits phytoplankton growth in the northeast Pacific subarctic. *Nature* (London) 331:341–343.

Martin, J.H., Knauer, G.A., Karl, D.M., and Broenkow, W.W. (1987). Vertex: Carbon cycling in the northeast Pacific. *Deep-Sea Res.* 34:267–285.

McCarthy, J.J. (1984). Measuring oceanic primary production. *Global Ocean Flux Study. Workshop Proc.* Nat. Acad. Press. Washington, DC, pp. 151–165.

McCarthy, J.J., and Goldman, J.C. (1979). Nitrogenous nutrition of marine phytoplankton in nutrient depleted waters. *Science* 203:670–672.

McElroy, M.B. (1983). Marine biological controls on atmospheric CO_2 and climate. *Nature* (London) 302:328–329.

Morel, A., Lazzara, L., and Gostan, J. (1987). Growth rate and quantum yield time response for a diatom to changing irradiances (energy and color). *Limnol. Oceanogr.* 32:1066–1084.

Pace, M.L., Knauer, G.A., Karl, D.M., and Martin, J.H. (1987). Primary production, new production and vertical flux in the eastern Pacific Ocean. *Nature* (London) 325:803–804.

Parsons, T.R., and Takahashi, M. (1973). *Biological Oceanographic Processes* (1st ed.). Pergamon Press, NY.

Pearman, G.I., and Hyson, P. (1986). Global transport and inter-reservoir exchange of carbon dioxide with particular reference to stable isotope distributions. *Atmos. Chem.* 4:81–124.

Perry, M.J. (1986). Assessing marine primary production from space. *BioScience* 36:461–467.

Peterson, B.J. (1980). Aquatic primary productivity and the ^{14}C-CO_2 method: A history of the productivity problem. *Ann. Rev. Ecol. Syst.* 11:359–385.
Platt, T. (1986). Primary production of the ocean water column as a function of surface light intensity: algorithms for remote sensing. *Deep-Sea Res.* 33:149–163.
Platt, T., and Harrison, W.G. (1986). Reconciliation of carbon and oxygen fluxes in the upper ocean. *Deep-Sea Res.* 33:273–276.
Platt, T., and Li, W.K.W. (eds.) (1986). Photosynthetic picoplankton. *Canad. Bull. Fish. Aquat. Sci.* 214.
Platt, T., Sathyendranath, S., Caverhill, C.M., and Lewis, M.R. (1988). Ocean primary production and available light: Further algorithms for remote sensing. *Deep-Sea Res.* 35:855–879.
Prezelin, B.B. (1981). Light reactions in photosynthesis. pp. 41–43. In T. Platt (ed.), *Physiological Bases of Phytoplankton Ecology. Canad. Bull. Fish. Aquat. Sci.* 210, Ottawa, Canada.
Reid, J.L., and Schulenberger, E. (1986). Oxygen saturation and carbon uptake near 28 N, 155 W. *Deep-Sea Res.* 33:267–271.
Riley, G.A. (1956). Oceanography of Long Island Sound, 1952–54. II. Physical oceanography. *Bull. Bingham Oceanogr. Coll.* 15:15–46.
Robinson, I.S. (1985). *Satellite Oceanography: An Introduction for Oceanographers and Remote-Sensing Scientists*. Wiley, NY.
Ryther, J.H. (1956). Photosynthesis in the oceans as a function of light intensity. *Limnol. Oceanogr.* 1:61–70.
Ryther, J.H. (1969). Photosynthesis and fish production in the sea. *Science* 166:72–76.
Sathyendranath, S., Lazzara, L., and Prieur, L. (1987). Variations in the spectral values of specific absorption of phytoplankton. *Limnol. Oceanogr.* 32:403–415.
Schulenberger, E., and Reid, J.L. (1981). The Pacific shallow oxygen minimum, deep chlorophyll maximum and primary productivity, reconsidered. *Deep-Sea Res.* 28:901–919.
Smith, E.L. (1936). Photosynthesis in relation to light and carbon dioxide. *Proc. Nat. Acad. Sci.* 22:504–511.
Smith, R.C., and Baker, K.S. (1978). The bio-optical state of ocean waters and remote sensing. *Limnol. Oceanogr.* 23:247–259.
Smith, P.E., and Eppley, R.W. (1982). Primary production and the anchovy population in the southern Californian bight: Comparison of time series. *Limnol. Oceanogr.* 27:1–17.
Smith, R.C., Eppley, R.W., and Baker, K.S. (1982). Oceanic chlorophyll concentrations as determined by satellite (Nimbus-7 Coastal Zone Color Scanner). *Marine Biol.* 66:269–279.
Steele, J.H. (1962). Environmental control of photosynthesis in the sea. *Limnol. Oceanogr.* 7:137–150.
Sturm, B. (1983). Selected topics of Coastal Zone Color Scanner (CZCS) data evaluation. pp. 137–167. In A.P. Cracknell (ed.), *Remote Sensing Applications in Marine science and Technology*. Reidel, Dordrecht.
Sundquist, E.T. (1985). Geological perspectives on carbon dioxide and the carbon cycle. pp. 5–59. In E.T. Sundquist and W.S. Broecker (eds.), *The Carbon Cycle and Atmospheric CO_2: Natural Variations Archaen to Present*. Amer. Geophys. Union, Washington, DC.
Sverdrup, H.U. (1953). On conditions for the vernal blooming of phytoplankton. *J. Int. Cons. Explor. Mer.* 18:287–295.
Takahashi, M., and Parsons, T.R. (1972). Maximization of the standing stock and primary productivity of marine phytoplankton under natural conditons. *Indian J. Mar. Sci.* 1:61–62.

Talling, J.F. (1957). The phytoplankton population as a compound photosynthetic system. *N. Phytol.* 56:133–149.

Talling, J.F. (1970). Generalized and specialized features of phytoplankton as a form of photosynthetic cover pp. 431–445. In *Prediction and Measurement of Photosynthetic Productivity. Proc. IBP/PP Tech. Meeting, Trebon, 14–21 Sept. 1969*. Pudoc, Wageningen.

Topliss, B.F. (1985). Optical measurements in the Sargasso Sea: Solar stimulated fluorescence. *Oceanol. Acta* 8:263–270.

Topliss, B.J., and Platt, T. (1986). Passive fluorescence and photosynthesis in the ocean: Implications for remote sensing. *Deep-Sea Res.* 33:849–864.

Venrick, E.L., McGowan, J.A., and Mantyla, A.W. (1973). Deep maxima of photosynthetic chlorophyll in the Pacific Ocean. *Fish. Bull.* 71:41–52.

Walsh, J.J. (1984). The role of ocean biota in accelerated ecological cycles: A temporal view. *BioScience* 34:499–507.

Yentsch, C.S., and Ryther, J.H. (1957). Short-term variations in phytoplankton chlorophyll and their significance. *Limnol. Oceanogr.* 2:140–142.

14. Analysis of Remotely Sensed Data

Jeremy F. Wallace and Norm Campbell

The prospects for obtaining new information on a global scale rest on suitable access to, and organization and processing of, immense volumes of remotely sensed and other data. This chapter addresses the issue of processing high-dimensional spectral data for extraction of information on surface conditions or processes. The second part of the chapter describes some statistical methods and developments relevant to the use of remotely sensed data for estimates of surface condition or classification.

Remotely Sensed Data and Information Content

Existing and future remotely sensed data present new opportunities, and also new problems, for research workers. The effective application of remote sensing to measurements of global processes will require that researchers come to terms with the nature of the data, and with the huge volumes of data that will be available. The challenge is considerable in such areas as database storage, integration, and processing for a required purpose.

Historical data records from existing instruments are enormous, and experiences with applications of these data indicate the magnitude of the global monitoring task. It is already the case that the technology for data collection far exceeds the capacity to collect relevant ground data, and to integrate and process the data. The instruments proposed for the Earth Observing System (EOS) and other platforms will supply data in volumes that are orders of magnitude greater than those now available. There is reason to hope that, buried within these masses of digits, there will be information of a scale and quality relevant to the understanding of global processes and climate. In fact, remote sensing provides the only prospect for obtaining measurements of some processes on the required scale.

If the information is to be obtained, it is essential that the data handling and analysis methods be suited to the nature of the data and the problems. Otherwise, we are likely to see repeated the disappointments of some of the early Landsat processing experiences. Spectacular broad-scale visual displays are readily produced from remotely sensed image data. However, extraction of reliable numerical estimates of surface processes and conditions is not straightforward. The Landsat Multi-Spectral Scanner (MSS) scanner was designed to monitor vegetation, and lists of indices for greenness and vegetative vigour have been proposed for MSS data. Yet in crop studies, none has proved to be an accurate indicator of crop performance over broad areas across seasons. Good correlations cannot be assumed with harder-to-measure processes such as photosynthetic activity. For the modeling and monitoring of biosphere processes, reliable numerical estimates are required. Estimates of variability are also required for the confident detection of trends or change.

High-Dimensional Data

The notable feature of remotely sensed spectral data is that the data are multidimensional; that is, the observation on a ground unit (pixel) is multivalued, the values being the reflectances in the available spectral bands. Each pass of the various instruments may record more values—a single pass of the High Resolution Imaging Spectrometer (HIRIS) will record 128 bands. Data of this spectral resolution may be considered a continuous spectral curve rather than discrete band measurements.

The observations may become very high dimensional indeed with the integration of data sources, and over a temporal sequence. It is this feature of the data, together with the scale of coverage, that provides the opportunity to derive measures which would otherwise be impossible and at the same time provides the challenge for data organization and analysis (see Figure 14.1).

Human perception is accustomed to two-dimensional representations such as maps, graphs, and scatterplots. Powers of discrimination and pattern detection are high for such figures. In higher dimensions, this is certainly not the case. It is conventional to present summaries of the data as two-dimensional figures, and tempting to believe that the chosen representation is meaningful and adequate. Temporal trends or spectral responses of pixels may be displayed in a limited way on graphs. Single-valued functions may be displayed on map figures as colors. Familiar representations are one-dimensional indices using color- or grey-level density

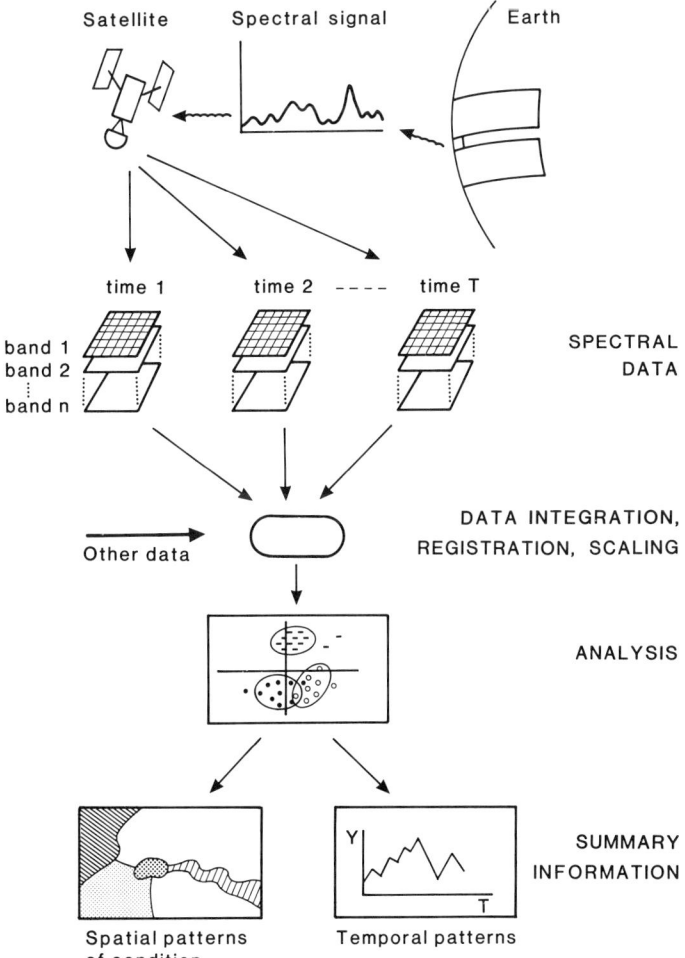

Figure 14.1. Schematic diagram of high-dimensional data—assembly, analysis, and reduction.

slicing, and classification maps of separable classes. A simple measure such as surface temperature at one fixed time can be well represented on a figure of this type. It is, however, unlikely that the full information in some high-dimensional spectral-temporal sequence can be so represented. Yet this naive assumption is not uncommon when the use of remote sensing is considered.

The Role of Analysis

The role of analysis in any application should be to examine the available data, to quantify the data's relevance to the problem, and to reduce the relevant information to an interpretable form, either numerical or pictorial. The role of the researcher is then to assemble data of a suitable scale and quality. The data set may include more than just remotely sensed data. These points are discussed further in the following, and one possible analysis strategy is outlined.

Qualitative or Quantitative Analysis?

There are two distinct approaches to data reduction and representation. The first relies on image interpretation and is largely qualitative or descriptive. The second approach, which includes the various classification methodologies, uses statistical methods to quantify the information in the data, and relationships with ground-measured quantities.

The qualitative approach has been more widely used in remote sensing applications to date. The display of standard false color images is a typical example of the approach. In general, the analyst displays some bands or indices that are believed to be relevant (e.g., theoretical vegetation indices or particular ratios of Thematic Mapper (TM) bands for geological information). Discernible differences and similarities on the images produced are assumed to be meaningful; the approach relies on the power of the human eye to perceive patterns and relationships on images. This image interpretation can be extremely effective in some contexts, such as the detection of geological structure. The normalized difference vegetation index (NDVI) images of Tucker et al. (1985) are an effective example in the ecological field. The images have a very high descriptive value because of the eye's ability to interpolate and find patterns. However, at a global scale, the predictive value of such an approach is generally poor or unknown; the relationships between the indices and the ground data on a pixel basis are often tenuous.

In the quantitative approach, the numerical data values on individual pixels are analyzed to describe and quantify relationships and clustering in the spectral space and their associations with ground data. The simplest examples of such methods are two-dimensional scatterplots of pixel band values. Various more sophisticated classification approaches have been put forward (e.g., Schowengerdt, 1983; Richards, 1984, 1986; Swain and Davis, 1978; Swain et al., 1981). The numerical classification approach, which is time consuming and often raises difficulties with data variation, has been criticized on these grounds. It is our view that where predictive measures are to be obtained from the data, there is no alternative to a numerical approach. If variation is a problem in the data, then it will certainly affect any standard index display. The essential advantage of quantitative methods is that the structure and variation in the data are revealed. If a standard index is adequate for the purpose, then an appropriate statistical analysis will establish this. If, however, the relationship is more complex, then a suitable classifier or combination or indices may be derived.

Statistical discrimination procedures allow assessment and comparison of data sets for applications. The essential notion is that the distribution of sample data may be used to attach confidence levels to spectrally separable classes and, where calibration data are available at least, to measurements extracted from the data. These techniques will be essential in the context of measuring or detecting subtle phenomena from the various high-dimensional data sets that are becoming available.

Database Storage

It is not intended here to discuss the organization of the primary archival database storage. This is a specialized and formidable task when the spatial, and especially temporal, dimensions are considered. However, the importance of this basic data organization cannot be overstated. It will not be sufficient to have records of single passes of various instruments available separately on tape or other media; this acts to discourage innovative uses of the data as a whole. Flexibility of storage and access are crucial to successful exploration of the data and subsequent applications. Research workers must have access to the data, the means to integrate data sources at any desired scale, and analysis procedures to establish whether the remotely sensed data are measuring the quantity of interest. The difficulties and solutions to the database problems of the First ISLSCP Field Experiment (FIFE) project (Sellers et al., 1988; this volume) may provide a model for the larger world. The structure for storage and distribution of the data on a national and international level will be significant in affecting user access.

Potential users now have the opportunity to consider the type and scale of data they might require. Experience in collaborative projects has shown that user access to image processing and analysis procedures is essential for gaining most information from the data. For research work and data exploration, it is essential that data be available for small areas at an affordable price.

Scale

The question of scale is critical at the data collection and processing stages. Expectations of what information the data may carry clearly depend on spatial, spectral and temporal resolution. For a lucid exposition with many references, see Townshend and Justice (1988).

The spatial resolutions of instruments vary considerably, and mixed pixels present problems at any scale. The measured spectral data are the integrated responses over a surface area. If the pixel is small and homogeneous, it may be reasonable to expect that the signature of some feature (e.g., vegetative vigor, leaf area index) will match predictions from laboratory spectra. This will not be the case when the pixel contains mixtures of classes in unknown proportions. Mixed pixels are to be expected when the pixel size is large or neighborhoods of pixels are scaled up to larger units. Some efforts have been made to deduce mixture information from spectral data in particular contexts (Graetz and Gentle, 1982; Pech et al., 1986; Jupp et al., 1986).

The "snapshot" nature of satellite imagery raises questions in measuring dynamic processes on the temporal scale. The data user is unlikely to have any control over the frequency or time of day of satellite overpasses or over the presence of clouds. For monitoring change and trends, multidate imagery is required. Processing of registered multitemporal imagery should give better results than separate processing of individual passes. In particular, multitemporal data should provide better information on mixed pixels where the components respond differently over time.

Information Value of the Data

It remains to be established by experiment whether remotely sensed data can measure the states or fluxes of interest. The FIFE project (Sellers et al., 1988; this volume) provides a major opportunity to relate exhaustively measured climate and atmospheric parameters to remotely sensed data from many instruments with a range of spatial and temporal resolutions. The analyses of these data, in attempting to define suitable temporal, spectral, and spatial scales to measure the processes, should be a model for the analysis of global data. They should also indicate which of the processes and states might be measurable with existing and planned instruments. The environment of this experiment is a fairly uniform grassland; it will be necessary to verify or recalculate the relationships in other environments, and to attempt to gain comparable information over the oceans.

Analysis of Existing Data

Existing data may also carry significant information. It has been mentioned elsewhere in this volume that there is an urgent need to attempt to use these data in current climate models. It is appropriate that analysis efforts now should be directed to these existing data and to data from experimental projects. The methods described in the section following may be directed to the discovery of indices or classifiers of interest for these data. Success implies confidence for use on a broader scale or extension to the new data sources. Landsat MSS data date from 1972 and could be examined to provide indications of surface changes since that date.

The applications of existing data provide indicators of approaches for new problems and data sets. The experience is cautionary for those with high expectations for the new data. Landsat MSS and TM sensors are designed to detect vegetation and soil information, and undoubtedly there is information in these data relevant to global ecology. Applications of existing data have generally not taken into account the full information in the spectral-temporal record, relying in the main on image interpretation or classifications of single-pass data. Extraction of information on global states and dynamic processes will require processing methods for the integration and selection of suitable data sources, and then reduction of the data to indices or classifiers that measure the desired quantities to an acceptable level of accuracy.

Powerful descriptive images have been produced, but there is a lack of convincing numerical results or predictions. Classification methods have proved too time consuming or complex, whereas theoretical and empirical indices failed to give reliable results on a broad scale. Failure to obtain the right data, and to understand the nature of the data, may often be responsible rather than any inadequacy of the data themselves. When appropriate methods are applied, the information in these data may be surprisingly good. In a recent vegetation-mapping study, for example, multitemporal MSS data gave adequate results, whereas single-image data proved inadequate (Hobbs et al., 1989). Where numerical estimates are required, there is no short cut; discrimination needs to be established and variation taken into account. It is possible that signal variation problems are exaggerated by high spatial resolution.

Preprocessing to larger pixel size and temporal averaging may provide data that are better behaved and more suitable for large-scale measurements. The Advanced Very High Resolution Radiometer (AVHRR)-derived NDVI images (Tucker et al., 1985) display the power of this approach for illustration of states on a continental scale. The theoretical association with vegetation vigor makes the images of change particularly useful. Yet it is not clear what the numerical value of this index actually tells us about the vegetation on the pixel. When it is known that some bare soils may give an NDVI value of 0.3 (J. Norman, personal communication), it is clear that the raw value of this index is of little use as a numerical measure for photosynthesis or any other vegetative function. Change in the NDVI may be better correlated with change in vegetative function, but this should be established before it is used as such, or a better classifier of the function must be sought.

In fact, as previously observed, no one-dimensional index can be expected to summarize the information in the multidimensional spectral data space. Adequate indices and classifiers may be found for different purposes, and researchers need access to the data and appropriate analysis methods to establish these measures for their particular problems. The following sections give details of procedures that can be applied to the general class of problems known as cover class mapping.

Statistical Methods for Cover Class Mapping Using Remotely Sensed Data

This section discusses some statistical considerations and developments in the use of remotely sensed spectral data for cover class mapping. This is the general class of problems where the numerical values of the data are used to identify states in the sampled pixels. The particular methodology outlined is based on experiences gained within the CSIRO Division of Mathematics and Statistics, which has an established research project on the analysis of remotely sensed image data. Other quantitative methodologies exist and have been applied; see, for example, Schowengerdt (1983), Richards (1984), and Swain and Davis (1978). Supervised classification algorithms incorporate procedures analogous to those described in the following. The general aim is to provide statistically based reduction of the data to useful summary information. It is desirable that the discrimination procedure describes the data structure and that the allocation routines take into account the data structure and variation.

In resource assessment and monitoring, the most common applications concern cover class mapping—either the assignment of a class label to each pixel, or the detection of particular classes within the image. Suitable discrimination and classification procedures can help to quantify the relevance of a data set to the problem. For example, the effect of adopting simplified discriminant indices or reducing the number of spectral bands can be calculated. It is our view that there is a lot of information to be gained from the application of relatively sophisticated statistical techniques to remotely sensed data in detailed area studies. The trend to make satellite image data available in smaller units is encouraging, as is the availability of low-cost image processing and analysis systems. In Western Australia, several collaborative studies have been carried out using software now running on the Commodore Amiga computer. These studies include the mapping of saline areas in agricultural districts and native vegetation communities (using multitemporal Landsat MSS data), the mapping and monitoring of forest clearing in catchment areas (TM data), and a preliminary investigation into the mapping of waterlogged soils (airborne MSS data).

Remotely Sensed Data

Remotely sensed data are typically digital counts arising from reflected (or emitted) energy in different regions of the spectrum. Each set of values corresponds with a nominal ground area. The sampling is usually on a regular grid with adjacent (or overlapping) ground units. For a statistician, these data may be considered a multivariate integer response with some special features. The data values are related spectrally owing to correlations of reflectance values in regions of the spectrum. Spatial correlation is also likely to be introduced by the characteristics of the sensors and atmospheric transmission. Moreover, the size of cover classes relative to pixel size often results in neighborhood information that can be used to improve the information on a central pixel.

Ground Data

Experience has shown that, while there are huge volumes of satellite data available, the collection of accurate ground data is a real problem. Even where aerial photographs and ground records are available, photointerpretation of classes and registration to a pixel grid are not trivial exercises. This problem is critical when mapping transient or dynamic phenomena, and should not be underestimated in the context of establishing relationships between remotely sensed data and climatic and biological processes.

A Strategy for Cover Class Mapping

A general strategy for cover class mapping has evolved from research and collaborative studies. As mentioned above, alternative approaches exist but the components of these are generally analogous to those described here. The process is concerned with class definition and allocation in the spectral data space. The results will be meaningful if the spectral classes correspond to useful ground cover classes. In principle, however, the stages described here can be carried out without reference to ground data, and ground class labels assigned only at the final stage.

The initial stage is to produce color composite displays of the data. From these and available ground information, training classes are selected. These training areas should be spectrally homogeneous and spatially contiguous. The training class data are used to estimate the location and dispersion of the class in the spectral space. For these class parameters to be well estimated, the number of pixels in each training class should be several times the number of bands used in the allocation procedure. It is better to choose too many rather than too few training areas.

Discriminant analysis [canonical variate analysis—see for example, Anderson, (1958)] can be used to provide information on the clustering and separation of classes in the spectral data space. Alternative indices may be derived and data sources compared. Ordination of the training class data is carried out and the means displayed on the first few canonical axes. The canonical roots indicate the overall degree of separation and the number of dimensions. It is possible here to direct the analysis to contrasts of particular interest, to test the effects of reducing the set of bands and of adopting simplified discriminant indices.

The results of the ordination are used to define spectral classes for the allocation procedure. Clustered training areas are grouped, with a subset chosen to represent the spectral class. Optimal selection of such a subset is a problem; at present, the choice is made using locations on the canonical variate plots, in conjunction with known ground data. It is also debated whether the spectral class is best represented by the data from the separate training areas or by the combined cluster of data.

The definition of spectral classes is followed by an allocation procedure. Pixels are assigned probabilities of class membership according to their relative and absolute closeness to the spectral classes (Campbell, 1984). These results may be displayed in various ways. It is only at this stage that ground labels need be assigned to the spectral classes. Several spectral classes may be needed to define a ground type, or different cover types may be inseparable in the spectral space.

The whole procedure allows for iterative refinement. Regions that are wrongly allocated or atypical of all known classes can be used to define further spectral classes, and the process repeated until the results are satisfactory.

Discriminant Analysis and Spectral Indices

Successful mapping of cover classes depends on the spectral data providing separation between classes. Ideally, we hope that the training site data will form discrete clusters corresponding to the cover classes of interest. The locations of these clusters provide the basis for subsequent classification and allocation procedures. In practice, the spectral data may afford good separation between some cover types (e.g., bushland and crop) and overlap between others (e.g., crop types). Canonical variate analysis is an ordination procedure that can be used to measure the spectral separation between sites and to display the locations of the training site data in spectral space.

The linear combination of bands that best discriminates between several classes is that which maximizes the ratio of between-class variation to within-class variation. This ratio is called the canonical root and the associated linear discriminator is called the canonical vector. This analysis is not new; for a general exposition see Anderson (1958). Estimates of means and variances and discriminant functions will be improved if robust analysis procedures are applied to downweight outlying pixel values (Campbell, 1982).

Successive canonical vectors are chosen to maximize this ratio, subject to the variate scores being uncorrelated with those already chosen. The associated canonical roots form a decreasing sequence; their sum measures the overall separation between classes and the individual values the proportion accounted for by the successive vectors. These values summarize the adequacy of the data to separate the classes and the essential dimensionality of the class separation. For example, in a recent vegetation-mapping study using multitemporal Landsat MSS data (eight bands) with 43 training classes, the first three canonical roots were, respectively, 34.6, 6.4, and 3.8 from a total of 45.6. These three dimensions account for 96% of the overall class separation (Hobbs et al., 1989).

It is often possible and instructive to produce simplified discriminant indices with little effect on class separation. The well-known greenness index of Kauth and Thomas (1976) is derived from Landsat MSS data with the coefficients -0.282, -0.660, 0.577, 0.388. A simple contrast of bands may do as well. The effect of reducing the set of bands can also be calculated. In the study mentioned, adequate discriminant information was retained using six bands with integer weights. Significant information was lost, however, when the data set was reduced to a single overpass. When this type of analysis is directed to particular class contrasts, insight into the nature of the spectral separation can be gained.

It is possible that a one-dimensional index will prove adequate to order the data for some applications. More commonly, the ordination reveals that higher dimensions are required to separate classes of interest and a classification approach is then appropriate.

Spectral-Temporal Indices

For monitoring purposes, the explanation and extension of results can be simplified if the discriminating vectors can be constrained (without significant loss of information) to a form in which the coefficients of the data bands are in the same proportion within each overpass. The interpretation then concerns the relative weightings for bands and times. In an unpublished study on the discrimination between crop and pasture using multitemporal MSS data, the optimal discriminator gave a canonical root of 5.97. A simplified spectral-temporal index retained most of the separation (canonical root 5.58). The greenness index performed poorly in comparison, giving a canonical root of 2.08. The optimal index, and even the discriminating bands, were shown to vary with local cultural conditions across farm boundaries (Campbell, unpublished data). The message from such analyses is clear—indices derived from one analysis may be quite inappropriate in a different context.

Allocation Procedures and Assessment of Results

Allocation results are generally presented in the form of a map of class labeled pixels. Accuracy assessments are difficult on a broad scale; on test areas accuracy is most often summarized by the proportion of pixels correctly labeled in comparison with a 'true' map. Schowengerdt (1983) gives a good introduction to allocation methods; see also Richards (1986). Using pixel data only, the maximum likelihood classifier based on multivariate Gaussian densities is generally reported to give the best results. In this procedure, the relative (posterior) probability of membership is calculated for each known class, and the pixel is labeled as belonging to the class for which this value is the greatest. Such a summary is adequate only if all classes are well separated and each pixel falls clearly into a known class. This is rarely the case in practice. It is also not always relevant to produce only a maximum likelihood map, and it may be of interest to locate pixels with even a low probability of membership in certain classes (e.g., saline land, erosion cells).

It is preferable to retain for each pixel two sets of indices: the relative probabilities of membership and the typicality probabilities of belonging to the known spectral classes. [For a detailed explanation of these measures, see Campbell (1984).] These values may be displayed in various ways to assess the completeness and adequacy of the allocation results.

In practice, three types of display have been found to be particularly useful. Assignment of class probabilities to the red, green, and blue guns allows equivocally labeled pixels to be readily identified as mixed colors in the display. For a single class, the confidence of allocation is well summarized by a display of the probability index in blue, and typicality in the other colors; pixels that have high probability and typicality for the class will be white in this display, and blue indicates that the pixel is close relative to the other known classes, but atypical of the training class data. The maximum likelihood class label display is modified by the application of a typicality threshold; pixels that are atypical of all known classes are then displayed in black. Side-by-side displays of the original image data and the allocation results allow ready identification of areas that are allocated equivocally or are atypical. These areas may be chosen as training sites for iterative refinement of the classification.

It should be clear that if the discrimination (i.e., separation) between classes in the spectral space is poor, then allocation is a futile exercise. It is necessary to seek more or better data. Multitemporal data have been found to be essential in collaborative studies for vegetation and crop mapping.

Neighborhood Models and Allocation

Where the size of cover classes is large relative to the ground pixel size, there is contextual information in neighboring pixels that may be used to improve the information on a central pixel, and hence overall classification accuracy. There has been considerable interest in recent years in methodologies for incorporation of this neighbor information. For a review, see for example Nagy (1984).

The general methodology is to iterate the allocation process. After the initial classification, the posterior probabilities are recalculated using local prior weights. The weights depend on the labels of the neighboring pixels, and the associations in the initial allocation; normally a three-by-three neighborhood is used.

Improvements to be expected from such methods will depend on the ground geometry of the spectral classes. Where large areas of spectrally close classes are found (e.g., fields of different crops), considerable overall improvement may result. The improvements also depend on the initial classification being reasonably accurate. Updating from a poor initial classification may actually reduce overall accuracy. In an example using MSS data for the classification of crop, pasture, and bush areas, incorporation of neighbor information improved overall accuracy of maximum-likelihood labels from 80.7% to 94.5%. Greater improvements have been reported by Di Zenzo et al. (1987), who compare alternative updating approaches.

General conditions for the applicability of these algorithms are not known. It may be difficult to establish that appropriate conditions for the use of these techniques do apply in monitoring large-scale processes. It may be found also that any gains in accuracy are small in relation to the precision required. Again, the appropriate analysis methods and data sources should be tested for the applications.

Conclusions

Suitable data organization and analysis tecnhiques are essential to the extraction of information from high-dimensional remotely sensed data. The volumes of data are tremendous. The primary database organization must allow for integration and registration of data sources and flexibility of user access to the data.

Data users face the problem of reduction of the data to useful numerical or pictorial summaries. They will require access to suitable candidate data sets and numerical processing methods to calculate and assess indices or classifiers of the data. Qualitative image interpretations of uncalibrated index displays have a role in indicating patterns and variations in the data and, by implication, variations in surface condition. However, the limitations of this approach in the predictive numerical context should be understood.

Where numerical estimates of processes are required on a continental scale, they should wait for the establishment by suitable analysis of optimal indices and error estimates. In any new application, it is appropriate initially to focus intensive analysis efforts on relatively small and well-known areas, examining possibly very complex data sets. By reduction and comparison, suitable data sets and classifiers may be selected, if they exist. Extension to broader scales may then be implemented. It is an implicit requirement of this process that research workers must have access to suitable statistical image processing systems, and to data sets integrated over suitable spatial and temporal scales.

References

Anderson, T.W. (1958). *An Inroduction to Multivariate Statistical Analysis.* Wiley, NY.
Campbell, N.A. (1982). Robust procedures in multivariate analysis. 1. Robust covariance estimation. *Appl. Statist.*, 29:231–237.
Campbell, N.A. (1984). Some aspects of allocation and discrimination. (pp. 177–192). In G.N. van Vark and W.W. Howells, (eds.), *Multivariate Statistical Methods in Physical Anthropology.* Reidel, Amsterdam, Netherlands.
Di Zenzo, S., Bernstein, R., Degloria, S.D., and Kolsky, H.G. (1987). Gaussian maximum likelihood and contextual classification algorithms for multicrop classification. *IEEE Trans. Geosci. Remote Sens.* GE-25(6):805–814.
Graetz, R.D., and Gentle, M.R. (1982). A study of the relationships between

reflectance characteristics in the Landsat wavebands and the composition and structure of an Australian semi-arid rangeland. *Photogramm. Eng. Remote Sens.* 48:1721–1732.

Hobbs, R.J., Wallace, J.F., and Campbell, N.A. (1989). Classification of native vegetation in the Western Australian wheatbelt using Landsat MSS data. *Vegetatio* 80:91–105.

Jupp, D.L.B., Walker, J., and Penridge, L.K. (1986). Interpretation of vegetation structure in Landsat MSS imagery: a case study in disturbed eucalypt woodlands. Part 2. *J. Envir. Manag.* 23:25–37.

Kauth, R.J., and Thomas, G.S (1976). The tasselled cap—A graphic description of the spectral temporal development of agricultural crops as seen by Landsat. *Proc. Symp. on Machine Processing of Remotely Sensed Data*, Purdue University, Lafayette, IN, pp. 4B41–4B51.

Nagy, G. (1984). Advances in information extraction techniques. *Remote Sens. Envir.* 15:167–175.

Pech, R.P., Graetz, R.D., and Davis, A.W. (1986). Reflectance modelling and the derivation of vegetation indices for an Australian semi-arid shrubland. *Int. J. Remote Sens.* 7(3):389–403.

Richards, J.A. (1984). Thematic mapping from multitemporal data using the principal components transformation. *Remote Sens. Envir.* 16:35–46.

Richards, J.A. (1986). *Remote Sensing Digital Image Analysis*. Springer-Verlag, Berlin.

Schowengerdt, R.A. (1983). *Techniques for Image Processing and Classification in Remote Sensing*. Academic Press. NY.

Sellers, P.J., Hall, F.G., Asrar, G., Strebel, D.E., and Murphy, R.E. (1988). The first ISLSCP field experiment (FIFE). *Bull. Amer. Meteorol. Soc.*, 69(1):22–27.

Swain, P.H., and Davis, S.M. (Eds.) (1978). *Remote Sensing: The Quantitative Approach*. McGraw-Hill, NY.

Swain, P.H., Vardeman, S.B., and Tilton, J.C. (1981). Contextual classification of multispectral image data. *Pattern Recogn.* 13(6):429–441.

Townshend, J.R.G., and Justice, C.O. (1988). Selecting the spatial resolution of satellite sensors required for global monitoring of land transformations. *Int. J. Remote Sens.* 9(2):187–236.

Tucker, C.J., Vanpraet, C.L., Sharman, M.J., and Van Ittersum, G. (1985). Satellite remote sensing of total herbaceous biomass production in the Senegalese Sahel: 1980–1984. *Remote Sens. Envir.* 17:233–250.

15. Remote Sensing of Biosphere Functioning: Concluding Remarks

Richard J. Hobbs and Harold A. Mooney

This volume was concerned with two distinct areas:
1. Remote sensing of metabolic processes and the determination of what drives changes in global metabolism
2. Remote sensing of changes in biosphere structure (vegetation and landscape dynamics).

Global climatic change is likely to result in changes in both metabolism and structure. Most of the early chapters centered on the use of remote sensing to estimate rates at the metabolic level, whereas the later chapters dealt with the detection of structural change. Some consideration was given to how these two levels could be linked, but it became apparent that both cannot be modeled simultaneously because of time-scale differences. There is a clear shift in domain from plant production to population and community dynamics. Modeling of metabolic responses assumes that vegetation boundaries remain the same (i.e., are unaffected by changing climatic conditions), whereas structural models assume that there will be boundary shifts. Clearly, both levels are required, since structural changes will feed back into metabolic processes (e.g., through changing carbon stores with changing vegetation boundaries). However, the conclusion is that attempts to produce linked models would prove ineffective.

Current Global Circulation Models (GCMs) involve two stages when dealing with vegetation. Stage 1 determines the present distribution of vegetation types (i.e., requires current vegetation structure), and stage 2 determines the physiological characteristics of each type. At present, the estimates of the distribution of major vegetation types and of current metabolism in these types are very crude and require considerable refinement before metabolic changes can be modeled. This points to a clear role for remote sensing in the development of more accurate estimates of global metabolism to aid in the prediction of future changes.

In the development of remote sensing techniques to evaluate biosphere function, there has not been an extensive utilization of experimental techniques, especially for forests. These approaches could, however, greatly clarify signal interpretation. For example, structural features of forest systems can be separated from chemical or environmental features by experimental manipulations such as irrigation and fertilization.

There is a need for more First ISLSCP Field Experiment (FIFE)–type experiments that integrate ground-based, aircraft, and satellite measurements. While such experiments are required in different systems, there is also some merit in continuing integrated measurements over the original site. This would allow detection of structural cf. metabolic changes.

There is a great need for long-term observations to monitor biospheric changes. Remote sensing technology presents a great opportunity to collect long-term data, but requires consistency of data collection and compatibility between successive generations of sensors. Long-term observations also require specific questions to specify the design of the system that will use the data. Long-term satellite data must also be matched with long-term ground data for continued truthing. This could be achieved by the accumulation of satellite data over Long Term Ecological Research Sites, or over the Biosphere Reserves proposed under International Geosphere-Biosphere Program.

Although there is continuing development of more sophisticated sensors and more data are becoming available, current technologies still have a lot to offer. It can be argued, in fact, that the available technology exceeds the scientific capability of interpreting and applying it. There are also probably many more applications for existing data than have been so far realized. The main problems in many cases therefore do not lie in data acquisition, but in data analysis and interpretation.

Index

A

Absorbed photosynthetically active radiation (APAR), 67–68, 149, 150, 182
Abundance, of vegetation, 11
Acid rain, 5
Advanced Very High Resolution Radiometer (AVHRR), 2, 33, 35, 54, 110, 194, 205, 207, 211, 253, 263, 265, 297
  estimates of terrestrial NPP using, 66–71
Aerial photography, 8, 10, 223, 232
Aerodynamic roughness, 16
Africa, Sahel, 17
Air Force Geophysics Laboratory, 35
Airborne Imaging Spectrometer (AIS), 92, 93, 94
Aircraft-based remote sensing, of trace gas fluxes, 162–164
Albedo, 16
Amazon Boundary Layer Experiment (ABLE), 159–160, 162, 254
Amazon River Basin, remote sensing, 249–266
Ammonia, 157
Animal distribution models, 238
Antecedent Precipitation Index (API), 57
Antecedent precipitation method, 115
Arboretum, see University of Wisconsin Arboretum
Atmosphere models, 20
Atmospheric-geometric correction, 175
Atmospheric water vapor divergence method, 117–118
Australia, 227, 231, 232, 234, 236, 240, 298
Australian Department of Industry, Technology and Commerce, 2

B

Background (soil) exposure fraction, 68
Backscattering coefficient, 55
Beer-Lambert's law, 142
Biosphere-atmosphere interactions modeling, 170–174
  role of remote sensing, 174–176
Biosphere-Atmosphere Scheme (BATS), 16, 124, 261, 262
Biosphere Reserves, 306
Blackhawk Island, 91–92, 94, 95
Blue shift, 138
Brazil, 253
Brightness temperature, 47

Bucket model, 113, 115
Budyko, 112

C

Calibration, 175
Canopies, 14, 17, 21
  biochemistry, evaluation, 135–151
  interception and evaporation, 75, 77
  radiometric characteristics, 17
  water flux, vegetation index measurements and, 119
Carbon in the Amazon River Experiment (CAMREX), 254, 255, 264, 266
Carbon dioxide, 170
  global atmospheric, 65
Carbon monoxide, 157
Carbon storage, 16
Cellulose, lignin and, 90
Centro de Energia Nuclear na Agricultura (CENA), 253, 254
Century model, 160–161
Chlorophyll, 136, 138, 150, 271, 272
Classification, of vegetation structure, 11–12
Classification image processing models, 20
Climate
  change in, 9, 14, 16
  decay rates of fresh litter and, 88
Coastal Zone Color Scanner (CZCS), 271, 272
Community composition and structure (CCS), 172
Computer-automated time compositing, 66
Context modeling, 10
Continuous model, 18
Continuous scene model, 18, 20
Cover, as variable in characterizing vegetation structure, 14
Cover class mapping, 297–303
Crop canopy geometry, 21
CSIRO, Division of Mathematics and Statistics, 298

D

Data analysis, 291–303
  cover class mapping, 297–303
  database storage, 295
  existing data, 296–297
  high-dimensional data, 292–293
  information content and, 291–292
  information value of data, 296
  qualitative or quantitative, 294–295
  role of, 293–294
  scale, 295–296
Decay, of fresh litter, 88; *see also* Decomposition
Decomposition, 75
  remote sensing of, 87–98
  fresh litter, 88–89
  nitrogen mineralization, 91–98
  soil organic matter, 90–91
Deforestation, 5, 14, 206, 211
  tropical river basins, 249
Departmento Nacional de Aguas e Energia Eletrica (DNAEE), 253
Derivative spectrometry, 143–145
Desertification, 14
Dielectric constant, 41–42
  of water, 55
Directional changes, in vegetation, 205–206
Discrete model, 18
Dissolved inorganic carbon (DIC), 283
Disturbance, 8, 10
  regimes, 8
Drought, 115
Drought-dominated regions (DDRs), 231, 232

E

Earth Observing System (EOS), 2, 60–61, 150, 164, 197, 198, 291
Earth System Sciences Committee, 203, 214
Ecosystem modeling, to estimate NPP, 71–78
Ecosystems, vegetation component, 7–10
Eddy correlation systems, 164
El Niño, 5
Emissivity, of soil, 44, 46
ERS-1, 224
European Center for Medium Range Weather Forecast (ECMWF), 259
Evapotranspiration, 73, 106, 112, 121, 260
  using rainfall to estimate, 113–116
  see also Water and energy exchange

F

First ISLSCP Field Experiment (FIFE), 58, 178–198, 249, 295, 296, 306
  analysis of data, 195–196
  coordinating measurement program, 187–189
  experiment design, 184–187
  experiment execution, 189–195
  implications for future, 196–198
  theoretical background, 178–184

Flood-dominated regions (FDRs), 231, 232
Foliar litter, decomposition of, 88
FOREST-BGC model, 75–77, 81–83
    comparison with NDVI, 78–81
Forest ecosystems, remote sensing of decomposition in, 87–98
Forest fires, 211
Four-dimensional data assimilation, 126
France, 33
French National Space Program, see SPOT
Fresh litter, decomposition of, 88–89
Fresnel equations, 44

G

General Circulation Models (GCMs), 33, 106, 109, 170, 172, 174, 253
Geographic Information System (GIS), 10, 263–264, 265
Geomorphic unit hydrograph (GUH) theory, 260
Geophysical Fluid Dynamics Laboratory (GFDL), 109
Geostationary Operational Environmental Satellites (GOES), 253
Global Biosphere Models (GBMs), 16
Global Circulation Models (GCMs), 16, 65–66, 163, 305
Global habitability, 6, 14
Global Troposphere Experiment (GTE), 162, 164
Global vegetation index (GVI), 67
Goddard Space Flight Center, 66, 189
GOES thermal channel, 108, 194, 197
Grasslands, 13, 14
Grazing, 238
Greenhouse effect, 5, 14, 170

H

Habitability, global, 6, 14
HAPEX-MOBILHY experiment, 33
High Resolution Imaging Spectrometer (HIRIS), 150–151, 263, 292
High-resolution imaging spectrometry, 71, 91
H-resolution model, 18, 20
Humus, decay of, 90, 91
Hydrologic Atmospheric Pilot Experiment, 33, 249

I

Ice, dielectric constant, 42
Illumination geometry, 18

Image processing models, 18–22
Imaging spectrometry, current and potential uses, 149–151
Institute of Hydrology (U.K.), 254
Instituto de Pesquisas Espaciais (INPE), 253, 254
Instituto Nacional de Pesquisas da Amazonia (INPA), 253, 254
Interannual variability, in vegetation, 205
International Atomic Energy Agency (IAEA), 266
    Amazonia I project, 254
International Biological Program, 73
International Geosphere-Biosphere Program (IGBP), 1, 82, 306
International Satellite Land Surface Climatology Project (ISLSCP), 178–198, 209, 249

J

Jet Propulsion Laboratory, 92

L

Landsat Multi-Spectral Scanner (MSS), 207, 209, 210, 223, 232–233, 234, 239, 240, 253, 263, 264, 265, 292, 296, 300, 302
Landsat series, 2, 17, 24, 25, 65, 68, 212, 213, 223, 233, 253, 292
Landsat Thematic Mapper (TM), 159, 207, 209, 223, 224, 233, 234, 239, 253, 263, 265, 266, 294, 296
Landscape processes, 221–244
    landscape change, 228–234
    landscape properties, 223–226
    process domains, 226–228
    using models, 235–244
Layered atmospheric absorption models, 20
Leaching, 75
Leaf area index (LAI), 22, 23, 24, 67–68, 211
    integrating into ecosystem models, 75–78
    satellite estimation of, 73–75
Leaf orientation, 16
Leaf stomatal closure, 71
Lettau, 113
Lignin, 88, 92
    cellulose and, 90
    in whole canopies, 93–98, 101–103
Ligno-cellulose index (LCI), 90
Litter, decomposition of, 87, 88–89

Litterfall, 75, 77
Lockheed Electra, 162
Long-Term Ecological Research (LTER), 82
Long-Term Ecological Research Sites, 306
L-resolution model, 18, 20
Lysimeters, 78

M

Manabe bucket model, 115
Manual of Remote Sensing (Colwell), 55
Marine photosynthesis, 269–286
    effects of climate change on, 283–284
    pigment concentration and, 273–283
    remote sensing, 271–273
Metal contamination, 269, 281
Methane, 157, 159
Microwave, 17, 224
    radiometry, 47–48
    sensing, of soil moisture, 41–61
Middle infrared (MIR), 223, 224
Mineralization, 75
Mixture model, 20
Moderate resolution Imaging Spectrometer (MODIS), 151, 263
Moisture availability, 108
Multispectral scanners, 223
Munday, 264

N

NAS (National Academy of Science), 66
NASA (National Aeronautics and Space Administration), 17, 33, 66, 91, 92
    AVIRIS, 263
    C-130, 33, 34, 189
    Global Troposphere Experiment (GTE), 162
    Goddard Space Flight Center, 66, 189
    H-1 helicopter, 191
NASA-Ames Research Center, 99
National Center for Atmospheric Research (NCAR), 33, 109, 124
    King Air, 189, 191
National Oceanic and Atmospheric Administration (NOAA), *see* NOAA
National Research Council of Canada (NRC), 191
National Science Foundation, 266
Near-infrared (IR) surface reflectances, 176, 208, 223, 224
Near-IR spectroscopy (NIRS), 139
Net primary productivity (NPP), 17, 65, 87
    estimating, 65–83
        ecosystem modeling, 71–78
        NDVI and forest BGC compared, 78–81
        using AVHRR/NDVI, 66–71
        validating regional estimates, 81–82
    nitrogen mineralization and, 97, 98
Nimbus 5, 57
Nimbus 6, 57
Nimbus 7, 57, 120, 271
Nitrogen
    fresh litter decomposition and, 88, 90
    in whole canopies, 101–103
Nitrogen mineralization, prediction by remote sensing, 91–98
Nitrous oxide, 157, 159, 160
NOAA, 2, 33
    –7 polar orbiter, 110
    Advanced Very High Resolution Radiometer (AVHRR), 110, 194, 205, 207
    global vegetation index (GVI), 67, 78
    satellites, 17, 54
Nonmethane hydrocarbons, 157
Normalized difference vegetation index (NDVI), 17, 22, 54, 66–71, 118, 119, 120, 149, 150, 209, 211, 294, 297

O

Oceans, 269
    remote sensing of, 271–273
    see also Marine photosynthesis
Optical wavelengths, 17
Organic mixtures, spectral analysis of, 142–149
Ormsby, J., 199
Ozone, 164
    tropospheric, 158

P

Palmer Drought Severity Index (PDSI), 115
Parameter-to-biophysical-quantity calculation, 175–176
Perpendicular vegetation index (PVI), 54
Phaeopigment, 272
Photosynthesis, 75, 77, 78
    marine, 259–286; *see also* Marine photosynthesis
Phytoplankton, 270, 271, 275–276, 279
Planck/Blackbody equation, 31–32
Planetary boundary layer (PBL), 162
Plant canopies, 13, 14, 17
    biochemistry, 135–151
Pollution, vegetation changes and, 206

Potometers, 78
Process domains, 226–228
Prognostic variables, 126
Projected foliage cover (PFC), 7, 11

Q

Qualitative analysis, vs. quantitative, 294–295
Quantitative analysis, vs. qualitative, 294–295

R

Radambrasil, 253
Radarsat satellites, 224
Radiance-to-parameter calculation, 175
Radiation
  interaction with plants, 13–14
  scene models and, 18
Radioactive isotope transport methods, 78
Radiosonde station data approach, 118
Rainfall approach, to water and energy exchange, 113–116
Reflectance, of radiation, 2
Reflectivity, of soil, 44, 46
Remote sensing
  Amazon River Basin, 249–266
  of canopy biochemistry, 135–151
  data analysis, 291–303
  landscape processes, 221–244
  of litter and soil organic matter decomposition, 87–98
  marine photosynthesis, 269–286
  present methods and limitations, 17–26
  problems in data collection and interpretation, 16–17
  of spatial and temporal dynamics of vegetation, 203–214
  of terrestrial primary productivity, 65–83
  trace gas fluxes and, 158–165
  water and energy exchange, 105–129
Resistance analogue concept, 261
Resolution, of discrete scene models, 18
Resource depletion, 6
Restrahlen bands, 40
Retranslocation, 88
Rothampstead grassland plots, 205
Runoff, 121, 264

S

Sahel, Africa, 17
Satellite-based remote sensing
  field experiments and, 169–198
  of trace gas flux, 164–165
Savannahs, 8, 14
Scanning Multichannel Microwave Radiometer (SMMR), 54
Scattering behavior, of terrain, 55–56
Scene inference, 18
Scene models, 18, 20, 25
Scour-transport-fill (STF) sequence, 239–241
Sea surface temperature, estimation, 32–33
Seasat, 57
Seasonal variations, in vegetation, 204
Sea-surface temperature (SST) field, 172
Sediment, availability, 236–238
Senescence, 88
Sensor look angle, 68
Sensor models, 20–22
Shadows, variations in, 25
Simple Biosphere Model (SiB), 16, 122, 261, 262
SIR A, 57
SIR B, 57
Skin-temperature method, 106–112
Skylab, 57
Snow melt, 75, 77
Soil
  dielectric constant, 41, 42
  dielectric properties, 43–44
  emissivity and reflectivity, 44, 46
  landform behavior and, 221
Soil erosion, 5, 14, 17
Soil moisture
  direct measurements of, 120–121
  see also Surface soil moisture and temperature measurement
Soil organic matter, decomposition of, 87, 90–91
Soil temperature, see Surface soil moisture and temperature measurement
Soil texture, nitrogen mineralization and, 97, 98
Soil water content and outflow, 75
Solar insolation, 31
Solar zenith angle, 68
Southern oscillation index (SOI), 284
Spatial patchiness, 25
Spatial resolution, 20, 21
Spectral analysis, of organic mixtures, 142–149
Spectral resolution, 20, 21
Spectral variations, of land surface emissivity, 39–41
Spectral-spatial scene models, 25

Spectrometry
  derivative, 143–145
  imaging, current and potential uses, 149–151
Spectrophotometry, 142
Split-window technique, 33, 41, 71
SPOT, 2, 197, 207, 223, 226, 233
Stomata, 16
Subtropics, 6, 7
Successional recovery, 8
Surface albedo, 31
Surface resistance factor, 70–71
Surface roughness, 49–52
Surface soil moisture and temperature measurement, 31–61
  microwave sensing, 41–61
  thermal infrared, 32–41
  see also Soil moisture
Synoptic scales, 250
Synthetic aperture radar (SAR), 55, 224, 225

T
Tasseled cap scene models, 25
Taxonomy, of vegetation, 8–9, 10
Temperature, global, 65–66
Terrestrial primary productivity, remote sensing of, 65–83
Thematic Mapper (TM), 159, 207, 209, 223, 224, 233, 239, 253, 263, 265, 266, 294, 296
Thematic Mapper Simulator data, 73
Thermal inertia method, 106, 108–109
Thermal Infrared Multispectral Scanner (TIMS), 33–41
Thermal infrared sensing, of surface temperature, 32–41
TIROS meteorological satellites, 66
Total ozone mapping system (TOMS), 164
Trace gas fluxes, 157–165
  aircraft-based remote sensing, 162–164
  classification-based estimates, 158–160
  ground-based approaches, 161
  remote sensing for driving variables of models, 160–161
  satellite-based flux measurements, 164
Transpiration, 75, 77, 78
Tropical Rainfall Mission (TRM), 116, 259
Tropical river basins, 249
Tropics, 6, 7
Tropospheric ozone, 158
T_s, 71

T_s/NDVI, 70
Twin Otter, 191

U
U.S. National Academy of Science (NAS), 66
U.S. National Aeronautics and Space Administration, see NASA
U.S. National Science Foundation, 2
Universal Soil Loss Equation (USLE), 237
University of Washington, Seattle, 254
University of Wisconsin, College of Agriculture and Life Sciences, 99
University of Wisconsin Arboretum, 91, 94, 97
University of Wyoming King Air, 189, 191

V
Vegetation
  community composition and structure (CCS), 172
  detecting change in, 211–214
  detecting differences in, 206–211
  dynamics, 8
  effects on microwave emission, 53–55
  patterns of change, 204–206
  physiognomy, 7–8
  spectral properties of, 136–142
  structure, 8, 11–16
  taxonomic composition, 8–9
Vegetation functional types (VFTs), 13
Vegetation index approach, to water and energy exchange, 118–120
Vegetation indices, 22
Vegetative cover, estimation of, 22–25
Venezuela, 260
Vertical distribution of biomass, 11

W
Water, dielectric constant, 41–42, 55
Water and energy exchange, 105–129
  atmospheric water vapor divergence method, 117–118
  combined method, 121–129
  direct measurements of soil moisture, 120–121
  rainfall approach, 113–116
  skin-temperature method, 106–112
  use of soil and terrain formation, 121
  use of spatial variability, 121
  vegetation index approach, 118–120
Water site availability, LAI and, 73
Water vapor continuum, 32
Wetness factor, 108